AKADEMIE DER WISSENSCHAFTEN DER DDR

Zentralinstitut für Physik der Erde (ZIPE)

Seismological Bulletin 1970 Station Moxa (MOX)

By

Johannes Stelzner, Dorothea Güth,
and Joachim Weyrauch

With 4 Figures

AKADEMIE-VERLAG · BERLIN
1976

Erschienen im Akademie-Verlag, 108 Berlin, Leipziger Straße 3—4

Lizenznummer: 202 · 100/548/76
Gesamtherstellung:
VEB Druckerei „Thomas Müntzer“, DDR-582 Bad Langensalza
Bestellnummer: 762 126 9 (2004/B/1970) · LSV 1437
Printed in GDR
EVP 38,—

PREFACE

The annual Seismological Bulletin 1970 for the Station Moxa (MOX) continues the series of publications about seismological records at the Central Station Moxa edited by the Central Earth Physics Institute of the Academy of Sciences of the German Democratic Republic.

The provisional analysis of the records of station Moxa was performed in the Seismological Service under the direction of JOHANNES STELZNER by JOACHIM WEYRAUCH and BRIGITTE HÄNSCH.

The annual Bulletin 1970 was prepared by JOHANNES STELZNER and DOROTHEA GÜTH with the technical assistance of URSULA DÖRING.

Control of the instruments of the station Moxa was carried out under the direction of CHRISTIAN TEUPSER.

There is added to the Bulletin a scientific contribution concerning „Distribution of Azimuth and Epizentral Distance of Earthquakes Relative to Moxa (MOX).“

H. KAUTZLEBEN
Director

TABLE OF CONTENTS

Preliminary notes for the interpretation of seismograms

In the Bulletin the international code is used:

1. Phase interpretation

Pg — direct longitudinal wave in near epicentral distances ($D < 10°$)

Pb, Pn — guided longitudinal head waves along the CONRAD- or MOHOROVIČIĆ-discontinuity ($D < 10°$)

P — direct longitudinal wave travelled through the earth mantle

PKIKP — direct longitudinal wave travelled through the inner core (travel-time branch DF)

PKHKP — direct longitudinal wave refracted in the intermediary zone between inner and outer core. Phase symbol according to BOLT [1] (travel-time branch GH)

PKP2 — direct longitudinal wave travelled through the — outer core only (travel-time branch AB)

PKP — first noticeable onset of longitudinal core phase not identified

PP, PPP — waves reflected on the earth surface with permanent longitudinal character

PKKP — core phase reflected once within the core at the outer core boundary

PKPPKP — longitudinal core phase reflected at the earth surface

Sg — direct transversal wave in near epicentral distances ($D < 10°$)

Sb, Sn — guided transversal head waves along the CONRAD- or MOHOROVIČIĆ-discontinuity ($D < 10°$)

S — direct transversal wave travelled through the earth mantle

SKS — direct wave travelled transversal through the mantle and longitudinal through the core

SS, SSS — waves reflected on the earth surface with permanent transversal character

SKKS — wave travelled transversal through the mantle, longitudinal through the core and reflected within the core at the outer core boundary

PcP, ScS, PcS, ScP — longitudinal and transversal waves with steady or changing character reflected at the outer core

PS, SP, PPS — longitudinal and transversal waves with changing character reflected at the surface of the earth

pP, sP, pPP, sPP, pPKIKP, sPKP2, pS — phases of deep-focus earthquakes of longitudinal or transversal waves with steady or changing character. p;s — reflected near the epicentre

pPKP, sPKP — phases of deep focus earthquakes of longitudinal core waves not exactly to be coordinated

SKP, PKS — core phases with different character before and after the direct transit of the core

SKSP — SKS wave with longitudinal character after the reflection at the surface of the earth

P1, P2, P3,...S1, S2,... — multiple onsets of body waves

Pn, Sn — teleseismic Pn and Sn waves in the epicentral distances $23° < D < 40°$ after BÅTH [2]

Pa, Sa — waves probably guided in the astenosphere channel or higher modes of surface waves

PL — leaking modes, normal dispersed train of waves of periods greater than about 10 s, beginning at or near the time of initial P-wave

X, Y, Z — remarkable phases of body waves, not to be identified

LmV, LmH — maximum of the vertical and horizontal component respectively of longperiodical surface waves. If there are several maxima with comparable proportions in A/T, the numeration was carried out in a temporal sequence e.g. Lm1H, Lm2H

The phase symbol is followed by the designation of the type of seismometer from which the time of onsets is taken.

A — seismograph with amplitude characteristic of type A (short-period)
B — seismograph with amplitude characteristic of type B (middle-period)
C — seismograph with amplitude characteristic of type C (long-period)

2. *Measurement of amplitudes and calculation of magnitudes*

All data of amplitudes and periods printed in the column „remarks" are always taken from the records of the same instruments, from which are taken the onset-times of the corresponding phases. The symbol of phase and component is followed by the symbol of the type of instruments in parenthesis e.g.: PV(A), PV(B), SH(B)

Data of amplitudes obtained from records of instruments of type A are given in units of length of nm (1 nm = 1 nanometre = 10^{-6} millimetre). Data of amplitudes obtained from instruments of type B and such obtained from instruments of type C are given in units of length μm (1 μm = 1 mikrometre = 10^{-3} millimetre) e. g.: PV: 1.25 s 38.6 nm, SH: 10 s 3.2 μm, LmH: 22 s 15 μm.

Magnitudes are determined from all those phases, for which calibrating functions are known and internationally used, i.e.

for maxima of body waves P(PH, PV), PP(PPH, PPV), and S(SH)-Q-functions from Gutenberg and Richter [3] — and

for maxima of surface waves (h $<$ 100 km) LmH, LmV — calibrating functions from Prague σ [4] —.

The station correction S was not yet taken into consideration.

MPV, MP1V, MP2V, MPPV — magnitude of vertical component V of the adequate body waves
MPH, MPPH, MSH — magnitude of horizontal component H of the adequate body waves
MLV, ML1V, ML2V — magnitude of the vertical component V of the maximum surface waves
MLH, ML1H, ML2H — magnitude of the horizontal component H of the maximum surface waves
MAG or MB — magnitude of vertical component V of the first onset of P-waves (body-waves) given by USCGS
MS resp. ML — magnitude of horizontal component H of maximum surface waves given by USCGS resp. BCIS

If there are several evaluations of amplitudes from different types of seismographs for the same wave, the symbol of magnitudes is followed by the symbol of instruments e.g.: MPV(A), MPV(B).

3. *Direction of body-wave onsets*

If the direction of motion at the beginning of a wave onset is clearly to be recognized, the sign + or — is placed before the phase symbol. It means:

in the Z component	+	ground motion upwards, compression
	—	ground motion downwards, dilatation
in the N component	+	ground motion to the north
	—	ground motion to the south
in the E component	+	ground motion to the east
	—	ground motion to the west

4. *Further abbreviations*

i — sharp beginning of phase motion (impetus)

e — gradual beginning of phase motion (emersio)

D — epicentral distances in degree (°), calculated according to geocentric coordinates, the maximum error of the own calculations amounts to ± 0,1°

Az — azimuth: clockwise measured angle between north direction in epicentre and the connecting line from epicentre to station Moxa

h — depth of focus in km, our data for depth of focus are based on travel-time curves for deep focus earthquakes after GUTENBERG and RICHTER [5]

H — origin time in GMT (Greenwich Mean Time)

USCGS — United States Coast and Geodetic Survey, Washington

BCIS — Bureau Central International de Séismologie, Strasbourg

ANUSSR — Academia Nauk USSR, Moscow

USAEC — United States Atomic Energy Commission, Washington, D.C.

ISC — International Seismological Centre, Edinburgh

USNOAA — United States National Oceanic and Atmospheric Administration, Boulder

HAN- — Niedersächsisches Landesamt für Bodenforschung Hannover-Buchholz, FRG

For abbreviations of seismological stations and other agencies in the international three letter code see the introductions to the Regional Catalogue of Earthquakes, Edinburgh and the Bulletins of the International Seismological Centre, Edinburgh.

Round brackets indicate uncertainties in interpretation of phase, time, depth of focus or epicentral distances respectively.

[1] BOLT, A.,The velocity of seismic waves near the earths center. Bull. Seism. Soc. Am. 54 (1964) 1, 191—208.

[2] BÅTH, M., Propagation of Sn and Pn teleseismic distances. Pure and Applied Geophysics 64 (1966/II) 19—30.

[3] GUTENBERG, B. and RICHTER, C. F., Magnitude and energy of earthquakes. Annali di Geofisica 9 (1956) 1, 1—15.

[4] KÁRNÍK, V., KONDORSKAJA, N. V. u. a., Standardization of the earthquake magnitude scale. Stud. Geophys. et Geodet., Prague 6 (1962) 41—48.

[5] GUTENBERG, B. and RICHTER, C. F., Materials for the study of deep-focus earthquakes. Bull. Seism. Soc. Am. 26 (1936) 4, 341—390.

Seismological Station Moxa (MOX)
of the Central Earth Physics Institute

Elevation above mean sea level:	455 m
Bedrock:	clay slate of the lower carboniferous formation
Geographic coordinates:	$\varphi = 50°38'46''$N $\lambda = 11°36'58''$ E
Address:	Central Earth Physics Institute (since February 1969) Seismological Service GDR-69, Jena, Burgweg 11 German Democratic Republic Telex: 058 86275 seis dd

Seismographs and their parameters 1970

T_s — seismometer free period
T_g — galvanometer free period
D_s — seismograph damping
D_g — galvanometer damping
V_0 — magnification factor
N — north-south component
E — east-west component
Z — vertical component
σ^2 — coupling coefficient

SKM — Seismograph Kirnos modified
SSJ — Seismic Station Apparatus Type Jena
VSJ — Vertical Seismograph Type Jena

Type of Seismograph	Comp.	$T_s/[s]$	$T_g/[s]$	D_s	D_g	V_e	σ^2
VSJ-II	Z	0.23	0.054	0.35	1.1	300000	
VSJ-II	Z	1.0	1.0	0·5	0.5	47200	0.56
A SKM-III	N	1.64	0.39	0.52	1.98	25700	
	E	1.64	0.39	0.51	1.93	25400	
	Z	1.64	0.39	0.51	1.99	24600	
B SSJ-I	N	20	1.13	0.50	8.87	109	0.074
		20	1.14	0.50	8.79	1110	0.074
	E	20	1.13	0.49	8.85	103	0.071
		20	1.16	0.49	8.61	1050	0.071
	Z	20	1.13	0.48	8.82	108	0.048
		20	1.24	0.48	8.05	1070	0.048
C SSJ-I/L	N	30	70.7	1.24	0.5	1500	0.15
	E	30	79.1	1.3	0.5	1200	0.087
	Z	30	77.2	1.3	0.5	1300	0.066
Strain/L	NS		85.4		0.70	65*)	
	EW		86.2		0.70	67*)	
(coupled)	NS+EW		86.2		0.70	42*)	
Wiechert 1200 kp	N	8.5		0.32		190	
	E	8.3		0.35		190	
Mainka 150 kp	N	20		0.36		50	
200 kp	E	20		0.47	4	40	

*) for wave velocity 5 km s^{-1}

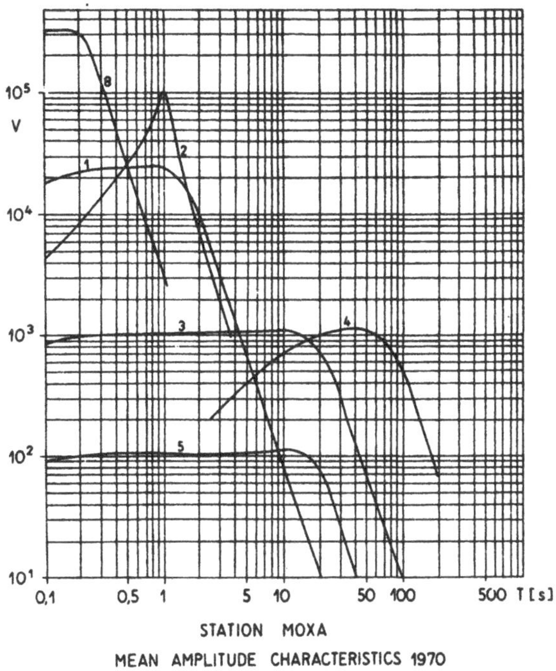

Fig. 1. Station Moxa, mean amplitude characteristics 1970

1 — Seismograph Kirnos Modified-III (SKM-III) (NS-, EW- and Z-component)

2 — Seismograph Type Jena II (Z-component)

3 — Seismic Station Apparatus Type Jena I/1000 (SSJ-I/1000) (NS-, EW- and Z-component)

4 — Seismic Station Apparatus Type Jena I/L (SSJ-I/L) (NS-, EW- and Z-component)

5 — Seismic Station Apparatus Jena I/100 (SSJ-I/100) (NS-, EW- and Z-component)

8 — Seismograph Type Jena II (Z-component)

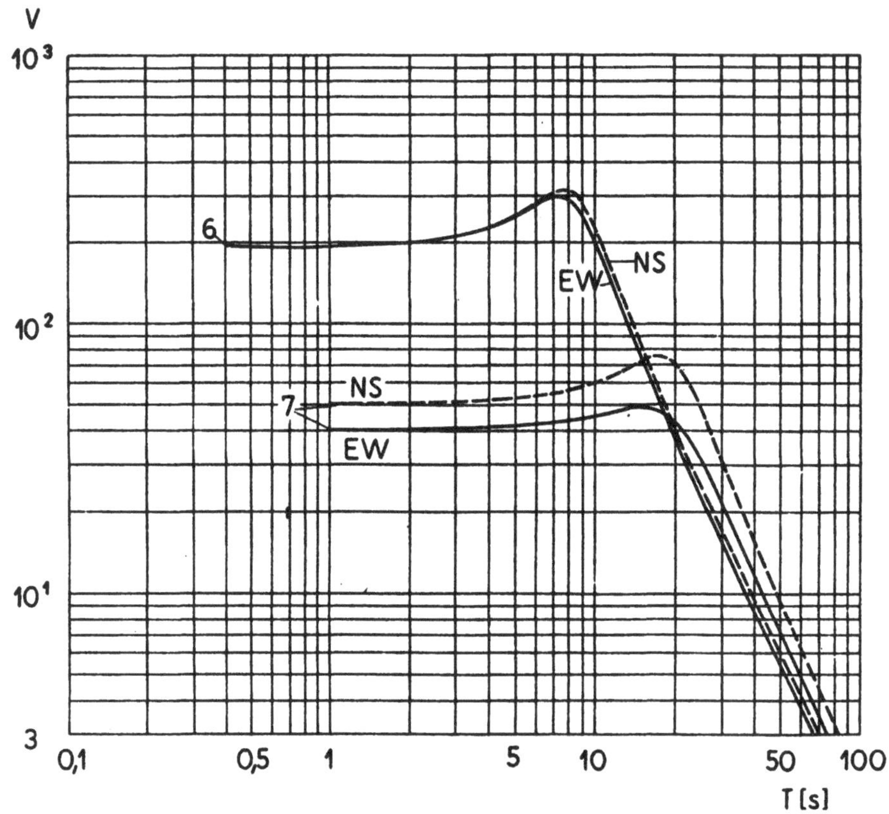

Fig. 2. Station Moxa, mean amplitude characteristic of mechanical seismographs 1970

6 — Wiechert Seismograph (NS- and EW-component)

7 — Mainka Seismograph (NS-and EW-component)

Seismological Recordings at Station Moxa 1970

Day	Phase		h m s	Remarks
1.	eP1	A	01 56 26.5	Costa Rica 8.60 N 83.53 W
	eiP2	A	56 33	H = 01 43 46.7 h = 48 km MB = 5.2
	eiP3	A	56 41.5	D = 86.67 Az = 39.4 (USCGS)
	e	A	58 35	P2V(A):1.8s 81.0nm
	eSKS	C	02 06 52	P3V(A):1.6s 55.0nm
	eS	C	07 05	
	eSS	C	13 00	
	LmH	B	37.6	
	LmV	B	44.5	
1.	+eiP	A	02 02 22.5	Ryukyu Islands 28.62 N 129.31 E
	e	A	02 26	H = 01 49 55.6 h = 39 km MB = 5.2
	e	A	02 33	D = 83.84 Az = 325.5 (USCGS)
	e	A	02 44	PV(A):1.0s 11.8nm MPV(A)=5.1
	LmH	B	37.6	
	LmV	B	44.5	
1.	iP	A	10 06 00.5	Kurile Islands 45.78 N 154.37 E
	ePcP	A	06 06.5	H = 09 53 59.9 h = normal MB = 5.3
	epP	A	06 11	D = 78.69 Az = 336.9 (USCGS)
	LmH	B	45	PV(A):1.3s 69.8nm MPV(A)=5.6
	LmV	B	47.4	PcPV(A):1.4s 37.2nm
				pPV(A):1.3s 43.6nm
				LmH(B):15s 0.69/um MLH(B)=5.1
				LmV(B):15s 0.48/um MLV(B)=5.0
1.	e(PKP2)	A	10 29 09	Loyalty Islands 21.06 S 168.88 E
	eX	A	29 16	H = 10 09 28.5 h = 35 km MB = 4.7
				D = 145.48 Az = 334.3 (USCGS)
				PKP2V(A):1.4s 11.6nm
				XV(A):1.8s 27.1nm
1.	ePKP	A	12 59 06.5	Loyalty Islands Region 21.10 S 169.01 E
				H = 12 39 28.2 h = 0 km
				D = 145.57 Az = 334 (ISC)
				MAG=4.5 (NOU)
1.	ePKIKP	A	17 30 52.5	Kermadec Islands 29.36 S 177.50 W
	ePKHKP	A	31 05	H = 17 11 00.6 h = 44 km MB = 5.4 (USCGS)

Day	Phase		h m s	Remarks
cont.				
1.	eiPKP2	A	17 31 25	D = 157.7
				PKIKPV(A):1.5s 79.2nm
				PKHKPV(A):1.5s 20.1nm
				PKP2V(A): 1.3s 34.9nm
1.	ePKIKP	A	22 27 41	<u>New Ireland</u> 4.04 S 153.92 E
	LmH	C	23 11	H = 22 09 28.6 h = 404 km MB = 5.2
	LmV	C	12	D = 123.82 Az = 332.1 (USCGS)
				PKIKPV(A):1.2s 16.3nm
2.	eP	A	01 30 33	<u>Leeward Islands</u> 16.06 N 59.69 W
				H = 01 19 46.1 h = normal MB = 5.1(USCGS)
2.	eP	A	07 34 14.5	<u>Rumania</u> 45.46 N 26.31 E
				H = 07 31 37.9 h = 134 km MB = 4.4
				D = 11.11 Az = 303.1 (USNOAA)
				PV(A):1.5s 20.1nm
2.	eP	A	19 46 58.5	<u>Yugoslavia</u> 44.75 N 17.09 E
	e	A	47 19	H = 19 45 16.5 h = 24 km MB = 4.4
	e	A	47 21.5	D = 6.96 Az = 329.9 (USCGS)
	eSn	A	48 18	PV(A):0.8s 15.4nm
	ei	A	48 27	LmH(B):12s 1.2/um
	eSg	A	49(02)	
	LmH	B	49.2	
	LmV	B	55.2	
2.	LmH	B	22 15.0	LmH:16s 0.7/um
	LmV	B	21.0	LmV:16s 0.5/um
3.	iP	AB	06 59 54	<u>Turkey-USSR Border Region</u>
	e	A	59 58.5	41.77 N 43.24 E
	iS	B	07 04 16	H = 06 54 49.4 h = 68 km MB = 5.1
	ei	B	04 29	D = 23.47 Az = 303.0 (USCGS)
	LmH	B	11.2	PV(A):1.8s 67.6nm MPV(A)=4.9
	LmV	B	13.2	LmH(B):11s 0.8/um MLH(B)=4.4
				LmV(B):20s 1.3/um MLV(B)=4.5

Day	Phase		h m s	Remarks
3.	e	A	13 42 13	Leeward Islands 15.83 N 59.58 W
				H = 13 31 17.7 h = 34 km MB = 4.9
				D = 66.01 Az = 41.3 (USCGS)
3.	LmH	B	15 56.6	LmH:18s 0.8 /um
	LmV	B	56.8	LmV:20s 0.6 /um
3.	e(P)	A	16 42 24	Western Iran 32.64 N 48.77 E
	LmH	B	57.8	H = 16 35 48.4 h = 32 km MB = 5.0
	LmV	B	17 00.0	D = 32.64 Az = 314.5 (USCGS)
				PV(A):1.5s 15.1nm MPV(A)=4.7
				LmH:14s 0.5 /um MLH(B)=4.4
				LmV:15s 0.9 /um MLV(B)=4.7
4.	eP	A	08 45 01	Northern Celebes 0.34 N 121.94 E
	ePP	A	49 17	H = 08 31 23.1 h = 168 km MB = 5.4
	eSS	C	09 03 40	D = 102.51 Az = 322.3 (USCGS)
	LmH	B	25	e(A) 49 22 e(A) 49 40.5
4.	-eP1	A	17 12 04.5	Yunnan Province, China 24.14 N 102.50 E
	iP2	A	12 09	H = 17 00 40.2 h = 31 km MB = 5.9
	iP3	A	12 12.5	D = 72.27 Az = 318.1 (USCGS)
	ePP	B	14 44	P1V(A):1.2s 26.4nm MP1V(A)=5.1
	e	B	16 48	P2V(A):1.6s 341.0nm MP2V(A)=6.1
	iS	B	21 28	P3V(A):2.1s 940.0nm MP3V(A)=6.5
	ePS	B	22 00	PV(B):5s 3.2 /um MPV(B) =6.7
	e	B	23 38	PPV(B):10s 3.5 /um MPPV(B)=6.6
	eSS	B	26 00	SH(B):14s 22.7 /um MSH(B) =7.2
	LmH	B	43.7	LmH:19s 292.0 /um MLH =7.6
	LmV	B	48.7	LmV:16s 172.0 /um MLV =7.5
4.	eP	A	17 44 03.5	Yunnan Province, China 24.20 N 102.47 E
				H = 17 32 40.2 h = normal MB = 5.2
				D = 72.20 Az = 318.0 (USCGS)
4.	eP	A	18 43 32	Yunnan Province, China 24.27 N 102.30 E
				H = 18 32 09.5 h = normal MB = 4.9
				D = 72.04 Az = 318.0 (USCGS)

Day	Phase		h m s	Remarks
4.	eiP1	A	21 55 57.5	Yunnan Province, China 24.15 N 102.48 E
	eP2	A	56 00	H = 21 44 32.8 h = normal MB = 5.0
				D = 72.25 Az = 318.1 (USCGS)
				P2V(A):1.2s 22.3nm MP2V(A)=5.2
5.	eP1	ABC	00 32 57	Philippine Islands 19.19 N 121.20 E
	eP2	A	33 02	H = 00 20 15.0 h = 53 km MB = 5.4
	eP3	A	33 16.5	D = 87.12 Az = 323.1 (USCGS)
	eiS	BC	43 20	P1V(A):1.8s 47.3nm MP1V(A)=5.5
	ePS	BC	44 25	P2V(A):1.9s 75.8nm MP2V(A)=5.6
	eSS	C	49 20	SH:15s 1.62 µm MSH =5.8
	LmH	B	01 15.6	LmH:18s 5.0 µm MLH =6.0
	LmV	B	16.9	LmV:17.5s 6.2 µm MLV =6.1
5.	iP	A	04 26 25.5	North of Svalbard 81.58 N 4.39 W
				H = 04 20 04.9 h = normal MB = 4.3
				D = 31.47 Az = 160.3 (USCGS)
				PV(A):1.0s 17.7nm MPV(A)=5.0
5.	eP	A	09 20 34	Leeward Islands 16.05 N 59.61 W
				H = 09 09 47.8 h = 20 km MB = 5.3
				D = 65.87 Az = 41.3 (USCGS)
				PV(A):2.4s 41.4nm MPV(A)=5.3
5.	eP	A	12 00 35	Yunnan Province, China 23.92 N 102.84 E
	LmH	B	31.0	H = 11 49 10.7 h = 33 km MB = 4.8
	LmV	B	37.4	D = 72.64 Az = 318 (ISC)
				MS =5.6 (USCGS)
				LmH:18s 2.9 µm MLH=5.6
				LmV:16s 1.4 µm MLV=5.3
5.	e(P)	A	14 33 13	Yunnan Province, China 23.93 N 102.20 E
				H = 14 21 35.3 h = normal MB = 4.8
				D = 72.24 Az = 318.0 (USCGS)
5.	eP	A	19 50 20	North Atlantic Ocean 35.96 N 10.46 W
				H = 19 45 30.4 h = normal MB = 4.2
				D = 21.65 Az = 40.4 (USCGS)

Day	Phase		h m s	Remarks
6.	ePKP	A	00 56 14	Loyalty Islands Region 21.7 S 170.1 E
				H = 00 36 27 h = 0 km (ISC)
				D = 146.6
				PV(A):1.1s 14.1nm
6.	ePKIKP	AB	05 54 59	Dentrecasteaux Islands Region
	ePP	AB	57 00	9.63 S 151.46 E
	e	A	55 03.5	H = 05 35 51.8 h = 8 km MB = 5.7
	e	A	55 07.5	D = 127.44 Az = 328.9 (USCGS)
	LmH	B	06 52.0	PKIKPV(A):1.5s 55.2nm
	LmV	B	52.0	LmH:20s 5.6/um MLH=6.3
				LmV:21s 4.6/um MLV=(6.1)
				e(A) 56 41 e(A) 56 49
6.	eP	A	06 09 30	Sakhalin Island 49.68 N 142.15 E
	e	A	09 35	H = 05 58 09.3 h = 37 km MB = 4.9
				D = 71.49 Az = 329.3 (USCGS)
				PV(A):0.9s 7.8nm MPV(A)=4.8
6.	eP	A	07 23 56	Sakhalin Island 49.61 N 142.33 E
				H = 07 12 38.5 h = normal MB = 4.8(USCGS)
				D = 71.5 traces
6.	eP	A	13 06 54.5	Leeward Islands 15.82 N 59.71 W
	e	A	06 58	H = 12 56 05.9 h = normal MB = 5.3
	eS	BC	15 40	D = 66.10 Az = 41.3 (USCGS)
	e(PSS)	BC	16 18	PV(A):1.2s 12.2nm MPV(A)=5.0
	LmH	B	33.0	LmH:18s 1.1/um MLH =5.1
	LmV	B	33.0	LmV:18s 1.5/um MLV =5.3
7.	eP	A	01 22 26.5	Yunnan Province, China 24.26 N 102.36 E
				H = 01 11 03.6 h = normal MB = 4.8
				D = 72.09 Az = 318.0 (USCGS)
7.	ePKP2	A	04 04 49	South of Fiji Islands 26.57 S 178.68 W
				H = 03 45 12.2 h = 359 km MB = 4.5
				D = 154.70 Az = 344.6 (USCGS)
				PKP2V(A):0.7s 13.4nm

Day	Phase		h m s	Remarks
7.	iPKP	A	05 12 46.7	New Hebrides Islands 18.85 S 169.18 E
				H = 04 53 42.7 h = 249 km MB = 4.9
				D = 143.61 Az = 335.8 (USCGS)
7.	iP	ABC	08 06 58.7	Leeward Islands 15.88 N 59.73 W
	iS	BC	15 48	H = 07 56 11.1 h = 25 km MB = 5.7
	eSSS	BC	23 20	D = 66.07 Az = 41.3 (USCGS)
	eLR	C	27 10	PV(A):2.1s 91.0nm MPV(A)=5.6
	LmH	B	35.0	SH:10s 1.3/um MSH =6.0
	LmV	B	34.6	LmH:16s 3.0/um MLH =5.6
				LmV:18s 4.1/um MLV =5.7
7.	ePKIKP	A	09 13 36.5	East New Guinea 5.40 S 146.96 E
				H = 08 55 04.7 h = 190 km MB = 5.0
				D = 121.54 Az = 328.3 (USCGS)
				PKIKPV(A):1.0s 13.8nm
7.	e(pP)	A	14 07 03	Southern Sumatra 1.85 S 100.40 E
				H = 13 53 52.4 h = 40 km MB = 5.2
				D = 90.64 Az = 320.5 (USCGS)
8.	ePKP	A	08 52 20	Tonga Islands 15.23 S 174.86 W
				H = 08 32 49.3 h = 64 km MB = 5.1
				D = 144.29 Az = 352.9 (USCGS)
8.	iPKIKP	ABC	17 32 16.2	South of Kermadec Islands
	ipPKIKP	AB	33 03.7	34.74 S 178.57 E
	epPKP2	A	33 52	H = 17 12 39.1 h = 179 km MB = 6.1
	ePP	C	36 40	D = 161.48 Az = 333.1 (USCGS)
	iPP	B	36 44	PKIKPV(A):2.0s 398.0nm
	e(pPP)	C	37 28	pPKIKPV(A):1.7s 309.0nm
	e(sPP)	C	38 00	
	ePPP	C	40 25	
	ePcPPKP2	C	42 24	
	eiSS	C	56 40	
	iPSS	C	58 11	
	F	C	19 35	

Day	Phase		h m s	Remarks
8.	eP	A	21 27 09	Hindu Kush 36.30 N 70.95 E H = 21 19 10.4 h = 135 km MB = 5.0 D = 44.20 Az = 308.2 (USCGS)
8.	eP	A	23 18 34.5	Honshu, Japan 40.87 N 140.30 E H = 23 06 36.3 h = 46 km MB = 4.9 D = 78.41 Az = 329.5 (USCGS)
9.	eP	A	05 59 55	Carlsberg Ridge 9.26 N 58.08 E H = 05 50 15.7 h = normal MB = 4.6 D = 56.21 Az = 326.3 (USCGS)
9.	eP	A	09 29 42	Iran-USSR Border Region 38.50 N 55.38 E H = 09 23 05.4 h = 23 km MB = 4.8 D = 32.98 Az = 306.0 (USCGS) PV(A):1.3s 30.6nm MPV(A)=5.1
9.	eP	A	19 35 52	Carlsberg Ridge 9.20 N 58.10 E H = 19 26 12.3 h = normal MB = 4.5 D = 56.27 Az = 326.3 (USCGS)
9.	-iPKP ePKP2	A A	20 17 26.5 17 30	Fiji Islands 15.20 S 178.67 W H = 19 58 42.7 h = 435 km MB = 5.1 D = 143.68 Az = 348.9 (USCGS)
9.	e(pP)	A	23 34 59	Sumbawa Island 9.28 S 117.35 E H = 23 16 21.7 h = 64 km MB = 5.7 D = 107.09 Az = 320.1 (USCGS)
10.	LmH LmV	C C	00 14.0 14.4	LmH:36s 1.4/um LmV:40s 1.2/um
10.	-iP epP	A A	04 32 45.5 32 50.5	Gulf of Alaska 59.45 N 144.97 W H = 04 21 42.7 h = normal MB = 4.9 D = 68.63 Az = 15.8 (USCGS) PV(A):1.2s 34.6nm MPV(A)=5.5 pPV(A):1.5s 55.1nm

Day	Phase		h m s	Remarks
10.	-eP	A	04 58 29.5	Tadzhik-Sinkiang Border Region
				38.14 N 73.68 E
				H = 04 50 28.6 h = 146 km MB = 5.0
				D = 44.81 Az = 307.0 (USCGS)
				PV(A):0.7s 9.6nm MPV(A)=4.8
10.	e(Sn)	A	12 11 30	Belgium 50.97 N 5.45 E
	eSg	A	11 42	H = 12 09 39 (BCIS)
				h = 0 km MAG=4.8 (ISC)
				D = 3.9
10.	+iP1	AB	12 20 48.5	Mindanao, Philippine Islands
	+iP2	A	20 50	6.83 N 126.74 E
	-iP3	A	20 53	H = 12 07 08.6 h = 73 km MB = 6.1
	ePP	B	25 00	D = 100.22 Az = 324.1 (USCGS)
	eiSKS	B	31 28	P1V(A):1.8s 47.3nm MP1V(A)=5.8
	eS	B	32 23	P2V(A):1.8s 154.0nm MP2V(A)=6.3
	eiPSS	B	39 40	P3V(A):1.4s 186.0nm MP3V(A)=6.5
	LmH	B	13 10.6	PV(B):16s 6.23µm MPV(B) =7.0
	LmV	B	10.9	PPV(B):12s 5.35µm MPPV(B) =6.8
				LmH:20.5s 145.0µm MLH= 7.5
				LmV:21s 228.0µm MLV= 7.7
10.	eP	A	14 13 19	Mindanao, Philippine Islands
				6.78 N 126.84 E
				H = 13 59 38.1 h = 67 km MB = 5.4
				D = 100.31 Az = 324.2 (USCGS)
10.	eP	A	14 30 10.5	Mindanao, Philippine Islands
				6.76 N 126.92 E
				H = 14 16 29.1 h = 62 km MB = 5.6
				D = 100.38 Az = 324.2 (USCGS)
				PV(A):1.3s 17.5nm MPV(A)=5.5
10.	eP	A	15 55 21	Ryukyu Islands 28.74 N 128.99 E
				H = 15 43 00.7 h = 94 km MB = 4.8
				D = 83.58 Az = 325.3 (USCGS)

Day	Phase		h m s	Remarks
11.	ePKHKP	A	01 35 59	Tonga Islands 20.42 S 174.89 W
	esPKP	A	36 09.5	H = 01 16 12.1 h = normal MB = 4.8
				D = 149.40 Az = 351.8 (USCGS)
				PKHKPV(A):1.7s 36.4nm
11.	eP	A	03 26 10	Chagos Archipelago 6.07 S 71.22 E
	ePcP	A	26 16	H = 03 14 23.6 h = normal MB = 4.9
				D = 76.15 Az = 325.6 (USCGS)
				PV(A):1.0s 7.9nm MPV(A)=4.8
11.	eP	A	05 15 44	Mindanao, Philippine Islands
	epP	A	16 01	7.32 N 126.82 E
				H = 05 02 05.8 h = 70 km MB = 5.5
				D = 99.86 Az = 324.2 (USCGS)
				PV(A):2.4s 90.0nm (MPV(A)=6.0)
11.	eiPKHKP	ABC	05 39 18.5	Loyalty Islands 22.63 S 171.49 E
	eiPKP2	A	39 23	H = 05 19 37.0 h = 43 km MB = 5.2
				D = 147.92 Az = 335.6 (USCGS)
				PKHKPV(A):1.8s 154.0nm
				PKP2V(A):1.5s 58.0nm
11.	ePKHKP	A	24 00 41.5	Tonga Islands 20.04 S 173.70 W
				H = 23 40 54.0 h = normal MB = 4.3
				D = 149.17 Az = 353.4 (USCGS)
12.	eP	A	02 09 08.5	Chile-Bolivia Border Region
				21.6 S 67.7 W
				H = 01 55 48 h = 133 km
				D = 99.89 Az = 39 (ISC)
12.	eP	A	04 38 47	Mindanao, Philippine Islands
				6.81 N 126.97 E
				H = 04 25 01.1 h = 70 km MAG=5.1
				D = 100.36 Az = 324 (ISC)

Day	Phase		h m s	Remarks
13.	e(PKP)	A	11 08 21.5	New Hebrides Islands 19.78 S 169.86 E H = 10 48 51.0 h = 34 km MB = 4.4 D = 144.71 Az = 335.9 (USCGS)
13.	eP	A	17 08 53	Philippine Islands 8.57 N 127.33 E H = 16 55 20.4 h = 80 km MB = 5.1(USCGS) D = 99.1
13.	eP	A	21 02 06	Mindanao, Philippine Islands 7.08 N 126.83 E
	LmH	C	50.6	H = 20 48 26.1 h = 65 km MB = 5.4
	LmV	C	50.5	D = 100.07 Az = 324.2 (USCGS) PV(A):1.5s 20.1nm MPV(A)=5.5 LmH(C):20s 0.5/um MLH(C)=5.0 LmV(C):20s 0.8/um MLV(C)=5.2
13.	ePKHKP	A	22 51 49	South of Fiji Islands 23.36 S 179.87 W
	epPKP2	A	54 00	H = 22 32 55.3 h = 531 km MB = 4.8 D = 151.33 Az = 344.7 (USCGS) PKHKPV(A):1.0s 9.8nm
14.	eP	A	02 35 43.5	Taiwan Region 23.97 N 122.36 E H = 02 23 14.5 h = normal MB = 4.9 D = 83.96 Az = 323.2 (USCGS) PV(A):0.6s 13.4nm MPV(A)=5.4
14.	LmH	B	03 30.8	Timor 10.34 S 123.15 E
	LmV	B	38.6	H = 02 22 34.6 h = 26 km MAG=5.1 (ISC) D = 111.5 LmH:18.5s 2.3/um MLH=5.8 LmV:16s 1.6/um MLV=5.7
14.	eP	A	10 25 45	Yunnan Province, China 24.35 N 102.21 E
	esP	A	25 52	H = 10 14 22.3 h = 39 km MB = 5.2
	LmH	B	57.3	D = 71.94 Az = 318.0 (USCGS)
15.	+iP	A	08 54 57	Kuril Islands 43.56 N 146.77 E H = 08 42 59.6 h = 47 km MB = 4.9 D = 78.39 Az = 332.7 (USCGS) PV(A):1.0s 15.8nm MPV(A)=5.1

Day	Phase		h m s	Remarks
15.	LmH	B	17 43.0	LmH:18s 1.3 µm
	LmV	B	43.2	LmV:18s 0.9 µm
16.	iPg	A	07 59 32	Czechoslowakia 49.72 N 13.44 E
	iSg	A	59 51	H = 7^h 59^m explosion of 13 tons (PRU)
				D = 1.5
16.	-iP	ABC	08 16 34.8	Southern Alaska 60.31 N 152.72 W
	-sP	B	17 04	H = 08 05 39.6 h = 91 km MB = 5.6(USCGS)
	+iS	BC	25 28	D = 68.6
	i	BC	25 32	PV(A):0.8s 107.5nm MPV(A)=6.1
	ePPS	C	26 20	SH(B):10s 2.9 µm MSH(B)=6.4
	eScS	B	26 22	
	eSS	BC	30 00	
16.	ePKHKP	A	11 05 11	Fiji Islands 21.13 S 179.08 W
				H = 10 46 30.1 h = 638 km MB = 4.7
				D = 149.37 Az = 346.6 (USCGS)
16.	eP	A	22 46 03	Crete 34.97 N 23.85 E
				H = 22 41 53.5 h = 72 km MAG=4.3
				D = 18.01 Az = 334 (ISC)
17.	iPg	A	03 06 39	Bernburg (GDR), rockburst
	e	A	06 51.5	51°46'N 11°46'E
	iSg	A	06 55	
17.	eP	A	08 00 56.5	Kuril Islands 49.55 N 158.53 E
				H = 07 49 02.6 h = normal MB = 4.6
				D = 76.25 Az = 339.0 (USCGS)
18.	+iP	A	00 31 58	Mariana Islands 21.38 N 146.65 E
	e	A	35 55	H = 00 18 23.9 h = 39 km MB = 5.7
	iPP	A	35 58.5	D = 98.07 Az = 333.0 (USCGS)
	ePPS	C	45 35	PV(A):2s 64.1nm MPV(A)=6.0
	iSS	C	50 38	PPV(A):1.8s 87.8nm MPPV(A)=5.9
	LmH	B	01 15.2	LmH:17s 2.4 µm MLH =5.8
	LmV	B	21.3	LmV:16s 1.7 µm MLV =5.6

Day	Phase		h m s	Remarks
18.	ePKP	A	04 30 26.5	Fiji Islands 15.60 S 179.90 W H = 04 10 42.0 h = 20 km MB = 5.1 D = 143.82 Az = 347.6 (USCGS)
18.	-iPKP	A	07 13 46	Fiji Region 21.2 S 177.9 W H = 06 54 05.7 h = 33 km MAG = 4.6 D = 149.7 Az = 348 (ISC) PKPV(A):0.8s 15.4nm
18.	eP	A	13 25 36.5	Kurile Islands 44.17 N 147.87 E H = 13 13 37.6 h = normal MB = 4.7 D = 78.20 Az = 333.3 (USCGS) PV(A):1.0s 11.8nm MPV(A)=4.9
19.	eP	A	00 39 27	Kirgiz SSR 41.07 N 69.27 E H = 00 31 52.6 h = 43 km MB = 5.0 D = 40.34 Az = 303.8 (USCGS)
19.	ePKP	A	09 41 59	New Hebrides Islands 19.30 S 169.13 E H = 09 22 38.0 h = 113 km MB = 4.3 D = 143.99 Az = 335.5 (USCGS) PKP(A):0.7s 15.3nm
19.	e(P)	A	13 08 23	Burma-India Border Region 27.03 N 96.96 E H = 12 57 28.4 h = 45 km MB = 4.6 D = 66.75 Az = 316.3 (USCGS)
20.	eiP epP	A A	00 50 09.5 50 20	Unimak Islands 53.78 N 163.53 W H = 00 38 24.3 h = normal MB = 5.1 D = 75.87 Az = 3.2 (USCGS) PV(A):1.3s 54.6nm MPV(A)=5.5
20.	eP	A	03 32 29.5	Santa Cruz Islands 12.51 S 167.03 E H = 03 13 35.0 h = 234 km MB = 5.1 D = 137.00 Az = 337.1 (USCGS)

Day	Phase		h m s	Remarks
20.	iP	A	06 44 30	Unimak Islands 53.69 N 163.61 W H = 06 32 43.5 h = normal MB = 4.1 D = 75.96 Az = 3.1 (USCGS) PV(A):0.5s 23.0nm MPV(A)=5.6
20.	iPKIKP	ABC	07 39 34	South of Fiji Islands 25.80 S 177.35 W
	isPKP	B	40 10	H = 07 19 51.2 h = 80 km MB = 6.5
	epPKP2	B	40 24	D = 154.25 Az = 346.8 (USCGS)
	eiPP	BZ	43 25	e(B) 50 38 e(B) 54 08
	iPP	BN	43 30	PKIKPV(A):1.5s 126.0nm
	eSKKS	B	50 10	
	eSS	BN	08 02 56	
	eSS	BE	03 00	
20.	eP	A	08 40 25	Near Coast of Nicaragua 11.48 N 86.35 W
	epP	A	40 48	H = 08 27 48.0 h = 46 km MB = 5.4 D = 86.22 Az = 39.2 (USCGS)
20.	+ePKIKP	A	08 57 19	South of Fiji Islands 25.96 S 177.39 W
	+ePKHKP	A	57 27.5	H = 08 37 37.6 h = 100 km MB = 5.1
	+iPKP2	A	57 42	D = 154.40 Az = 346.7 (USCGS)
	epPKHKP	A	57 54	
	+ePPKP2	A	58 09.5	
20.	ePKP2	A	14 06 17.5	South of Fiji Islands 25.9 S 177.2 W H = 13 46 07.6 h = 85 km MB = 4.5(USCGS) D = 154.3 PKP2V(A):1.3s 10.9nm
20.	ePKIKP	A	15 31 29.5	Solomon Islands 5.42 S 154.32 E
	ePP	A	33 23	H = 15 12 51.0 h = 196 km MB = 5.4 D = 125.21 Az = 331.8 (USCGS) PPV(A):1.8s 81.0nm
20.	ePKHKP	A	17 27 22.5	South of Fiji Islands 26.03 S 177.23 W
	ePKP2	A	27 35	H = 17 07 31.2 h = 105 km MB = 5.4
	eipPKP	A	27 42	D = 154.50 Az = 346.9 (USCGS) PKHKPV(A):2.2s 65.4nm

Day	Phase		h m s	Remarks
20.	iP	ABC	17 45 01	Hokkaido, Japan 42.52 N 142.97 E
	isP	B	45 11	H = 17 33 05.4 h = 46 km MB = 6.3
	ei	A	45 14	D = 77.98 Az = 330.7 (USCGS)
	iS	B	54 52	PV(A):2.0s 770.0nm MPV(A)=6.5
	ePKPPKP	A	18 12 01	LmH:16.5s 33.5/um MLH =6.8
	LmH	B	18.6	LmV:13s 19.1/um MLV =6.7
	LmV	B	24.6	
21.	ePKP2	A	00 45 54	South of Fiji Islands 26.21 S 177.02 W H = 00 25 48.0 h = 95 km MB = 4.3 D = 154.72 Az = 347.1 (USCGS)
21.	iP	A	00 57 10	Unimak Island Region 54.01 N 163.56 W H = 00 45 21.0 h = 0 km MAG=4.3 D = 75.64 Az = 3 (ISC)
21.	ePKP2	A	03 19 27	South of Fiji Islands 26.25 S 177.14 W H = 02 59 25.8 h = 112 km MB = 4.5 D = 154.73 Az = 346.9 (USCGS)
21.	e(P)	A	10 44 42.5	Northern Norway 66.5 N 11.9 E H = 10 39 50 D = 15.90 Az = 181 (UPP)
21.	e(pP)	A	13 11 32.5	Hokkaido, Japan 42.52 N 143.01 E
	e	A	11 46.5	H = 12 59 32.8 h = 51 km MB = 4.7 D = 77.99 Az = 330.8 (USCGS)
21.	eP	ABC	18 05 23.5	Off Coast of Mexico 7.02 N 104.30 W
	eiPP	BC	09 35	H = 17 51 38.5 h = normal MB = 6.2
	eiSKS	BC	16 12	D = 100.52 Az = 35.6 (USCGS)
	eiPS	BC	18 14	PV(A):2.4s 166.0nm MPV(A)=6.2
	eiPPS	B	19 11	PV(B):8s 1.48/um MPV(B)=6.6
	ePKKP	A	21 46	PPV(B):12s 4.15/um MPPV(B)=6.7
	e	A	22 12	LmH:24s 39.2/um MLH =6.8
	eSS	C	23 45	LmV:23s 33.8/um MLV =6.8
	i	BC	24 00	e(A) 05 30 e(B,C) 07 28 e(B,C) 08 20
	LQ	C	33 33	
	LmH	B	45.0	
	LmV	B	45.2	

Day	Phase		h m s	Remarks
21.	eP	A	21 48 19	Dodecanese Islands 36.46 N 26.70 E
				H = 21 44 17.8 h = 140 km
				D = 17.86 Az = 327.3 (USCGS)
22.	ePKHKP	A	00 04 03	South of Fiji Islands 26.23 S 177.11 W
	ePKP2	A	04 15	H = 23 44 09.8 h = 95 km MB = 5.0
				D = 154.71 Az = 346.9 (USCGS)
				PKHKPV(A):0.8s 42.3nm
				PKP2V(A):1.5s 45.2nm
22.	ePKP	A	03 46 59	South of Fiji 25.6 S 176.8 W
				H = 03 27 10 h = 71 km
				D = 154.19 Az = 348 (ISC)
				PKPV(A):0.6s 9.6nm
22.	+eP	AC	04 07 27.5	Rat Islands, Aleutian Is.
	e	A	10 38.5	51.18 N 177.28 E
	eS	C	17 20	H = 03 55 32.7 h = 38 km MB = 5.3
	eScS	C	17 48	D = 77.82 Az = 350.7 (USCGS)
	eSS	C	22 48	PV(A):1.6s 49.5nm MPV(A)=5.4
	LmH	B	49.5	LmH:16s 2.0/um MLH =5.5
	LmV	B	48.0	LmV:16s 1.4/um MLV =5.4
22.	iPn	A	15 26 01.9	Federal Rep. Germany 48.33 N 9.09 E
	iPg	AB	26 08	H = 15 25 17.0 h = 16 km MB = 4.5
	iSn	B	26 36	D = 2.84 Az = 34.4 (USCGS)
	iSg	B	26 45	LmH:6.0s 90.0/um MLH=5.2
	i	B	26 48	LmV:5.0s 43.8/um
	LmH	B	27.0	LmH:5.5s 66.0/um MLH=5.1 (Wiechert)
	LmV	B	27.1	M (nach Iida) = 5.5 bei τ = 34 sec.
				Swabian Jura Region/FRG
				48°21'N 0°08.5'E (STU)
22.	ePn	A	15 46 42	Swabian Jura Region/FRG 48.3 N 9.0 E
	ePg	A	46 53	H = 15 45 56 (BCIS)
	iSg	A	47 28.5	D = 2.9 Az = 35 (ISC)

Day	Phase		h m s	Remarks
23.	eiP	A	03 43 14.5	Unimak Island 53.76 N 163.58 W
	epP	A	43 23.5	H = 03 31 29.0 h = normal MB = 5.1
	esP	A	43 26	D = 75.89 Az = 3.2 (USCGS)
				PV(A):1.4s 76.7nm MPV(A)=5.6
23.	eP	A	03 56 01	Unimak Island 53.86 N 163.56 W
				H = 03 44 16.2 h = normal MB = 4.0
				D = 75.79 Az = 3.2 (USCGS)
23.	eP	A	07 07 32	Unimak Island 53.81 N 163.52 W
				H = 06 55 45.9 h = 16 km MB = 4.4
				D = 75.83 Az = 3.2 (USCGS)
23.	ePKIKP	A	09 51 28.5	South of Fiji 26.49 S 177.0 W
	ePKP2	A	51 46	H = 09 31 38 h = 61 km MAG=4.9
				D = 154.99 Az = 347 (ISC)
23.	eiP	A	22 34 08	Kurile Islands 49.78 N 154.87 E
	epP	A	34 27	H = 22 22 38.7 h = 130 km MB = 5.3
				D = 75.14 Az = 336.8 (USCGS)
				PV(A):1.4s 102.0nm
24.	eiSg	A	04 18 29	Federal Rep. Germany 48.3 N 9.0 E
				H = 04 16 55 (BCIS)
				D = 2.90 Az = 35 (ISC)
24.	ePKP2	A	08 43 30	South of Fiji 26.0 S 177.0 W
				H = 08 23 18 h = 39 km MAG=4.6
				D = 154.52 Az = 347 (ISC)
24.	eP	A	15 47 34	Southern Greece 37.29 N 23.43 E
				H = 15 43 55.2 h = 110 km MB = 4.3
				D = 15.80 Az = 331.4 (USCGS)
24.	eP	A	19 12 50	Luzon, Philippine Islands
	e	A	13 01.5	17.40 N 122.13 E
	epP	A	13 06.5	H = 19 00 00.3 h = 63 km MB = 5.1
	LmV	B	56.3	D = 89.08 Az = 323.4 (USCGS)

Day	Phase		h m s	Remarks
cont. 24.	LmH	B	20 00.2	pPV(A):1.8s 50.6nm MPV(A)=5.5 LmH:16s 1.2/um MLH =5.4 LmV:12s 0.9/um MLV =5.4
24.	ePn	A	21 15 18.5	Federal Rep. Germany 48.0 N 9.1 E
	ePg	A	15 22.5	H = 21 14 26 h = 0 km
	eSn	A	15 44	D = 3.09 Az = 31 (ISC)
	eSg	A	15 56	
24.	ePKP	A	22 15 05	New Hebrides Islands 19.12 S 168.87 E H = 21 55 39.0 h = 70 km MB = 4.7 D = 143.73 Az = 335.4 (USCGS) PKPV(A):1.2s 12.2nm
24.	ePKP	A	23 57 20	Fiji Islands 15.29 S 177.11 W
	ePKP2	A	57 22.5	H = 23 38 34.4 h = 410 km MB = 4.5 D = 144.04 Az = 350.5 (USCGS)
26.	eiP	A	00 49 35	Near East Coast of Kamchatka 54.20 N 160.33 E H = 00 38 11.5 h = normal MB = 4.9 D = 72.28 Az = 339.7 (USCGS) PV(A):1.1s 22.2nm MPV(A)=5.2
26.	ePKIKP	A	10 20 40	Santa Cruz Islands 12.58 S 166.37 E
	iPP	B	23 20	H = 10 01 20.5 h = 50 km MB = 5.7
	LmH	B	11 27.0	D = 136.81 Az = 336.6 (USCGS)
	LmV	B	27.4	LmH:18s 6.2/um MLH=6.4 LmV:17.5s 5.8/um
26.	ePKHKP	A	14 34 36	Fiji Islands 20.79 S 178.59 W H = 14 15 50.6 h = 570 km MB = 4.4 D = 149.14 Az = 347.3 (USCGS) PKHKPV(A):0.9s 9.7nm
26.	eP	A	16 46 18	Hindu Kush Region 36.51 N 70.44 E
	e	A	46 45.5	H = 16 38 31.7 h = 224 km MB = 4.7
	eipP	A	47 05	D = 43.75 Az = 308.0 (USCGS)

Day	Phase		h m s	Remarks
26.	eP	A	19 22 17.5	Unimak Island 53.86 N 163.55 W H = 19 10 32.7 h = normal MB = 3.9 D = 75.79 Az = 3.2 (USCGS)
26.	eP	A	24 03 26.5	Fox Islands Aleutian Is. 53.84 N 165.50 W H = 23 51 44.1 h = 51 km MB = 4.4 D = 75.86 Az = 1.9 (USCGS)
27.	eP	A	07 18 46	Mongolia 45.78 N 94.20 E H = 09 09 31.2 h = normal MB = 4.5 D = 52.53 Az = 307.3 (USCGS)
27.	ePKIKP	A	09 22 07	Santa Cruz Islands 10.87 S 165.90 E H = 09 02 51.8 h = 50 km MB = 5.5 D = 135.07 Az = 337.0 (USCGS)
27.	+eP ePcP	A A	09 41 54 41 59	Northern Colombia 7.48 N 72.15 W H = 09 29 43.1 h = 22 km MB = 5.7 D = 80.31 Az = 39.9 (USCGS) PV(A):1.8s 47.3nm MPV(A)=5.2
27.	eP	A	11 00 02	Tsinghai Province, China 34.90 N 101.32 E H = 10 49 31.4 h = normal MB = 5.1 D = 63.78 Az = 314.8 (USCGS) PV(A):1.2s 12.2nm MPV(A)=5.0
27.	eP	A	11 28 36	Near East Coast of Kamchatka 57.72 N 163.58 E H = 11 17 29.4 h = 41 km MB = 5.1 D = 69.56 Az = 341.4 (USCGS) PV(A):1.2s 22.4nm MPV(A)=5.2
28.	eP	A	08 51 46.5	Off East Coast of Honshu, Japan 40.23 N 143.60 E H = 08 39 36.1 h = 20 km MB = 4.8 D = 80.81 Az = 331.3 (USCGS) PV(A):1.1s 12.1nm MPV(A)=4.8

Day	Phase		h m s	Remarks
28.	-iPKIKP	ABC	23 24 38.5	Fiji Islands 20.68 S 178.85 W
	+iPKHKP	A	24 43	H = 23 06 01.7 h = 608 km MB = 5.6
	iPKP2	A	24 48.5	D = 148.97 Az = 347.0 (USCGS)
	epPKIKP	BC	26 55	PKIKPV(A):2.0s 316.0nm
	esPKIKP	BC	27 55	PKHKPV(A):1.4s 815.0nm
	ePSKS	BC	38 25	PKP2V(A) :1.0s 252.0nm
	eSPP	BC	40 10	e(C) 31 18 e(B,C) 45 30 e(C) 51 30
	ePPS	BC	41 40	
	eSS	BC	46 40	
29.	ePKP	A	03 12 34.5	Tonga Islands 19.92 S 173.77 W
	eiPKP2	A	12 37.5	H = 02 52 50.0 h = 32 km MB = 5.2
	ei	A	12 39	D = 149.05 Az = 353.3 (USCGS)
	epPKP	B	12 46	PKP2V(A):2.1s 96.2nm
	esPKP	A	12 49	LmH(C):18s 0.7/um MLH(C)=5.4
	e	BC	13 05	LmV(C):18s 0.7/um
	LmH	C	04 20.5	
	LmV	C	26.5	
29.	eP	A	06 15 39	Near East Coast of Honshu, Japan
	ePcP	A	15 47	35.91 N 140.41 E
	e	A	18 39.5	H = 06 03 21.7 h = 70 km MB = 5.1
				D = 82.74 Az = 330.0 (USCGS)
				PV(A):1.0s 27.6nm MPV(A)=5.2
29.	+iP	A	07 10 46.5	Easthern Kazakh SSR 49.80 N 78.21 E
	e	A	12 19	H = 07 02 57.5 h = 0 km MB = 5.6
				D = 41.30 Az = 297.7 (USCGS)
				PV(A):0.7s 161.0nm MPV(A)=5.9
				Underground explosion (UPP)
29.	eiP	A	11 12 09.5	Sicily 38.76 N 14.88 E
				H = 11 09 24.1 h = 280 km MB = 4.7
				D = 12.11 Az = 350.1 (USCGS)
				PV(A):1.3s 74.2nm MPV(A)=4.8
29.	e	A	12 09 44	Tonga Islands 20.07 S 173.53 W
				H = 11 49 47.8 h = 50 km MB = 4.7
				D = 149.22 Az = 353.6 (USCGS)

Day	Phase		h m s	Remarks
29.	ePKHKP	A	13 16 44	Fiji Islands 20.53 S 176.83 W
				H = 12 57 28.3 h = 285 km MB = 4.9
				D = 149.21 Az = 349.5 (USCGS)
				PKHKPV(A):0.8s 15.4nm
30.	ePKIKP	BC	08 47 20	New Hebrides Islands 14.59 S 167.33 E
	epPKIKP	BC	48 12	H = 08 28 22.7 h = 172 km MB = 5.7
	eiPP	BC	50 22	D = 139.02 Az = 336.5 (USCGS)
	eiSKP	BC	50 50	i(B) 47 30 e(B,C) 51 45 e(B,C) 52 o8
				e (B,C) 53 05
30.	eP	A	17 22 21	Kurile Islands 43.52 N 147.01 E
				H = 17 10 23 h = 48 km MAG=4.9
				D = 78.50 Az = 333 (ISC)
31.	eP	A	01 30 04	Greece 38.93 N 21.23 E
				H = 01 26 52.6 h = normal MB = 4.3
				D = 13.54 Az = 333.0 (USCGS)
31.	eP	A	04 28 38	North Atlantic Ridge 52.32 N 31.73 W
				H = 04 23 00.2 h = normal MB = 4.5
				D = 26.75 Az = 76.1 (USCGS)
				PV(A):0.8s 9.6nm MPV(A)=4.4
31.	eP	A	08 09 24.5	Philippine Islands 6.85 N 127.08 E
				H = 07 55 41.8 h = 63 km MB = 5.2
				D = 100.39 Az = 324.2 (USCGS)
31.	eP	A	11 54 16.5	Northern Sumatra 4.14 N 96.09 E
	epP	A	54 31	H = 11 41 53.6 h = 56 km MB = 5.3
				D = 83.31 Az = 320.4 (USCGS)
				pPV(A):2.4s 83.0nm MPV(A)=5.3
31.	eP	AB	16 40 59	North Atlantic Ocean 53.88 N 35.48 W
	Pm	A	41 03	H = 16 35 03.9 h = normal MB = 5.1
	ePP	B	41 38	D = 28.59 Az = 77.0 (USCGS)
	eS	BC	45 44	PmV(A):2.2s 76.3nm MPmV(A)=5.0
	e	C	48 25	LmH:16s 4.3 um MLH =5.2

Day	Phase		h m s	Remarks
cont. 31.	LmH	B	16 52.0	LmV:15.5s 4.5/um MLV=5.3
	LmV	B	54.0	e(C) 41 32
31.	eP	A	19 05 22.5	<u>Ascension Island</u> 7.13 S 12.91 W H = 18 55 08.5 h = normal MB = 4.9 D = 61.33 Az = 17.5 (USCGS) PV(A):1.4s 27.9nm MPV(A)=5.2
31.	ePKP2	A	20 57 21.5	<u>Kermadec Islands</u> 28.02 S 176.93 W H = 20 37 00.8 h = normal MB = 4.8 D = 156.49 Az = 346.3 (USCGS)
31.	eP	A	21 20 53	<u>Kuril Islands Region</u> 49.68 N 158.99 E
	esP	A	21 10	H = 21 09 06.6 h = 40 km MB = 5.1 D = 76.23 Az = 339.3 (USCGS) PV(A):1.2s 12.2nm MPV(A)=4.8
31.	eP	A	22 36 05	<u>Honduras</u> 15.73 N 88.75 W H = 22 23 37.1 h = 46 km MB = 5.0 D = 84.41 Az = 39.0 (USCGS)

Day	Phase		h m s	Remarks
1.	eP	A	03 09 23.5	Cyprus 34.43 N 32.62 E
	eX	A	09 28	H = 03 04 26.8 h = 33 km MB = 4.4
				D = 22.29 Az = 323.0 (USCGS)
				PV(A):1.2s 18.3nm MPV(A)=4.4
				XV(A):1.2s 19.2nm
1.	ePn	A	16 28 50	Federal Rep. Germany 48.3 N 9.0 E
	e	A	28 56	H = 16 28 05 (BCIS)
	e	A	28 58	D = 2.8
	ePg	A	29 02	
	iSn	A	29 24.4	
	e	A	29 31.5	
	eiSg	A	29 35	
2.	eP	A	15 49 26	Kurile Islands 43.65 N 145.96 E
	e	A	49 28	H = 15 37 31.5 h = 60 km MB = 5.1
	e	A	49 37	D = 78.03 Az = 332.3 (USCGS)
	e	A	49 44	
2.	+iP	A	17 34 08.7	Kurile Islands 43.51 N 147.48 E
	eipP	A	37 21.5	H = 17 22 08.0 h = normal MB = 5.5
	LmH	B	18 09.6	D = 78.66 Az = 333.1 (USCGS)
	LmV	B	13.7	PV(A):1.2s 93.5nm MPV(A)=5.7
				pPV(A):1.5s 115.0nm
				LmH(B):16s 2.5/um MLH(B)=5.6
				LmV(B):15s 1.8/um MLV(B)=5.6
2.	eiP	A	17 38 20	Kurile Islands 43.57 N 147.53 E
	epP	A	38 32	H = 17 26 18.7 h = normal MB = 5.1
	e	A	38 46	D = 78.63 Az = 333.1 (USCGS)
				PV(A):1.2s 38.6nm MPV(A)=5.4
2.	+iP	A	18 01 52.2	Kurile Islands 43.49 N 147.42 E
	eipP	A	02 05	H = 17 49 51.9 h = normal MB = 5.5
	LmH	B	37.2	D = 78.67 Az = 333.1 (USCGS)
	LmV	B	48.9	PV(A):1.3s 183.5nm MPV(A)=6.0
				pPV(A):1.5s 226.0nm
				LmH(B):16s 4.3/um MLH(B)=5.9
				LmV(B):12s 3.0/um MLV(B)=5.9

Day	Phase		h m s	Remarks
2.	eiP	A	18 17 51.5	<u>Kurile Islands</u> 43.29 N 147.54 E
	epP	A	18 04	H = 18 05 49.6 h = normal MB = 5.2
				D = 78.88 Az = 333.2 (USCGS)
				PV(A):1.2s 50.8nm MPV(A)=5.4
				pPV(A):1.5s 57.8nm
3.	eP	A	06 09 49	<u>Kurile Islands</u> 43.55 N 147.42 E
				H = 05 57 47.9 h = normal MB = 4.7
				D = 78.60 Az = 333.1 (USCGS)
				PV(A):1.3s 21.8nm MPV(A)=5.1
3.	eP	A	10 13 11	<u>Kurile Islands</u> 43.38 N 147.53 E
				H = 10 01 12.8 h = 60 km MB = 4.3
				D = 78.80 Az = 333.2 (USCGS)
3.	ePKP	A	19 14 36	<u>New Hebrides Islands</u> 17.53 S 167.69 E
				H = 18 55 06.2 h = 39 km MB = 5.2
				D = 141.83 Az = 335.3 (USCGS)
3.	eP	A	19 29 18	<u>Kurile Islands</u> 43.65 N 147.73 E
	e	A	29 29.5	H = 19 17 16.2 h = 26 km MB = 5.3
	LmH	B	20 08.9	D = 78.62 Az = 333.2 (USCGS)
	LmV	B	08.9	PV(A):1.3s 52.4nm MPV(A)=5.4
				LmH(B):13s 1.7 µm MLH(B)=5.6
				LmV(B):13s 1.6 µm MLV(B)=5.6
4.	eP1	A	05 21 51	<u>Off Coast of Guerrero, Mexico</u>
	eP2	A	21 56.5	15.53 N 99.48 W
	ePP	B	25 22	H = 05 08 48.0 h = 21 km MB = 6.0
	e	A	25 30.5	D = 90.90 Az = 36.4 (USCGS)
	eSKS	B	32 28	P2V(A):1.8s 74.4nm MP2V(A)=5.7
	iS	B	33 00	PPV(B):7s 1.67 µm MPPV(B)=6.6
	eiSS	B	38 56	LmH(B):18.5s 13.4 µm MLH(B)=6.4
	LmH	B	06 05.3	LmV(B):17.0s 12.6 µm MLV(B)=6.4
	LmV	B	09.3	

Day	Phase		h m s	Remarks
4.	iP	A	10 30 03	<u>Near East Coast of Honshu, Japan</u>
				36.53 N 140.63 E
				H = 10 17 46.2 h = 59 km MB = 4.8
				D = 82.29 Az = 330.1 (USCGS)
				PV(A):0.7s 11.5nm MPV(A)=4.9
4.	eP	A	13 19 13	<u>Kurile Islands</u> 43.54 N 147.78 E
	epP	A	19 25	H = 13 07 12.1 h = normal MB = 5.1
	esP	A	19 31.5	D = 78.73 Az = 333.3 (USCGS)
	LmH	P	58.9	PV(A):1.2s 32.5nm MPV(A)=5.3
	LmV	B	58.9	pPV(A):1.3s 39.3nm
				sPV(A):1.4s 44.1nm
				LmH(B):13.5s 1.6 μm MLH(B)=5.5
				LmV(B):13.5s 1.2 μm MLV(B)=5.4
4.	eiP	A	17 12 18	<u>Southern Nevada</u> 37.10 N 116.03 W
				H = 17 00 00.0 h = 0 km MB = 5.6
				D = 81.24 Az = 30.7 (USCGS)
				PV(A):1.5s 65.3nm MPV(A)=5.4
				Nevada test site "Grape B"
				37°05'51.3" N 116°01'35.4" W (USAEC)
4.	ePKHKP	A	23 05 38	<u>Loyalty Islands</u> 22.85 S 171.39 E
				H = 22 45 58.2 h = 57 km MB = 5.2
				D = 148.08 Az = 335.4 (USCGS)
				PKHKPV(A):1.1s 16.2nm
5.	+iP	A	03 51 26	<u>Yunnan Province, China</u>
	e	A	51 33	24.34 N 102.33 E
	LmH	B	04 21.6	H = 03 40 03.1 h = normal MB = 5.2
	LmV	B	28.2	D = 72.01 Az = 318.0 (USCGS)
				PV(A):1.6s 60.5nm MPV(A)=5.5
				LmH(B):18s 2.0 μm MLH(B)=5.4
				LmV(B):14s 1.3 μm MLV(B)=5.4
5.	ePg	A	04 11 37.5	rockburst near <u>Bernburg, GDR</u>
	iPg	A	11 39.5	51°46' N 11°46' E
	eSg	A	11 55	D = 1.0 (MOX)
	iSg	A	11 56	

Day	Phase		h m s	Remarks
5.	eP1	AB	12 58 32	<u>Kurile Islands</u> 47.02 N 154.20 E
	iP2	A	58 33.4	H = 12 46 38.2 h = normal MB = 5.5
	eS	BC	13 08 24	D = 77.50 Az = 336.7 (USCGS)
	LmH	B	32.7	P1V(B):7s 1.39 µm MP1V(B)=6.2
	LmV	B	42.6	P1V(A):0.8s 26.9nm MP1V(A)=5.4
				P2V(A):2.2s 453.0nm MP2V(A)=6.2
				SH(B):9.5s 1.29 µm MSH(B)=6.0
				LmH(B):17.5s 6.7 µm MLH(B)=6.0
				LmV(B):15s 3.1 µm MLV(B)=5.8
5.	eP	A	13 51 44	<u>Kurile Islands</u> 47.07 N 154.16 E
				H = 13 39 49.4 h = normal MB = 4.7
				D = 77.44 Az = 336.6 (USCGS)
				PV:1.0s 15.7nm MPV(A)=5.1
5.	+iP1	A	14 53 34.5	<u>Kurile Islands</u> 47.07 N 154.10 E
	iP2	A	53 35.7	H = 14 41 41.0 h = normal MB = 5.4
				D = 77.43 Az = 336.6 (USCGS)
				P1V(A):0.9s 23.4nm MP1V(A)=5.3
				P2V(A):1.0s 94.5nm MP2V(A)=5.9
5.	e(P)	A	16 08 55	<u>South of Alaska</u> 52.93 N 164.85 W
				H = 15 57 03.4 h = normal MB = 4.3
				D = 76.76 Az = 2.3 (USCGS)
				PV(A):0.6s 9.6nm MPV(A)=5.1
5.	eP1	A	22 19 12.5	<u>Luzon, Philippine Islands</u>
	iP2	ABC	19 16.6	12.60 N 122.12 E
	e	BC	21 32	H = 22 05 58.3 h = 11 km MB = 6.0
	eiPP	BC	22 56	D = 92.90 Az = 323.3 (USCGS)
	eiSKS	BC	29 44	P1V(A):2.0s 154.0nm MP1V(A)=6.1
	eiS	C	30 18	P2V(A):1.6s 192.0nm MP2V(A)=6.3
	eiSS	C	36 30	P2V(B):7s 2.77 µm MP2V(B)=6.8
	LmH	B	58.8	PPV(B):11s 3.0 µm MPPV(B)=6.6
	LmV	B	23 05.9	SH(C):14s 8.1 µm MSH(C)=6.7
				LmH(B):21.5s 62.0 µm MLH(B)=7.0
				LmV(B):16s 39.2 µm MLV(B)=7.0

Day	Phase		h m s	Remarks
5.	eP	A	22 58 55.5	<u>Luzon, Philippine Islands</u> 12.72 N 122.12 E H = 22 45 39.9 h = 20 km MB = 5.0 D = 92.80 Az = 323.3 (USCGS) PV(A):1.6s 22.0nm MPV(A)=5.3
6.	eiP1	A	00 23 13.5	<u>Off East Coast of Kamchatka</u> 54.64 N 163.56 E H = 00 11 49.6 h = 43 km MB = 5.6 D = 72.49 Az = 341.7 (USCGS) P1V(A):1.0s 47.2nm MP1V=5.6 P2V(A):1.4s 86.0nm MP2V=5.7
	eP2	A	23 18.5	
	epP	A	23 26.5	
6.	eP	A	02 30 44	<u>Mindoro, Philippine Islands</u> 12.55 N 121.85 E H = 02 17 30.1 h = 18 km MB = 5.4 D = 92.78 Az = 323.3 (USCGS) LmH(B):18s 1.8/um MLH(B)=5.6 LmV(B):17s 1.3/um MLV(B)=5.5
	e	A	30 47.5	
	e	A	30 53	
	e	A	31 05	
	eS	C	41 40	
	LmH	B	03 14.2	
	LmV	B	17.2	
6.	ePKP	A	09 00 53	<u>Tonga Islands</u> 16.00 S 175.18 W H = 08 41 50.5 h = 301 km MB = 4.6 D = 145.01 Az = 352.5 (USCGS) PKPV(A):1.6s 27.5nm
6.	+iP	A	14 09 08.5	<u>Kurile Islands</u> 43.93 N 147.91 E H = 13 57 14.3 h = 78 km MB = 5.0 D = 78.43 Az = 333.3 (USCGS) PV(A):1.0s 35.4nm MPV(A)=5.2
6.	eP	A	22 22 07	<u>Yunnan Province, China</u> 23.10 N 100.78 E H = 22 10 41.6 h = normal MB = 5.4 D = 71.98 Az = 318.0 (USCGS) PV(A):1.6s 55.0nm MPV(A)=5.4 LmH(B):16s 5.9/um MLH(B)=6.0 LmV(B):12s 3.8/um MLV(B)=5.9
	e	C	31 30	
	eSS	C	36 16	
	eSSS	C	39 32	
	LmH	B	56.1	
	LmV	B	56.7	

Day	Phase		h m s	Remarks
6.	eP	A	22 44 04.5	Yunnan Province, China 22.98 N 100.83 E H = 22 32 41.3 h = normal MB = 4.8 D = 72.09 Az = 318.0 (USCGS) PV(A):1.2s 14.2nm MPV(A)=5.0
7.	eiP	AB	10 12 58	Kurile Islands 47.22 N 154.10 E
	eS	B	22 58	H = 10 01 05.4 h = normal MB = 5.4
	LmH	B	47.2	D = 77.29 Az = 336.6 (USCGS)
	LmV	B	57.0	PV(A):1.8s 122.0nm MPV(A)=5.7 LmH(B):17.5s 6.1/um MLH(B)=6.0 LmV(B):14s 3.0/um MLV(B)=5.8
7.	-iP	A	10 15 48	Kurile Islands 48.2 N 152.2 E H = 10 03 59.1 h = 0 km MB = 5.5 D = 75.89 Az = 325 (ISC) PV(A):1.4s 69.7nm MPV(A)=5.6
7.	eP	A	10 28 38	Kurile Islands 47.25 N 153.99 E H = 10 16 44.3 h = normal MB = 4.5 D = 77.23 Az = 336.5 (USCGS)
7.	eiP	A	12 19 08.5	Kurile Islands 47.02 N 154.07 E H = 12 07 14.4 h = normal MB = 5.0 D = 77.46 Az = 336.6 (USCGS) PV(A):0.9s 31.1nm MPV(A)=5.4
7.	eP	AB	12 19 28	Kurile Islands 47.27 N 154.03 E
	LmH	B	53.5	H = 12 07 35.8 h = normal MB = 5.5
	LmV	B	13 01.2	D = 77.23 Az = 336.5 (USCGS) PV(A):1.2s 102.0nm MPV(A)=5.8 LmH(B):18s 6.2/um MLH(B)=6.0 LmV(B):15s 2.7/um MLV(B)=5.7
7.	eP	A	12 31 43	Kurile Islands 47.02 N 153.76 E H = 12 19 47.8 h = normal MB = 4.4 D = 77.38 Az = 336.4 (USCGS) PV(A):0.6s 15.3nm MPV(A)=5.3

Day	Phase		h m s	Remarks
7.	+eP	A	15 05 50.5	Kurile Islands 47.1 N 154.1 E H = 14 53 56.1 h = normal MB = 4.7 (USCGS) D = 77.4 PV(A):1.0s 23.6nm MPV(A)=5.3
7.	eP e	A A	15 07 52 07 54	Kurile Islands 47.07 N 154.29 E H = 14 55 58.1 h = normal MB = 4.9 D = 77.48 Az = 336.7 (USCGS) PV(A):0.8s 9.6nm MPV(A)=5.0
7.	+iP +eX	A A	21 27 52 28 05.5	Kurile Islands 47.12 N 154.11 E H = 21 15 59.0 h = normal MB = 5.0 D = 77.38 Az = 336.6 (USCGS) PV(A):1.6s 82.4nm MPV(A)=5.6 XV(A):1.8s 81.4nm
7.	e(P)	A	21 36 54	Kurile Islands 47.26 N 154.05 E H = 21 24 59.9 h = normal MB = 4.6 D = 77.24 Az = 336.5 (USCGS)
7.	+iP	A	23 48 47	Kurile Islands 47.32 N 154.02 E H = 23 36 53.3 h = normal MB = 5.0 D = 77.18 Az = 336.5 (USCGS) PV(A):1.2s 28.5nm MPV(A)=5.3
8.	ePKP	A	07 18 20	Fiji Region 18.11 S 177.52 W H = 06 58 44.2 h = 33 km MB = 4.5 D = 146.73 Az = 349 (ISC)
8.	eP	A	08 08 24	Hokkaido, Japan 41.97 N 140.57 E H = 07 56 25.2 h = 8 km MB = 5.1 D = 77.56 Az = 329.5 (USCGS)
8.	e(P)	A	11 22 08	Iceland 65.09 N 17.05 W H = 11 17 28.4 h = normal MB = 4.0 D = 20.74 Az = 120.4 (USCGS) PV(A):0.8s 7.7nm MPV(A)=4.1

Day	Phase		h m s	Remarks
8.	eiP	A	19 29 16.5	Andreanof Islands, Aleutian Is. 50.58 N 172.37 W H = 19 17 11.0 h = 15 km MB = 5.0 D = 79.10 Az = 357.4 (USCGS) PV(A):0.8s 25.0nm MPV(A)=5.3
8.	eiP	A	22 20 13.5	Nicobar Islands 6.70 N 93.55 E H = 22 08 07.3 h = normal MB = 5.2 D = 79.74 Az = 320.2 (USCGS) PV:1.2s 24.4nm MPV(A)=5.0
9.	eP	A	02 07 13.5	Greece-Albania Border Region 39.99 N 20.80 E H = 02 04 17.1 h = 53 km MB = 4.1 D = 12.45 Az = 331.9 (USCGS)
9.	e	A	06 19 07	Northern Italy 44.12 N 11.85 E H = 06 17 10.3 h = 0 km D = 6.53 Az = 359 (ISC)
9.	e	A	06 44 32	Central Italy 43.6 N 12.6 E H = 06 41 22 h = 10 km D = 7.11 Az = 355 (ISC)
9.	e	A	07 40 51	Northern Italy 44.31 N 11.71 E
	ei	A	41 03.5	H = 07 39 12.7 h = normal MB = 4.4
	ePg	A	41 16	D = 6.35 Az = 359.5 (USCGS)
	ei	A	41 22.5	PV(A):0.5s 11.5nm
	iSn	A	41 58	
9.	ePn	A	08 52 49.5	Northern Italy 44.31 N 11.58 E
	ePg	A	53 18	H = 08 51 15.6 h = normal MB = 4.2
	eSn	A	54 00	D = 6.34 Az = 0.2 (USCGS)
	eSg	A	54 49	PV(A):0.6s 9.6nm
9.	e	A	09 54 52	Central Italy 43.5 N 10.5 E H = 09 52 55 D = 7.19 Az = 6 (ISC)

Day	Phase	h m s	Remarks
10.	-eiPKIKP A	02 47 36	New Hebrides Islands 15.34 S 167.57 E H = 02 28 22.7 h = 141 km MB = 5.0 D = 139.79 Az = 336.3 (USCGS) PKIKPV(A):1.4s 25.6nm
10.	eP A	10 52 55	Near East Coast of Honshu, Japan 36.18 N 139.99 E H = 10 40 39.0 h = 67 km MB = 4.9 D = 82.33 Az = 329.8 (USCGS)
10.	eP A	17 17 55	Greenland Sea 73.30 N 6.25 E H = 17 12 53.4 h = normal MB = 4.7 D = 22.86 Az = 171.2 (USCGS) PV(A):1.8s 33.8nm MPV(A)=4.5
10.	ePKP A	18 04 08	Fiji Islands 17.25 S 178.55 W H = 17 45 29.2 h = 553 km MB = 4.8 D = 145.70 Az = 348.5 (USCGS) PKPV(A):2.0s 47.0nm
10.	ePKIKP A epPKIKP A	19 52 28 53 17.5	Banda Sea 5.92 S 130.69 E H = 19 34 05.7 h = 136 km MB = 5.7 D = 112.75 Az = 322.9 (USCGS) PKIKPV(A):1.8s 16.9nm pPKIKPV(A):1.6s 33.0nm
11.	eiPKHKP A +iPKP2 A +iX A i A LmH C LmV C	02 19 43 19 48 19 54 19 59 03 29.4 32.4	Tonga Islands 20.88 S 174.32 W H = 01 59 53.5 h = normal MB = 5.1 D = 149.93 Az = 352.5 (USCGS) PKHKPV(A):1.2s 48.7nm PKP2V(A):1.5s 93.0nm XV(A):1.5s 115.5nm LmH(C):20s 0.4/um MLH(C)=5.2 LmV(C):20s 0.3/um
11.	eX A LmH B LmV B	19 04 53 10.5 12.7	Southern Greece 37.62 N 22.65 E H = 19 01 17.5 h = 66 km MB = 5.0 D = 15.22 Az = 332.4 (USCGS) XV(A):0.7s 42.2nm

Day	Phase		h m s	Remarks
11.	e	A	21 11 51	Turkey 40.8 N 34.2 E
	e	A	12 04	H = 21 07 22 (BCIS)
	LmH	B	21.2	D = 18.5
	LmV	B	21.7	
12.	eP	A	02 01 24.5	Nepal 29.36 N 81.64 E
				H = 01 51 51.4 h = 44 km MB = 5.4
				D = 55.53 Az = 313.5
				PV(A):1.0s 27.6nm MPV(A)=5.2
13.	eP	A	03 25 45	Volcano Islands 24.48 N 141.07 E
	LmH	B	04 07.3	H = 03 12 52.4 h = 187 km MB = 5.6
	LmV	B	10.0	D = 92.92 Az = 330.5 (USCGS)
				PV(A):1.2s 16.3nm MPV(A)=5.3
				LmH(B):16.5 1.8/um
13.	e	A	12 28 18	Taiwan Region 21.15 N 121.13 E
	e	A	28 24	H = 12 15 18.4 h = 71 km MB = 4.8 (ISC)
				D = 85.6
13.	eSg	A	14 54 57	Federal Rep. Germany 48.3 N 9.0 E
				H = 14 53 27 (BCIS)
				D = 2.9
13.	iP	AB	15 56 17.8	Java Sea 5.95 S 113.03 E
	eiPP	ABC	16 00 36.5	H = 15 43 28.7 h = 636 km MB = 5.8
	esPP	BC	03 40	D = 101.81 Az = 320.4 (USCGS)
	eiSKS	BC	05 48	PV(A):1.6s 82.4nm MPV(A)=6.0
	eiS	BC	07 02	PV(B):4.5s 1.2/um MPV(B)=6.7
	esSS	BC	11 00	PPV(A):1.2s 93.5nm MPPV(A)=5.9
	eisSP	BC	12 37	PPV(B):5.5s 2.0/um MPPV(B)=6.5
	esSS	BC	17 55	SH(B):9.5s 1.7/um MSH(B)=6.1
	LmH	B	16 47.5	LmH(B):16s 2.2/um
	LmV	B	48.6	LmV(B):14s 1.5/um
13.	eP	A	16 51(31)	Off East Coast of Kamchatka
	e	A	51 35	54.70 N 163.54 E
	e	A	51 42.5	H = 16 40 05.2 h = normal MB = 4.7
				D = 72.43 Az = 341.7 (USCGS)

Day	Phase		h m s	Remarks
13.	eiSg	A	17 09 50.5	Federal Rep. Germany 48.3 N 9.0 E H = 17 08 17 (BCIS) D = 2.9
13.	eP	A	17 39 38.5	Southern Peru 15.94 S 71.67 W H = 17 26 16.9 h = 141 km MB = 5.4 D = 97.98 Az = 39.7 (USCGS)
14.	eP	A	11 30 41	Peru 9.88 S 75.57 W
	e	A	30 48	H = 11 17 16.1 h = 35 km MB = 5.9
	e	A	34 46	D = 95.78 Az = 39.7 (USCGS)
	e(SP)	BC	43 15	PV(A):1.4s 23.2nm MPV(A)=5.5
	LmH	B	12 11.5	LmH(B):20s 2.4 um MLH(B)=5.6
	LmV	B	11.5	LmV(B):20s 1.8 um MLV(B)=5.6
15.	eSg	A	01 10 21	Federal Rep. Germany 48.3 N 9.0 E H = 01 08 47 (BCIS) D = 2.9
15.	eP	A	07 13 36.5	Off Coast of Central America 12.86 N 90.21 W H = 07 00 54.0 h = 70 km MB = 4.5 D = 87.52 Az = 38.6 (USCGS)
15.	+eP	A	09 06 00	Sea of Okhotsk 48.29 N 146.48 E H = 08 55 09.0 h = 444 km MB = 4.8 D = 74.10 Az = 332.0 (USCGS) PV(A):1.5s 32.7nm MPV(A)=4.7
15.	eSg	A	12 05 34	Federal Rep. Germany 48.3 N 9.0 E H = 12 04 00 (BCIS) D = 2.9
15.	eP	A	12 50 21	Northern Celebes 0.01 S 122.94 E
	eX	A	54 29	H = 12 36 36.7 h = 154 km MB = 5.7
	ePP	A	54 43	D = 103.40 Az = 322.4 (USCGS)
	e	A	55 14	PV(A):1.2s 14.2nm MPV(A)=5.6
	ePKKP	A	13 06 15	XV(A):1.5s 22.6nm
				PPV(A):1.3s 45.8nm MPPV(A)=5.7

Day	Phase		h m s	Remarks
16.	eiSg	A	10 40 23	Federal Rep. Germany 48.3 N 9.0 E
				H = 10 38 57 (BCIS)
				D = 2.9
16.	ePP	A	16 13 11.5	Northern Celebes 1.12 N 120.21 E
	LmH	B	17 03.7	H = 15 55 04.9 h = 50 km MB = 5.4
	LmV	B	04.0	D = 100.83 Az = 322.1 (USCGS)
				PPV(A):1.4s 23.3nm MPPV(A)=5.4
				LmH(B):15s 0.6/um MLH(B)=5.2
				LmV(B):15s 0.6/um MLV(B)=5.2
16.	ePKIKP	A	21 55 07	South of Fiji Islands 25.15 S 178.29 E
	+iPKHKP	A	55 15.8	H = 21 36 22.5 h = 582 km MB = 5.3 (USCGS)
	-iPKP2	A	55 29.5	D = 152.3
	epPKIKP	A	57 26	PKIKPV(A):1.5s 50.2nm
				PKHKPV(A):1.4s 74.5nm
				PKP2V(A):1.6s 110.0nm
17.	eP	A	00 19 31	Greece-Albania Border Region
	LmH	C	25.6	39.38 N 20.50 E
	LmV	C	25.7	H = 00 16 25.4 h = 23 km MB = 4.7
				D = 12.89 Az = 333.8 (USCGS)
				PV(A):0.7s 19.2nm
17.	eP	A	03 05 19	Turkey 38.63 N 43.23 E
	eS	BC	10 06	H = 02 59 54.9 h = 35 km MB = 4.9
	e	C	10 12	D = 25.30 Az = 308.7 (USCGS)
	LmH	B	16.6	LmH(B):14s 0.7/um MLH(B)=4.3
	LmV	B	18.1	LmV(B):12s 0.7/um MLV(B)=4.6
17.	eP	A	04 54 26	Greece-Albania Border Region
	e	A	54 38	38.8 N 21.5 E
	LmH	B	05 01.0	H = 04 51 12 (BCIS)
	LmV	B	01.0	D = 13.8
17.	+iP	AB	05 59 29.5	Mindanao, Philippine Islands
	LmH	B	06 47.3	9.85 N 125.97 E

Day	Phase		h m s	Remarks
cont. 17.	LmV	B	06 48.7	H = 05 46 02.4 h = 72 km MB = 5.9 D = 97.34 Az = 324.2 (USCGS) PV(A):1.3s 56.7nm MPV(A)=5.9 LmH(B):13s 0.7/um MLH(B)=5.3 LmV(B):18s 0.6/um MLV(B)=5.1
17.	eP	A	07 34 36	Southern Italy 39.74 N 16.09 E
	iX	A	34 42	H = 07 32 00.9 h = 262 km MB = 4.4 D = 11.35 Az = 345.4 (USCGS) H = 07 32 2.6 39.83 N 15.92 E h = 273 km MB = 4.1 D = 11.23 Az = 346 (ISC) XV(A):0.8s 44.2nm
17.	e	A	09 10 54.5	Haiti Region 18.4 N 73.5 W H = 08 59 37.8 h = 77 km D = 72.95 Az = 42 (ISC)
17.	e(P)	A	16 22 22	Turkey 38.60 N 43.11 E H = 16 16 47.7 h = 9 km MB = 4.7 D = 25.24 Az = 308.8 (USCGS)
17.	ePKP	A	17 44 41.5	Loyalty Islands Region 22.24 S 170.44 E
	+iX	A	44 43	H = 17 25 01.0 h = 46 km MB = 4.9 D = 147.17 Az = 334.9 (USCGS) XV(A):1.7s 60.6nm
17.	ePKP	ABC	19 34 02	Loyalty Islands Region 22.20 S 170.51 E
	e	A	34 10	H = 19 14 21.3 h = 40 km MB = 5.6
	e	BC	34 24	D = 147.16 Az = 335.0 (USCGS)
	e	BC	34 55	PKPV(A):1.3s 30.6nm
	LmH	C	20 48.6	LmH(C):20s 0.4/um MLH(C)=5.1
	LmV	B	51.6	LmV(B):18s 0.7/um
18.	e(P)	A	02 19 18.5	Vancouver Island 50.25 N 129.79 W H = 02 07 40.3 h = 28 km MB = 4.7 D = 74.23 Az = 24.4 (USCGS)

Day	Phase		h m s	Remarks
18.	ePKIKP	A	07 18 23	Fiji Islands 19.53 S 177.84 W
	iPKHKP	A	18 27	H = 06 59 47.2 h = 598 km MB = 4.5
				D = 148.06 Az = 348.6 (USCGS)
				PKHKPV(A):1.2s 32.5nm
18.	eP	A	13 07 46.5	Rat Islands, Aleutian Is.
				52.06 N 175.50 E
				H = 12 56 00.0 h = 59 km MB = 5.0
				D = 76.77 Az = 349.5 (USCGS)
				PV(A):1.2s 20.3nm MPV(A)=5.1
18.	ePKP2	A	13 39 55	Loyalty Islands 22.29 S 170.56 E
				H = 13 20 08.1 h = normal
				D = 147.26 Az = 335.0 (USCGS)
18.	-iPKIKP	ABC	15 42 48	Fiji Islands 20.75 S 176.88 W
	iPKHKP	A	42 53	H = 15 23 33.7 h = 259 km MB = 5.8
	eipPKP	AB	43 56	D = 149.42 Az = 349.3 (USCGS)
	+sPKP	C	44 20	PKIKPV(A):1.5s 226.0nm
	+i	B	44 22	PKHKPV(A):1.7s 1408.0nm
	e	BC	45 58	
	epPP	C	47 40	
	eSKSP	BC	56 26	
	eSS	C	16 05 22	
	e	C	06 16	
	e	C	06 52	
18.	ePKIKP	A	16 43 13.5	South of Fiji Islands 22.94 S 176.21 W
	eiPKHKP	A	43 20	H = 16 23 36.2 h = 100 km MB = 5.2
	ePKP2	A	43 29	D = 151.68 Az = 349.5 (USCGS)
				PKIKPV(A):2.0s 51.3nm
				PKHKPV(A):1.4s 88.5nm
				PKP2V(A):1.6s 82.5nm
18.	eP	A	19 53 22	Dodecanese Islands 36.47 N 27.80 E
				H = 19 49 10.4 h = normal MB = 4.3
				D = 18.34 Az = 325.7 (USCGS)
				PV(A):1.6s 22.0nm MPV(A)=4.1

Day	Phase		h m s	Remarks
19.	eP	A	01 57 35	South Indian Ocean 36.21 S 53.11 E
				H = 01 44 18.3 h = normal MB = 5.4
				D = 93.89 Az = 335.0 (USCGS)
				PV(A):1.2s 26.4nm MPV(A)=5.4
19.	eP	A	01 58 38.5	Kurile Islands 47.27 N 153.90 E
				H = 01 46 45.7 h = normal MB = 4.7 (USCGS)
				D = 77.0
				PV(A):1.0s 21.6nm MPV(A)=5.2
19.	eP1	A	07 20 39.5	Eastern India 27.40 N 93.99 E
	iP2	A	20 42	H = 07 10 01.8 h = 18 km MB = 5.5
	eiP3	A	20 46	D = 64.66 Az = 315.7 (USCGS)
	eS	C	29 15	P2V(A):1.9s 197.0nm MP2V(A)=6.0
	eSS	C	33 32	P3V(A):1.6s 110.0nm MP3V(A)=5.8
	LmH	B	51.4	LmH(B):11.5s 2.9/um MLH(B)=5.7
	LmV	B	52.3	LmV(B):11s 3.2/um MLV(B)=5.8
19.	ePKIKP	A	11 07 31	Kermadec Islands 30.20 S 178.02 W
	ePKHKP	A	07 44	H = 10 47 34.4 h = 18 km MB = 5.6
	iPKP2	A	08 05.5	D = 158.33 Az = 343.2 (USCGS)
	eX	A	08 15	PKIKPV(A):1.8s 47.3nm
				PKHKPV(A):1.8s 27.0nm
				PKP2V(A):1.2s 77.2nm
				XV(A):1.5s 50.2nm
19.	eP	A	11 24 11.5	North of Severnaya Zemlya
	iX	A	24 17	83.48 N 115.05 E
				H = 11 16 23.5 h = normal MB = 4.8
				D = 41.48 Az = 290.8 (USCGS)
				PV(A):1.0s 19.7nm MPV(A)=4.8
				XV(A):1.1s 28.2nm
19.	ePKIKP	A	23 13 49.5	West New Guinea 4.50 S 140.11 E
	ePP	A	15 12	H = 22 55 03.5 h = 26 km MB = 5.7
				D = 117.08 Az = 326.0 (USCGS)
				PPV(A):1.7s 36.4nm MPPV(A)=5.8

Day	Phase		h m s	Remarks
20.	+iP	A	10 10 23	Dodecanese Islands 36.56 N 27.33 E H = 10 06 11.8 h = normal MB = 4.6 D = 18.05 Az = 326.2 (USCGS) PV(A):0.7s 19.2nm MPV(A)=4.3
20.	ePKHKP	A	10 53 15.5	Tonga Islands 21.17 S 174.33 W
	ePKP2	A	53 23	H = 10 33 25.4 h = normal MB = 4.9 D = 150.21 Az = 352.4 (USCGS) PKP2V(A):2.0s 76.9nm
20.	eP	A	13 00 05	Federal Rep. Germany 48.3 N 9.0 E H = 12 56 40.7 h = 0 km D = 14.19 Az= 332 (ISC)
20.	ePn	A	13 10 02	Federal Rep. Germany 48.3 N 9.0 E
	e	A	10 11	H = 13 09 17 (BCIS)
	iPg	A	10 15	D = 2.9
	eiSg	A	10 48	
20.	eP	A	20 23 43	Dodecanese Islands 36.56 N 27.18 E H = 20 19 30.3 h = 9 km MB = 4.8 D = 17.98 Az = 326.4 (USCGS)
21.	e	A	07 15 42	Ural Mountains Region 59.61 N 59.24 E
	e	A	21 10.5	H = 07 09 19.6 h = 1 km MB = 4.3 D = 28.09 Az = 272.8 (USCGS)
21.	ePn	A	11 15 38	explosion 50°35.2' N 14°03.2' E (PRU)
	iPg	A	15 41	D = 1.5
	e	A	15 57.5	
	iSg	A	15 59	
21.	e	A	17 10 49.5	Timor 8.62 S 124.06 E
	e	A	11 07.5	H = 16 52 00.4 h = 75 km MB = 5.3 D = 110.80 Az = 321.0 (USCGS)
21.	ePKP	A	22 12 03	Samoa Region 15.84 S 172.00 W H = 21 52 28.7 h = 33 km MB = 4.5 D = 145.16 Az = 356 (ISC)

Day	Phase		h m s	Remarks
22.	e	A	15 52 49.8	Crete 35.38 N 25.30 E
				H = 15 48 31.4 h = 42 km MB = 4.6
				D = 18.20 Az = 331.2 (USCGS)
22.	e	A	15 56 35	Crete 35.41 N 25.08 E
				H = 15 52 16.8 h = 37 km MB = 4.7
				D = 18.09 Az = 331.5 (USCGS)
22.	eP	A	20 19 33.3	Kurile Islands 45.70 N 151.58 E
				H = 20 07 36.5 h = 44 km MB = 4.1
				D = 77.97 Az = 335.2 (USCGS)
22.	eiP	A	22 41 39.3	Southwestern Ryukyu Is.
				24.95 N 124.02 E
				H = 22 29 18.8 h = 114 km MB = 5.1
				D = 84.08 Az = 323.7 (USCGS)
				PV(A):0.7s 15.3nm MPV(A)=5.0
22.	+iP	AB	23 46 08.3	Jan Mayen Island 71.12 N 8.63 W
	+i	A	46 14.3	H = 23 41 10.8 h = normal MB = 5.2
	eS	BC	50 14	D = 22.54 Az = 144.9 (USCGS)
	e	C	51 50	PV(A):1.6s 264.0nm MPV(A)=5.5
	LmH	B	56.8	LmH(B):13.5s 4.1/um MLH(B)=5.0
	LmV	B	57.7	LmV(B):13s 2.7/um MLV(B)=5.0
23.	eP	A	11 29 57.5	Southern Iran 27.82 N 54.53 E
	eS	C	36 15	H = 11 22 26.2 h = 20 km MB = 5.5
	e	C	39 01	D = 39.56 Az = 317.1 (USCGS)
	LmH	B	49.7	LmH(B):14s 4.6/um MLH(B)=5.5
	LmV	B	50.4	LmV(B):16s 4.0/um MLV(B)=5.5
23.	eP	A	13 01 04.5	South of Alaska 55.00 N 156.88 W
	epP	A	01 15.5	H = 12 49 29.1 h = normal MB = 5.1
				D = 74.29 Az = 7.6 (USCGS)
23.	ePg	A	14 44 04	France 49.08 N 6.46 E
	e	A	44 43	H = 14 42 53 (BCIS)
	eSg	A	44 49	D = 3.5

Day	Phase		h m s	Remarks
23.	ePKP	A	17 59 53	Fiji Islands 17.62 S 178.53 W
	ei	A	59 54	H = 17 41 16.7 h = 579 km MB = 5.1
	epPKP	A	18 02 07	D = 146.07 Az = 348.4 (USCGS)
23.	LmH	B	21 45.6	Probably Philippine Islands Region (USCGS)
	LmV	B	45.6	LmH(B):15s 1.1/um
24.	eP	A	00 44 49	North Atlantic Ocean 53.4 N 35.29 W
				H = 00 38 56 h = 32 km MB = 4.6 (ISC)
				D = 28.6
24.	+iPKHKP	A	00 56 15	Tonga Islands 22.46 S 174.42 W
	ePKP2	A	56 25	H = 00 36 24.3 h = 50 km MB = 5.3
				D = 151.47 Az = 351.9 (USCGS)
24.	-eP	AB	02 18 34	Szechwan Province, China 30.58 N 103.03 E
	eS	C	27 45	H = 02 07 36.8 h = normal MB = 5.9
	eSS	C	32 15	D = 67.87 Az = 316.6 (USCGS)
	eSSS	BC	35 08	PV(B):4.5s 1.4/um MPV(B)=6.5
	LmH	B	50.9	LmH(B):16s 4.8/um MLH(B)=5.8
	LmV	B	51.3	LmV(B):12s 5.1/um MLV(B)=6.0
24.	eP	A	06 52 34.5	Andaman Islands 13.40 N 93.31 E
				H = 06 40 58.5 h = 36 km MB = 5.1
				D = 74.51 Az = 319.2 (USCGS)
24.	eP1	AB	08 16 44	Gulf of Alaska 59.57 N 143.87 W
	+iP2	A	16 48.7	H = 08 05 39.6 h = 15 km MB = 5.0
	ei	A	17 32	D = 68.36 Az = 16.5 (USCGS)
	eS	BC	26 00	P1V(A):1.0s 11.8nm MP1V(A)=5.1
	e	BC	26 15	P2V(A):1.1s 24.2nm MP2V(A)=5.3
	eSS	C	31 20	LmV(B):14s 3.7/um MLV(B)=5.8
	LmV	B	53.9	LmH(B):13.5s 2.5/um MLH(B)=5.6
	LmH	B	54.2	
24.	-iPKIKP	A	15 27 37.8	Salomon Islands 7.07 S 155.61 E
	e	A	27 45.5	H = 15 08 55.5 h = 42 km MB = 5.3
				D = 127.26 Az = 331.9 (USCGS)
				PKIKPV(A):1.1s 26.2nm

Day	Phase		h m s	Remarks
24.	+iP	A	17 43 40.9	Unimak Island 53.81 N 163.09 W
				H = 17 31 55.7 h = 25 km MB = 4.4
				D = 75.82 Az = 3.5 (USCGS)
				PV(A):0.7s 15.3nm MPV(A)=5.2
25.	eP	A	08 09 05.7	Near East Coast of Honshu, Japan
	LmH	C	44.0	39.96 N 142.76 E
				H = 07 56 59.3 h = 57 km MB = 5.0
				D = 80.14 Az = 330.9 (USCGS)
				PV(A):1.2s 20.3nm MPV(A)=4.9
25.	+eP	A	10 33 25.2	Taiwan Region 24.06 N 122.25 E
	e	A	34 00	H = 10 20 59.4 h = 49 km MB = 5.2
	e	A	36 36.2	D = 83.82 Az = 323.2 (USCGS)
	e	A	36 39.7	PV(A):1.8s 74.3nm MPV(A)=5.6
	eS	BC	43 45	LmV(B):14s 5.1/um MLV(B)=6.1
	eSS	C	49 30	LmH(B):15.5s 4.9/um MLH(B)=6.0
	LmV	B	11 15.1	
	LmH	B	15.2	
25.	eP	A	14 40 56.6	Southern Nevada 37.04 N 116.00 W
				H = 14 28 38.0 h = 0 km MB = 5.2
				D = 81.29 Az = 30.7 (USCGS)
				Nevada test site "Cummarin"
				37°02'12.0" N 115°59'58.5" W (USAEC)
25.	eP	A	23 19 26.5	North Atlantic Ridge 28.81 N 43.27 W
	e	A	19 40.5	H = 23 11 01.9 h = normal MB = 4.7
	eS	C	26 20	D = 46.29 Az = 46.1 (USCGS)
	LmH	C	35.5	
26.	eP	A	03 17 06.5	Peru 13.65 S 74.01 W
				H = 03 03 40.7 h = 95 km MB = 5.4 (USCGS)
				D = 97.6
				PV(A):1.8s 30.4nm MPV(A)=5.6
26.	eP	A	06 10 18	Fox Islands, Aleutian Islands
	epP	A	10 28.5	51.21 N 170.00 W
				H = 05 58 18.5 h = normal MB = 4.4
				D = 78.51 Az = 359.0 (USCGS)

Day	Phase		h m s	Remarks
26.	iPn	A	14 45 39.5	Adelebsen/Solling, Fed.Rep.Germany
	iPg	A	45 42	51°37.00' N 9°45.00' E
	iSn	A	45 59.5	explosion, yield 5.850 to (Hannover)
	iSg	A	46 02.5	D = 1.5
26.	eP	A	15 42 18.5	Southern Nevada 37.12 N 116.07 W
				H = 15 30 00.0 h = 0 km MB = 5.3
				D = 81.24 Az = 30.7 (USCGS)
				Nevada test site "Yannigan"
				37°06'58.9" N 116°03'59.2" W (USAEC)
				PV(A):1.2s 24.4nm MPV(A)=5.1
26.	eP	A	16 03 09.5	Baguio 13.61 N 120.59 E
	ePP	AB	06 53.5	H = 15 50 11.0 h = 74 km MB = 5.3
	eS	C	14(00)	D = 91.20 Az = 323.0 (USCGS)
	eSS	C	20(15)	PV(A):1.4s 55.8nm MPV(A)=5.7
	LmH	B	45.9	LmH(B):16s 1.1 μm MLH(B)=5.4
	LmV	B	50.8	LmV(B):16s 0.6 μm MLV(B)=5.1
26.	eP	A	19 40 09.5	Nepal 27.67 N 85.92 E
	LmH	C	20 03.5	H = 19 30 07.7 h = normal MB = 5.2
	LmV	C	08.4	D = 59.40 Az = 314.6 (USCGS)
				PV(A):1.2s 28.4nm MPV(A)=5.2
				LmH(C):24s 0.5 μm MLH(C)=4.6
				LmV(C):14s 0.3 μm MLV(C)=4.6
26.	+iP	ABC	23 18 00.5	Kurile Islands 43.49 N 147.71 E
	+ipP	ABC	18 07.5	H = 23 06 00.0 h = 36 km MB = 5.4
	eS	C	27 52	D = 78.76 Az = 333.2 (USCGS)
	eSS	C	33 35	PV(A):1.4s 65.1nm MPV=5.5
	LmH	B	50.5	pPV(A):1.4s 172.0nm
	LmV	B	57.2	LmH(B):20s 6.4 μm MLH(B)=6.0
				LmV(B):17s 5.7 μm MLV(B)=5.7
26.	eP	A	23 31 25.5	Nepal 27.3 N 85.9 E
				H = 23 21 20 h = 31 km MB = 4.7
				D = 59.68 Az = 315 (ISC)

Day	Phase		h m s	Remarks
26.	+iP	AB	23 41 05.5	<u>Kurile Islands</u> 43.65 N 147.82 E
	ipP	A	41 18	H = 23 29 02.7 h = 15 km MB = 5.6
	LmH	B	24 13.9	D = 78.65 Az = 333.3 (USCGS)
	LmV	B	20.9	PV(A):1.2 83.3nm MPV(A)=5.7
				pPV(A):1.1s 92.6nm
				LmH(B):19.5s 9.0 μm MLH(B)=6.1
				LmV(B):15s 4.5 μm MLV(B)=6.0
26.	+iP	ABC	23 41 24.5	<u>Kurile Islands</u>
	ipP	A	41 38	H = 23 29 22 (UPP)
				PV(A):1.6s 159.5nm
				pPV(A):1.2s 81.2nm
26.	eP	A	24 01 18.5	<u>Kurile Islands</u> 43.48 N 147.60 E
				H = 23 49 17.1 h = 35 km MB = 4.6
				D = 78.73 Az = 333.2 (USCGS)
27.	eP	A	01 47 10.5	<u>Kurile Islands</u> 43.40 N 147.61 E
				H = 01 35 07.2 h = 20 km MB = 4.9
				D = 78.80 Az = 333.2 (USCGS)
				PV(A):1.2s 16.3nm MPV(A)=4.9
27.	eP	A	01 57 13.5	<u>Kurile Islands</u> 43.44 N 147.70 E
	e	A	57 25	H = 01 45 10.8 h = 15 km MB = 5.2
	e	A	57 45	D = 78.80 Az = 333.2 (USCGS)
				PV(A):1.5s 60.3nm MPV(A)=5.4
27.	eP	A	02 06 42	<u>Kurile Islands</u> 43.37 N 147.56 E
				H = 01 54 41.0 h = normal MB = 4.5
				D = 78.81 Az = 333.2 (USCGS)
				PV(A):1.0s 9.8nm MPV(A)=4.8
27.	eP	A	02 11 46.5	<u>Kurile Islands</u> 43.42 N 147.63 E
	epP	A	11 59	H = 01 59 44.8 h = 23 km MB = 4.6
	e	A	12 13	D = 78.79 Az = 333.2 (USCGS)
				PV(A):1.4s 18.6nm MPV(A)=4.9
27.	eP	A	02 12 31	<u>Kurile Islands</u>
	epP	A	12 44	H = 02 00 29 (UPP)

Day	Phase		h m s	Remarks
27.	eP	A	03 02 56.5	<u>Kurile Islands</u> 43.35 N 147.68 E H = 02 50 55.6 h = 38 km MB = 5.0 D = 78.87 Az = 333.2 (USCGS) PV(A):1.6s 38.5nm MPV(A)=5.2 XV(A):1.5s 55.3nm LmH(B):21s 2.3/um MLH(B)=5.5 LmV(B):15s 0.7/um MLV(B)=5.2
	eX	AB	03 16.5	
	LmH	B	35.4	
	LmV	B	41.9	
27.	ePKHKP	A	03 27 28	<u>South of Fiji Islands</u> 24.13 S 179.77 E H = 03 08 33.0 h = 545 km MB = 4.8 D = 151.97 Az = 343.8 (USCGS) PKHKP(A):1.2s 24.4nm
27.	eP	A	03 53 16	<u>Kurile Islands</u> 43.37 N 147.66 E H = 03 41 15.2 h = 37 km MB = 4.7 D = 78.85 Az = 333.2 (USCGS) PV(A):1.3s 17.5nm MPV(A)=5.0
27.	ePKHKP	A	05 05 03.5	<u>Tonga Islands</u> 22.16 S 174.86 W H = 04 45 12.4 h = 40 km MB = 4.7 D = 151.12 Az = 351.5 (USCGS) PKHKPV(A):0.7s 13.4nm
	ePKP2	A	07 14.5	
27.	eP	A	06 05 31.5	<u>Kurile Islands</u> 43.41 N 147.59 E H = 05 53 28.6 h = normal MB = 4.8 D = 78.79 Az = 333.2 (USCGS) PV(A):1.3s 17.5nm MPV(A)=4.9
	epP	A	05 44.5	
27.	e	A	07 18 50	<u>Kurile Islands</u> 4.33 N 147.71 E H = 05 53 25 h = 11 km MB = 4.9 D = 78.90 Az = 333 (ISC)
27.	-eP	ABC	07 20 02.5	<u>Andreanof Islands, Aleutian Is.</u> 50.13 N 179.56 W H = 07 07 58.1 h = 20 km MB = 6.0 D = 79.15 Az = 352.8 (USCGS) LmH(B):20s 5.1/um MLH(B)=5.9 LmV(B):16.5s 2.8/um MLV(B)=5.7
	ePP	BC	23 06	
	ePPP	BC	24 56	
	ePPPP	BC	26 28	
	eiS	C	30 02	
	e	B	30 04	
	ePS	B	30 48	

Day	Phase		h m s	Remarks
cont. 27.	eiSP	B	07 31 00	
	eSS	C	35 08	
	e	B	35 32	
	e(PKPPKP)	A	46 47	
	LmH	B	56.7	
	LmV	B	08 02.0	
27.	eP	A	07 29 56	Andreanof Islands, Aleutian Is. 50.12 N 179.78 W H = 07 17 51.7 h = 25 km MB = 5.0 D = 79.14 Az = 352.6 (USCGS)
27.	eP	A	08 02 32	Andreanof Islands, Aleutian Is. 50.02 N 179.66 W H = 07 50 28.7 h = 35 km MB = 4.8 (USCGS) D = 79.2
27.	eP	A	08 30 06	Andreanof Islands, Aleutian Is. 50.08 N 179.60 W H = 08 18 02.0 h = 18 km MB = 4.6 (USCGS) D = 79.2
27.	eP1	A	09 48 05	Kurile Islands 43.09 N 147.72 E
	eP2	A	48 08	H = 09 36 00.8 h = 21 km MB = 5.0
	eipP	A	48 19	D = 79.11 Az = 333.3 (USCGS) P2V(A):1.6s 27.5nm MP2V(A)=5.0 pPV(A):1.2s 24.4nm
27.	eP	A	09 51 24	Northwest of Kurile Islands 50.99 N 155.52 E H = 09 40 00.9 h = 130 km MB = 5.2 D = 74.19 Az = 337.1 (USCGS) PV(A):1.5s 22.6nm MPV(A)=4.7
27.	eP	A	10 01 23	Kurile Islands 43.38 N 147.54 E
	epP	A	01 35	H = 09 49 21.7 h = normal MB = 4.7 D = 78.80 Az = 333.2 (USCGS) PV(A):1.4s 14.0nm MPV(A)=4.8 pPV(A):1.2s 20.3nm

Day	Phase		h m s	Remarks
27.	eP	A	10 03 06	Kurile Islands 43.51 N 147.66 E
			03 17	H = 09 51 05.2 h = 37 km MB = 5.1
				D = 78.72 Az = 333.2 (USCGS)
				PV(A):1.3s 15.3nm MPV(A)=4.9
27.	eP	A	10 03 24	Kurile Islands
				H = 09 51 25 (UPP)
				PV(A):1.2s 16.2nm
27.	eP	A	10 17 50	Andreanof Islands, Aleutian Is.
				50.04 N 179.83 W
				H = 10 05 45.7 h = 24 km MB = 4.5
				D = 79.22 Az = 352.6 (USCGS)
27.	eP	A	10 25 54	Near West Coast of Honshu, Japan
	e	A	26 40	36.99 N 137.55 E
				H = 10 13 42.0 h = 25 km MB = 5.1
				D = 80.62 Az = 328.5 (USCGS)
				PV(A):0.7s 11.5nm MPV(A)=5.0
27.	eP	A	10 29 50	Kurile Islands 43.35 N 147.73 E
				H = 10 17 45.7 h = normal MB = 4.9 (USCGS)
				D = 78.8
27.	eP diff	A	11 01 52	Flores Sea 7.98 S 119.76 E
				H = 10 47 53.6 h = 183 km MB = 5.5
				D = 107.62 Az = 321 (ISC)
27.	ePKIKP	A	12 02 48	New Hebrides Islands 15.02 S 167.34 E
				H = 11 43 35.3 h = 130 km MB = 5.0
				D = 139.41 Az = 336.3 (USCGS)
27.	eP	A	13 12 43	South of Honshu, Japan 31.79 N 141.63 E
	eS	BC	23 16	H = 12 59 57.1 h = 13 km MB = 5.1
	e	C	23 20	D = 86.81 Az = 330.8 (USCGS)
	eSS	BC	29 00	LmH(B):14.5s 1.9 μm MLH(B)=5.6
	LmH	B	59.5	LmV(B):13s 1.6 μm MLV(B)=5.6
	LmV	B	59.6	

Day	Phase		h m s	Remarks
27.	eiPn	A	13 12 43	explosion
	ePg	A	13 06	D = ca. 5.8
	iSn	A	13 48	
	eSg	A	14 21	
27.	ePg	A	16 00 44	explosion, yield 29 to
	iSg	A	01 06.4	49°35.7' N 13°40.2' E
	e	A	01 0.7	(PRU)
				D = 1.7
27.	eP	A	17 13 54	<u>Kurile Islands</u> 43.74 N 147.96 E
	epP	A	13 05	H = 17 01 52.2 h = normal MB = 4.8
				D = 78.62 Az = 333.4 (USCGS)
28.	-iPKP	AC	05 17 47.4	<u>Samoa Islands</u> 16.30 S 171.93 W
	e	A	17 58	H = 04 58 10.4 h = normal MB = 5.5
				D = 145.62 Az = 356.0 (USCGS)
28.	-iP1	AB	11 04 06.8	<u>Andreanof Islands, Aleutian Is.</u>
	iP2	AB	04 10	52.71 N 175.10 W
	P2m	A	04 13	H = 10 52 31.2 h = 162 km MB = 6.1
	i	B	04 14	D = 76.86 Az = 355.6 (USCGS)
	ipP	B	04 52	P2mV(A):1.7s 2270.0nm MP2mV(A)=6.6
	iPP	B	07 07	LmH(B):16.5s 16.0 μm
	ePPP	A	08 57.3	LmV(B):14.5s 10.8 μm
	iPPP	B	09 00	
	eS	A	13 42	
	-iS	B	13 46	
	e	B	14 50	
	-isS	B	14 57	
	iSS	C	18 30	
	eSSS	C	24 00	
	e	A	11 30 55	
	ePKPPKP	A	31 02	
	e	A	31 15	
	LmH	B	45.3	
	LmV	B	45.5	

Day	Phase		h m s	Remarks
28.	e	A	14 04 10	El Salvador 13.1 N 89.3 W
				H = 13 51 23 h = 33 km MB = 4.2
				D = 86.78 Az = 39 (ISC)
28.	eP	A	20 06 26.5	Southern Iran 27.83 N 56.32 E
	e	A	06 30	H = 19 58 48.1 h = 35 km MB = 5.5
	e	A	07 14.5	D = 40.64 Az = 316.6 (USCGS)
	eS	C	12 28	PV(A):1.4s 23.2nm MPV(A)=4.7
	LmH	B	23.6	LmH(B):20s 1.2 µm MLH(B)=4.8
	LmV	B	30.4	LmV(B):11s 0.7 µm MLV(B)=4.9

Day	Phase		h m s	Remarks
1.	eP	A	20 20 00	Iran 33.96 N 58.91 E H = 20 12 44.8 h = 39 km MB = 5.2 D = 38.03 Az = 310.6 (USCGS)
1.	eP	A	21 26 54.5	Central Mid-Atlantic Ridge 0.81 N 26.79 W H = 21 16 53.9 h = normal MB = 4.7 D = 59.36 Az = 27.4 (USCGS)
2.	ePKP e	A A	01 16 20 16 24	New Hebrides Islands 15.29 S 167.60 E H = 00 57 12.6 h = 151 km MB = 4.9 D = 139.76 Az = 336.4 (USCGS) PKIKPV(A):1.3s 13.1nm
2.	ePKHKP ePKP2	A A	02 10 50 11 01	Tonga Islands 22.53 S 174.40 W H = 01 50 57.7 h = normal MB = 4.9 D = 151.54 Az = 351.9 (USCGS) PKHKPV(A):0.9s 23.4nm
2.	e	A	02 18 25	Probably Tonga Region (ISC)
2.	e	A	03 03 18	Probably Tonga Region (ISC)
2.	eP	A	05 09 33.5	Fox Islands, Aleutian Is. 52.78 N 167.45 W H = 04 57 42.4 h = normal MB = 4.3 D = 76.94 Az = 0.6 (USCGS) PV(A):1.5s 15.1nm MPV(A)=4.9
2.	eP	A	11 16 37.5	North of Ascension Island 3.17 S 12.20 W H = 11 06 45.1 h = normal MB = 5.0 D = 57.37 Az = 17.8 (USCGS) PV(A):1.4s 25.6nm MPV(A)=5.2

Day	Phase		h m s	Remarks
2.	ePKIKP	A	15 45 09.5	South of Fiji Islands 22.04 S 179.85 W
	+iPKHKP	A	45 15	H = 15 26 31.8 h = 623 km MB = 4.4
	ePKP2	A	45 24.5	D = 150.07 Az = 345.3 (USCGS)
				PKHKPV(A):1.0s 27.6nm
3.	eP	A	15 03 33	South of Honshu, Japan 32.28 N 140.71 E
				H = 14 50 58.9 h = 62 km MB = 4.7
				D = 95.91 Az = 330 (ISC)
3.	eP	A	16 39 03	Greece 38.26 N 20.59 E
				H = 16 35 44.8 h = 39 km MB = 4.5
				D = 13.93 Az = 335.6 (USCGS)
				PV(A):0.7s 7.7nm
4.	e	A	01 06 02	Probably South of Fiji Islands
4.	iP	A	01 30 45	Mindoro, Philippine Islands
	epP	A	30 57.5	13.45 N 120.38 E
				H = 01 17 44.5 h = 61 km MB = 5.2
				D = 91.20 Az = 322.9 (USCGS)
				PV(A):1.4s 32.6nm MPV(A)=5.5
4.	eP	A	01 55 56.5	Crete 34.47 N 26.36 E
				H = 01 51 30.1 h = 41 km MB = 4.5
				D = 19.42 Az = 330.8 (USCGS)
				PV(A):0.8s 11.5nm MPV(A)=4.2
4.	eP	A	03 44 43	South of Mariana Islands 12.13 N 143.71 E
	e	A	48 05	H = 03 30 35.4 h = normal MB = 6.2
	e	A	48 56	D = 104.85 Az = 330.7 (USCGS)
	ePP	BC	49 16	PV(A):2.0s 42.7nm MPV(A)=6.0
	eSKS	BC	55 18	LmH(B):16s 6.6 μm MLH(B)=6.1
	ePS	BC	58 10	LmV(B):18s 7.4 μm MLV(B)=6.2
	eSS	BC	04 03 50	
	LmH	B	34.1	
	LmV	B	37.1	

Day	Phase		h m s	Remarks
4.	ePKIKP	ABC	06 50 30.5	<u>Fiji Islands</u> 19.76 S 178.64 W
	-iPKHKP	A	50 35	H = 06 31 56.2 h = 624 km MB = 5.5 (USCGS)
	+iX	A	50 36	D = 148.3
	iPKP2	AC	50 40.5	XV(A):1.4s 214.0nm
	ipPKP	C	52 52	
	+i	A	52 56	
	esPKP	C	53 52	
	ePKKP	B	59 28	
	e	C	07 03 06	
	ePSKS	C	04 16	
	eSPP	C	06 00	
	e	B	06 18	
	eSS	B	12 16	
	eSS	C	12 18	
	+i	B	12 28	
	esSS	C	16 18	
4.	ePKP2	A	08 43 20.5	<u>Fiji Islands</u> 19.89 S 178.61 W
				H = 08 24 37.5 h = 625 km MB = 4.6
				D = 148.26 Az = 347.6 (USCGS)
4.	LmH	B	15 30.2	Probably <u>Philippine Islands</u> (USCGS)
	LmV	B	30.3	LmH(B):18s 3.6 μm
				LmV(B):18s 3.4 μm
5.	eP	A	05 00 53.5	<u>North Atlantic Ocean</u> 53.92 N 19.71 W
	i	A	00 55	H = 04 56 25.1 h = normal MB = 4.6
	eS	BC	04 32	D = 19.36 Az = 86.9 (USCGS)
	LmH	B	08.3	PV(A):1.2s 20.3nm MPV(A)=4.2
	LmV	B	08.9	
5.	iPg	A	15 05 30.5	<u>Dorheim/Hessen</u>, explosion
	eP	A	05 32.5	50°57.87' N 09°13.11' E
	eSg	A	05 51	H = 15 05 01.11 yield 7.0 to (Hannover-Buchholz)
				D = 1.5

Day	Phase		h m s	Remarks
6.	ePn	A	01 55 32.5	Federal Rep. Germany 47.8 N 11.1 E
	ePg	A	55 39	H = 01 54 47 (BCIS)
	eSn	A	56 07	D = 2.9
	eSg	A	56 20	
6.	ePKP2	A	22 00 31	Fiji Islands 20.98 S 176.43 W
				H = 21 41 08.2 h = 293 km MB = 4.4
				D = 149.72 Az = 349.8 (USCGS)
7.	ePn	A	08 02 14	Switzerland 46.3 N 7.5 E
	e	A	02 28.5	H = 08 01 07 (BCIS)
	ePg	A	02 43	D = 5.1
	i	A	02 46	
	i	A	02 51	
	e	A	03 07	
	eSg	A	03 52.5	
7.	eP	A	12 06 57.5	Central Mid-Atlantic Ridge
				0.47 S 19.67 W
				H = 11 57 09.4 h = normal MB = 4.6
				D = 57.47 Az = 23.1 (USCGS)
				PV(A):1.6s 30.2nm MPV(A)=5.1
7.	ePKHKP	A	17 19 49	Tonga Islands 22.57 S 175.17 W
				H = 16 59 57.9 h = normal MB = 5.1
				D = 151.48 Az = 350.9 (USCGS)
				PKHKPV(A):1.6s 27.5nm
7.	ePKHKP	A	17 31 45	Tonga Islands 22.43 S 174.41 W
	ePKP2	A	31 57	H = 17 11 52.3 h = normal MB = 5.2
				D = 151.45 Az = 352.0 (USCGS)
				PKHKPV(A):1.3s 35.0nm
7.	eP	A	19 04 39	Near East Coast of Honshu, Japan
				37.06 N 141.38 E
				H = 18 52 23.5 h = 60 km MB = 4.7
				D = 82.13 Az = 330.4 (USCGS)

Day	Phase		h m s	Remarks
8.	e	A	03 28 56	Probably <u>Northwest Russia</u>, explosion 67.6 N 32.8 E H = 03 23 42 (UPP) D = 20.0
8.	eP ePP	A A	12 05 22.5 09 40	<u>Molucca Passage</u> 1.80 N 126.60 E H = 11 51 21.8 h = 50 km MB = 5.6 (USCGS) D = 104.2 PV(A):1.0s 7.9nm MPV(A)=5.5
8.	ePKP	A	18 20 09.5	<u>Fiji Islands</u> 17.88 S 178.60 W H = 18 01 39.5 h = 640 km MB = 4.8 D = 146.30 Az = 348.3 (USCGS) PKPV(A):1.3s 13.1nm
8.	eP	A	19 03 04	<u>Alberta Province, Canada</u> 55.04 N 116.62 W H = 18 52 13.9 h = 9 km MB = 5.1 D = 66.19 Az = 33.1 (USCGS)
8.	eP	A	19 12 24.5	<u>Fox Islands, Aleutian Is.</u> 52.24 N 169.65 W H = 19 00 31.0 h = 32 km MB = 4.4 (USCGS) D = 77.3 PV(A):1.3s 13.1nm MPV(A)=4.8
9.	eP eS LmH LmV	AB BC B B	01 02 17 12 24 43.7 44.0	<u>Off East Coast of Honshu, Japan</u> 39.57 N 143.43 E H = 00 50 03.2 h = 14 km MB = 5.3 (USCGS) D = 80.7 PV(A):1.7s 57.5nm MPV(A)=5.3 LmH(B):14s 3.6 μm MLH(B)=5.9 LmV(B):13s 2.4 μm MLV(B)=5.7
9.	-eiPKIKP +iX -i ePP	ABC A B B	16 20 39 20 42 21 13 24 00	<u>New Hebrides Islands</u> 19.01 S 168.60 E H = 16 01 10.5 h = 41 km MB = 6.1 (USCGS) D = 143.6 XV(A):1.3s 698.0nm

Day	Phase		h m s	Remarks
cont. 9.	eSKKS	B	16 30 40	LmH(B):20s 19.3/um MLH(B)=6.8
	ePKKS	B	32 40	LmV(B):20s 11.9/um
	e	BC	34 18	
	ePPS	C	36 26	
	e	A	39 12	
	eSS	BC	42 30	
	LmH	B	17 28.8	
	LmV	B	28.8	
9.	ePKIKP	A	16 46 32.5	<u>New Hebrides Islands</u>
				H = 16 27 03 (UPP)
				PKIKPV(A):1.3s 54.6nm
9.	+eiPKIKP	A	18 50 25.5	<u>New Hebrides Islands</u> 19.09 S 168.47 E
				H = 18 30 55.7 h = 33 km MB = 5.3
				D = 143.55 Az = 335.1 (USCGS)
				PKIKPV(A):1.5s 62.8nm
10.	+iP	ABC	05 10 22.7	<u>Kurile Islands</u> 44.80 N 148.87 E
	eiS	BC	20 08	H = 04 58 26.3 h = 40 km MB = 6.0 (USCGS)
	LmH	B	43.3	D = 77.8
	LmV	B	49.4	PV(A):1.6s 960.0nm MPV(A)=6.7
				PV(B):4s 5.0/um MPV(B)=7.0
				SH(B):6.5s 2.8/um MSH(B)=6.5
				LmH(B):19s 10.2/um MLH(B)=6.2
				LmV(B):16s 6.7/um MLV(B)=6.1
10.	eiP	A	05 31 02	<u>Burma</u> 26.80 N 97.05 E
				H = 05 20 10.3 h = normal MB = 5.4 (USCGS)
				D = 67.0
10.	eP	A	06 25 07	<u>Luzon, Philippine Islands</u>
	epP	A	25 11	12.60 N 122.13 E
	LmH	B	07 04.7	H = 06 11 56.4 h = 30 km MB = 5.7 (USCGS)
	LmV	B	16.3	D = 92.9
				PV(A):1.3s 32.8nm MPV(A)=5.6
				pPV(A):1.7s 84.8nm
				LmH(B):21s 1.9/um MLH(B)=5.5
				LmV(B):13s 0.9/um MLV(B)=5.5

Day	Phase		h m s	Remarks
10.	ePKP2	A	07 40 10.5	Tonga Islands 19.77 S 174.90 W
				H = 07 20 18.3 h = normal MB = 4.9 (USCGS)
				D = 148.6
10.	-iP	A	08 31 11	Mid-Indian-Rise 15.36 S 67.24 E
	ei	A	31 15.5	H = 08 18 53.8 h = normal MB = 5.3
	e	A	31 21	D = 81.73 Az = 327.9 (USCGS)
11.	-iP	ABC	22 49 54.5	Kodiak Island 57.47 N 153.92 W
	+ipP	AB	50 05	H = 22 38 34.6 h = 29 km MB = 6.0 (USCGS)
	e	A	59 09	D = 71.5 h = 39 km
	eS	AB	59 13	PV(A):1.6s 830.0nm MPV(A)=6.5
	eSS	BC	23 03 50	PV(B):3.5s 2.4 µm MPV(B)=6.8
	e	C	08 40	pPV(A):1.8s 858.0nm
	LmH	B	20.6	SH(B):10s 4.2 µm MSH(B)=6.6
	LmV	B	20.6	LmH(B):23s 5.3 µm MLH(B)=5.7
				LmV(B):23s 3.8 µm MLV(B)=5.6
12.	ePKHKP	A	16 49 20	Fiji Islands 20.68 S 179.19 W
				H = 16 30 40.14 h = 615 km MB = 4.7
				D = 148.90 Az = 346.6 (USCGS)
				PKHKPV(A):0.8s 26.9nm
12.	eP	A	18 21 21	Yunnan Province, China 24.19 N 102.79 E
				H = 18 09 53.6 h = normal MB = 5.2
				D = 72.40 Az = 318.1 (USCGS)
				PV(A):1.5s 32.2nm MPV(A)=5.3
13.	iPg	A	15 05 20	Hilders/Rhön 50°32.71'N 10°02.44'E
	i	A	05 20.8	H = 15 05 00.6 yield 21.8 to (Hannover)
	iSg	A	05 35	D = 1.0
13.	ePKP	A	18 30 03	New Hebrides Islands 19.02 S 168.39 E
				H = 18 10 30.3 h = 17 km MB = 4.8
				D = 143.45 Az = 335.1 (USCGS)
				PKPV(A):1.5s 20.1nm

Day	Phase		h m s	Remarks
13.	eiP	A	18 35 39	Burma-India-Border-Region
				24.90 N 93.94 E
				H = 18 24 54.0 h = 62 km MB = 4.9
				D = 66.42 Az = 316.5 (USCGS)
				PV(A):1.0s 15.8nm MPV(A)=4.9
14.	eP1	ABC	01 57 19.5	Turkey-Iran Border Region
	eP2	A	57 22	38.60 N 44.71 E
	eS	BC	02 01 52	H = 01 51 44.4 h = 23 km MB = 5.3
	LmH	B	10.8	D = 26.23 Az = 308.1 (USCGS)
	LmV	B	10.8	P1V(B):10.5s 0.6/um MP1V(B)=5.0
				P2V(A):1.5s 22.6nm MP2V(A)=4.6
				SH(B):16s 1.8/um MSH(B)=5.3
				LmH(B):14s 2.7/um MLH(B)=4.9
				LmV(B):13.5s 4.0/um MLV(B)=5.2
14.	eP	A	07 42 13.5	North Atlantic Ridge 28.31 N 43.79 W
	eS	BC	49 09	H = 07 33 43.2 h = normal MB = 5.2
	LmH	B	59.0	D = 46.96 Az = 45.8 (USCGS)
	LmV	B	59.0	PV(A):2.0s 59.8nm MPV(A)=5.3
				SH(B):14s 0.5/um MSH(B)=5.1
				LmH(B):18s 0.8/um MLH(B)=4.7
				LmV(B):16s 1.0/um MLV(B)=5.1
14.	eP	A	08 23 46	North Atlantic Ridge 28.39 N 43.84 W
				H = 08 15 16.3 h = normal MB = 4.9
				D = 46.94 Az = 45.9 (USCGS)
				PV(A):2.0s 42.7nm MPV(A)=5.1
14.	eP	A	13 03 23	Kurile Islands 45.52 N 150.94 E
	e	A	03 27	H = 12 51 26.5 h = normal MB = 5.3
	LmH	B	41.0	D = 77.95 Az = 334.9 (USCGS)
	LmV	B	41.0	PV(A):1.2s 48.7nm MPV(A)=5.4
				LmH(B):20s 1.2/um MLH(B)=5.2
				LmV(B):20s 0.8/um MLV(B)=5.1
14.	e	A	15 51 26	Pyrenees 42.46 N 1.90 E
	e	A	51 29.5	H = 15 48 10.2 h = normal MB = 4.3 (USCGS)
	eSg	A	53 56	D = 10.6
	LmV	B	55.0	LmV(B):4.5s 0.9/um

Day	Phase		h m s	Remarks
14.	ePKHKP	A	21 08 25	Fiji Islands 19.61 S 178.19 W
				H = 20 49 46.1 h = 610 km MB = 5.1
				D = 148.07 Az = 348.2 (USCGS)
				PKHKP(A):1.3s 65.5nm
15.	e	A	05 42 46	Ryukyu Islands 26.36 N 129.48 E
	LmH	B	06 18.5	H = 05 29 56.1 h = 37 km MB = 5.3
	LmV	B	25.2	D = 85.78 Az = 325.6 (USCGS)
				LmH(B):18s 0.9 µm MLH(B)=5.2
				LmV(B):15s 1.1 µm MLV(B)=5.4
15.	eP diff	A	12 53 22	Chile-Argentinia Border Region
	ePKP	A	57 31.5	29.65 S 69.50 W
	ePP	ABC	57 46	H = 12 39 17.8 h = 119 km MB = 6.0
	epP	B	58 14	D = 107.10 Az = 41.2 (USCGS)
	e	A	58 18	PV(A):1.8s 30.4nm MPV(A)=6.1
	esPP	B	58 26	LmH(B):19s 0.4 µm
	e	AC	58 30	LmV(B):19s 0.4 µm
	eSKKS	BC	13 04 50	
	e	BC	06 16	
	e	BC	07 40	
	ePKKP	A	09 01.5	
	epPKKP	A	09 34	
	eSS	BC	13 38	
	LmH	B	42.0	
	LmV	B	42.0	
15.	ePKP2	A	17 52 46	Fiji Region 21.4 S 179.62 W
	e	A	52 51.5	H = 17 32 57.5 h = 33 km MB = 4.9 (ISC)
				D = 149.8
16.	eP	A	03 56 38.5	Tibet 33.86 N 86.30 E
				H = 03 47 06.4 h = 52 km MB = 4.9
				D = 55.42 Az = 311.8 (USCGS)
16.	e	A	06 29 58	Pyrenees 42.4 N 1.7 E
	eSg	A	32 24	H = 06 26 43 (BCIS)
				D = 10.8

Day	Phase		h m s	Remarks
16.	ePKP	A	16 44 56.5	New Hebrides Islands 19.22 S 168.50 E
	e	A	45 01	H = 16 25 22.1 h = 8 km MB = 4.9
	LmH	B	17 42.2	D = 143.67 Az = 335.0 (USCGS)
	LmV	B	18 00.2	PKPV(A):1.5s 22.6nm
				LmH(B):21s 1.0/um MLH=5.5
				LmV(B):17s 0.5/um
17.	ePKP2	A	01 37 54	South of Fiji Islands 26.27 S 177.07 W
				H = 01 17 47.7 h = 85 km MB = 4.9
				D = 154.76 Az = 347.0 (USCGS)
				PV(A):1.2s 18.3nm
17.	e	A	17 03 45	Yugoslavia 41.21 N 20.96 E
	e	A	03 48	H = 17 00 55.1 h = normal MB = 4.6
	LmH	B	07.5	D = 11.45 Az = 328.6 (USCGS)
	LmV	B	08.4	
17.	eP	A	17 19 22.5	Kyushu, Japan 31.43 N 130.64 E
				H = 17 07 14.1 h = 129 km MB = 5.1
				D = 82.18 Az = 325.8 (USCGS)
				PV(A):1.8s 40.6nm MPV(A)=4.9
17.	eX	A	22 11 29	Gulf of Alaska 59.20 N 147.90 W
	LmH	C	37	H = 22 00 12.4 h = 47 km MB = 5.1
				D = 69.25 Az = 13.8 (USCGS)
				XV(A):1.1s 52.4nm
17.	eP	A	23 27 07.5	Iran 33.88 N 59.72 E
	LmH	B	47.0	H = 23 19 42.3 h = 19 km MB = 5.0
	LmV	B	50.5	D = 38.60 Az = 310.6 (USCGS)
				PV(A):1.3s 43.7nm MPV(A)=5.0
18.	eP	A	17 23 25	Cyprus 34.49 N 32.50 E
	LmH	C	39.7	H = 17 18 29.0 h = 33 km MB = 4.5
	LmV	C	42.0	D = 22.18 Az = 323.0 (USCGS)
				PV(A):1.2s 32.5nm MPV(A)=4.6

Day	Phase		h m s	Remarks
19.	eP	A	15 33 01	Andreanof Islands, Aleutian Is.
				50.02 N 179.65 W
				H = 15 20 54.6 h = 15 km MB = 5.4
				D = 79.25 Az = 352.7 (USCGS)
				PV(A):1.0s 27.6nm MPV(A)=5.3
19.	eP1	ABC	23 45 23.5	Near Islands, Aleutian Is.
	iP2	A	45 26	51.32 N 173.76 E
	iP3	A	45 28.5	H = 23 33 29.1 h = 16 km MB = 5.8
	ePP	B	48 16	D = 77.28 Az = 348.5 (USCGS)
	eS	BC	55 16	P1V(A):1.4s 51.2nm MP1V(A)=5.3
	eSoS	C	55 44	P1V(B):11s 6.0/um MP1V(B)=6.6
	e	BC	58 08	P2V(A):1.3s 132.0nm MP2V(A)=5.8
	eSS	BC	24 00 12	P3V(A):1.2s 212.0nm MP3V(A)=6.1
	e	A	12 13	PPV(B):12s 2.5/um MPPV(B)=6.2
	e	A	12 18	LmH(B):15s 10.2/um MLH(B)=6.3
	e	A	12 27	LmV(B):14.5s 6.1/um MLV(B)=6.1
	LmH	B	32.9	
	LmV	B	35.4	
20.	eP	A	06 04 58	Kodiak Island 56.73 N 151.68 W
				H = 05 53 33.3 h = 18 km MB = 4.4
				D = 72.11 Az = 11.1 (USCGS)
				PV(A):1.1s 16.1nm MPV(A)=5.1
20.	eP	A	11 21 18	Mindanao, Philippine Islands
	LmH	C	12 09.3	5.00 N 125.22 E
	LmV	B	10.3	H = 11 07 35.6 h = 65 km MB = 5.7
				D = 100.79 Az = 323.6 (USCGS)
				PV(A):1.2s 16.3nm MPV(A)=5.6
				LmH(C):20.5s 1.1/um
				LmV(B):16s 0.5/um
20.	ePKP	A	11 37 26.5	New Hebrides Islands 18.57 S 169.10 E
				H = 11 18 22.7 h = 247 km MB = 4.7
				D = 143.32 Az = 335.9 (USCGS)
				PKPV(A):1.3s 32.8nm

Day	Phase		h m s	Remarks
20.	e(PKP)	A	22 53 08	Off Coast of Southern Chile
	LmV	B	23 48.4	44.99 S 80.29 W
	LmH	B	48.8	H = 22 34 16.9 h = normal MB = 5.1
				D = 123.94 Az = 50.1 (USCGS)
				LmV(B):18s 1.6/um MLV(B)=5.7
				LmH(B):17s 1.4/um MLH(B)=5.7
21.	e	A	02 52 45	Gulf of California 25.61 N 109.67 W
	LmV	B	03 33.0	H = 02 39 46.6 h = 5 km MB = 5.0
	LmH	B	33.2	D = 88.11 Az = 33.0 (USCGS)
				LmV(B):16s 0.9/um MLV(B)=5.3
				LmH(B):16s 0.8/um MLH(B)=5.2
21.	eiPKHKP	A	04 32 04	Fiji Islands 20.23 S 176.75 W
	ePKP2	A	32 09.5	H = 04 12 55.2 h = 350 km MB = 4.6
				D = 148.93 Az = 349.7 (USCGS)
				PKHKPV(A):1.7s 66.7nm
21.	e	A	10 23 17	Volcano Islands 24.05 N 142.70 E
	LmH	B	11 06.8	H = 10 09 55.4 h = normal MB = 5.4
				D = 94.02 Az = 331.2 (USCGS)
				LmH(B):15s 0.8/um MLH(B)=5.4
21.	e(PKP)	A	15 08 06.5	New Hebrides 19.1 S 168.7 E
				H = 14 48 26 h = 0 km MB = 4.7
				D = 78.70 Az = 334 (ISC)
				PKPV(A):1.2s 14.2nm
21.	eP	A	17 45 54	Kurile Islands 44.17 N 149.09 E
				H = 17 33 54.8 h = 45 km MB = 4.6
				D = 78.59 Az = 334.0 (USCGS)
21.	iPn	A	20 41 04	Federal Rep. Germany 48.46 N 9.15 E
	iPg	A	41 14.5	H = 20 40 21.3 h = 14 km MB = 4.1
	iSg	A	41 51	D = 2.72 Az = 35.3 (USCGS)
	LmH	B	42.1	LmH(B):6s 3.6/um MLH(B)=3.8
	LmV	B	42.1	LmV(B):5s 2.1/um

Day	Phase	h m s	Remarks
21.	ePKHKP A	22 52 12.5	Tonga Islands 21.04 S 174.00 W
			H = 22 32 23.7 h = normal MB = 5.0
			D = 150.12 Az = 352.8 (USCGS)
			PKHKPV(A):1.2s 20.3nm
22.	eiPKHKP A	08 44 23.5	Fiji Islands 21.59 S 176.12 W
	e A	44 34	H = 08 24 52.9 h = 190 km MB = 4.7
			D = 150.37 Az = 350.0 (USCGS)
			PKHKPV(A):1.3s 26.2nm
22.	eP A	10 07 53	Morocco 34.3 N 3.7 W
			H = 10 03 21 (BCIS)
			D = 19.7
22.	eP A	10 53 03	Szechwan Province, China
			31.69 N 103.81 E
			H = 10 42 05.8 h = 16 km MB = 5.0
			D = 67.52 Az = 316.5 (USCGS)
			PV(A):1.2s 14.2nm MPV(A)=5.1
22.	ePKIKP A	23 21 15.5	West of Macquarie Island
	e A	21 32	53.13 S 140.83 E
			H = 23 01 33.2 h = normal MB = 5.0
			D = 149.09 Az = 286.2 (USCGS)
			PKIKPV(A):1.8s 27.0nm
23.	eP1 AB	00 32 43.5	Honshu, Japan 40.14 N 140.19 E
	iP2 A	32 45	H = 00 20 54.7 h = 146 km MB = 5.7
	-ipP ABC	33 21	D = 79.00 Az = 329.5 (USCGS)
	e A	35 30	P2V(A):1.5s 77.9nm MPV(A)=5.2
	iS ABC	42 29	SH(B):8s 1.5/um MSH(B)=6.1
	eSS C	47 35	LmH(B):16s 2.1/um
	eSSS C	50 58	LmV(B):11s 0.7/um
	LmH B	01 05.3	
	LmV B	15.0	

Day	Phase		h m s	Remarks
23.	+eP	ABC	02 02 38	<u>India</u> 21.68 N 73.02 E
	e	A	02 44	H = 01 52 59.3 h = 3 km MB = 5.4
	e	A	02 50.5	D = 55.49 Az = 317.3 (USCGS)
	e	A	02 55.5	PV(A):1.2s 85.4nm MPV(A)=5.7
	eS	C	10 25	LmV(C):12s 0.7/um MLV(C)=5.0
	eSS	C	14 35	
	LmH	C	28.0	
	LmV	C	33.0	
23.	eiP	ABC	12 27 02.5	<u>Ryukyu Islands</u> 29.81 N 129.26 E
	i	A	27 14	H = 12 14 53.5 h = 148 km MB = 5.8
	ePP	BC	30 16	D = 82.84 Az = 325.4 (USCGS)
	e	BC	31 02	PV(A):1.6s 314.0nm MPV(A)=5.8
	eS	C	37 10	PV(B):5s 1.5/um MPV(B)=6.0
	ePS	B	38 19.5	LmH(B):16s 1.0/um
	ePS	C	38 20.	LmV(B):13s 0.7/um
	LmH	B	13 10.3	
	LmV	B	11.1	
23.	iP	A	20 59 10	<u>Greece-Albania Border Region</u>
	LmH	B	21 05.4	39.12 N 20.48 E
	LmV	B	05.5	H = 20 56 00.7 h = 10 km MB = 4.9
				D = 13.12 Az = 334.4 (USCGS)
				PV(A):0.8s 28.8nm MPV(A)=5.0
				LmH(B):12s 2.7/um
				LmV(B):12s 2.9/um
23.	eP	A	23 17 18	<u>Southern Nevada</u> 37.08 N 116.02 W
				H = 23 05 00.0 h = 0 km MB = 5.5
				D = 81.26 Az = 30.7 (USCGS)
				37°05'10.4" N 116°01'15.9" W
				Nevada test site "Shaper" (USAEC)
				PV(A):1.2s 28.4nm MPV(A)=5.4
24.	ePKIKP1	ABC	10 54 15.5	<u>Western Australia</u> 21.98 S 126.68 E
	ePKIKP2	A	54 18	H = 10 35 22.1 h = normal MB = 6.2
	ePP	ABC	55 54	D = 122.50 Az = 316.9 (USCGS)
	ePKKP	A	11 04 11	PKIKP1V(A):1.7s 60.6nm

Day	Phase		h m s	Remarks
cont. 24.	ePcPPKP	A	11 08 09	PKIKP2V(A):1.6s 98.8nm
	e	B	11 35	PKKPV(A):2.0s 42.7nm
	LmH	B	48.5	LmH(B):19s 1.2/um MLH(B)=5.6
	LmV	B	12 04.5	LmV(B):19s 1.1/um MLV(B)=5.5
24.	ePKHKP	A	13 24 25	Fiji Islands 20.68 S 178.73 W
	ePKP2	A	24 31.5	H = 13 05 43.8 h = 610 km MB = 4.1
				D = 149.00 Az = 347.2 (USCGS)
				PKHKPV(A):1.2s 20.4nm
24.	eP	A	23 58 28	Easthern China 26.20 N 104.87 E
				H = 23 47 00.7 h = 8 km MB = 4.9
				D = 72.18 Az = 318.1 (USCGS)
				PV(A):1.8s 20.3nm MPV(A)=5.0
25.	eP	A	12 32 38.5	Kodiak Island 56.67 N 152.08 W
	e	A	32 44	H = 12 21 11.7 h = 11 km MB = 4.9
				D = 72.21 Az = 10.8 (USCGS)
				PV(A):0.6s 15.3nm
25.	eP	A	17 19 04	Mediterranean Sea 34.8 N 25.0 E
				H = 17 14 48 (BCIS)
				D = 18.5
26.	eP	A	15 33 00	Andreanof Islands, Aleutian Is.
				50.71 N 175.00 W
				H = 15 20 53.7 h = 4 km MB = 4.7 (USCGS)
				D = 78.5
26.	+iP	AB	19 12 17.5	Southern Nevada 37.30 N 116.53 W
	+iPP	AB	15 23	H = 19 00 00.2 h = 0 km MB = 6.5 (USCGS)
	LmV	B	50.4	37°18'01.7" N 116°32'02.8" W
	LmH	B	50.5	Nevada test site "Handley" (USAEC)
				D = 81.3
				PV(A):1.3s 299.0nm MPV(A)=6.3
				LmV(B):15s 3.2/um MLV(B)=5.8
				LmH(B):17s 2.5/um MLH(B)=5.4

Day	Phase		h m s	Remarks
27.	eX	A	03 19 21	Norwegian Sea 72.85 N 3.8 E
				H = 03 14 13.2 h = 0 km (ISC)
				D = 22.3
				XV(A):1.2s 14.2nm
27.	eP	A	04 39 18.5	Near West Coast of Colombia
	e	A	39 24	5.57 N 77.63 W
				H = 04 26 42.3 h = 28 km MB = 5.2 (USCGS)
				D = 84.3
				PV:1.3s 39.3nm MPV(A)=5.5
27.	ePKP	A	04 53 37	New Hebrides Islands 19.11 S 168.50 E
				H = 04 34 03.4 h = 16 km MB = 4.4
				D = 143.57 Az = 335.1 (USCGS)
				PKPV(A):1.3s 39.3nm
27.	eP	A	05 10 45.5	Eastern Kazakh SSR 49.76 N 78.01 E
	e	A	12 28	H = 05 02 56.8 h = 0 km MB = 5.2 (USCGS)
				D = 41.3
				Probably underground explosion
				PV(A):0.9s 19,5nm MPV(A)=4.8
27.	ePKP	A	05 26 40	New Hebrides Islands 19.09 S 168.51 E
				H = 05 07 06.6 h = 11 km (USCGS)
				D = 143.6
27.	ePn	A	14 05 14	explosion yield 18.2 to (PRU)
	iPg	A	05 15	49°27.7' N 13°22.6' E
	eSn	A	05 37	D = 1.6
	iSg	A	05 37.5	
27.	eP1	A	18 50 37	Northern Celebes 0.37 N 119.35 E
	eP2	A	50 48.5	H = 18 36 45.8 h = 8 km MB = 6.2 (USCGS)
	eP3	A	50 52.5	D = 99.3
	ePP	B	54 52	P1V(A):1.6s 27.5nm MP1V(A)=5.5
	iSKS	BC	19 01 15	P2V(A):1.3s 30.5nm MP2V(A)=5.7
	eS	B	02 12	P3V(A):1.8s 94.6nm MP3V(A)=6.0
	eSP	BC	03 40	PPV(B):12s 3.1 μm MPPV(B)=6.6

Day	Phase		h m s	Remarks
cont. 27.	iPPS	B	19 04 20	LmH(B):20s 37.0 /um MLH(B)=6.9
	e	A	07 21.5	LmV(B):17s 19.3 /um MLV(B)=6.7
	eiSS	BC	09 32	
	LmH	B	39.8	
	LmV	B	49.5	
27.	e	A	23 52 22	<u>Dodekanese Islands</u> 36.03 N 28.36 E
				H = 23 47 51.3 h = 0 km (ISC)
				D = 18.9
28.	eP	A	05 29 31.5	<u>South Indian Ocean</u> 35.27 S 53.99 E
				H = 05 16 16.7 h = 21 km MB = 5.1 (USCGS)
				D = 93.3
28.	ePKIKP	A	08 04 55.5	<u>Solomon Islands</u> 6.26 S 154.63 E
	e	A	05 02	H = 07 45 59.9 h = 64 km MB = 5.9 (USCGS)
	ePP	A	06 52	D = 126.2
	eSKKP	A	17 54	PKIKPV(A):1.5s 30.2nm
	eSS	C	24 05	SKKPV(A):2.0s 51.3nm
	eSSS	C	28 55	
	LmH	B	51.4	
	LmV	B	05.6	
28.	eP	A	09 54 25.5	<u>Lake Baikal Region</u> 52.23 N 105.80 E
	e	A	54 33.5	H = 09 44 57.8 h = normal MB = 5.2
	e	A	55 30	D = 54.59 Az = 308.8 (USCGS)
	i	BC	10 14 58	LmH(B):14.5s 10.7 /um MLH(B)=6.1
	LmH	B	20.7	LmV(B):12s 9.1 /um MLV(B)=6.1
	LmV	B	21.0	
28.	ePKP	A	10 17 41	<u>Loyalty Islands</u> 22.36 S 170.36 E
				H = 09 57 52.8 h = 4 km (USCGS)
				D = 147.3
28.	e	A	16 44 59	<u>Yugoslavia</u> 43.3 N 19.0 E
	e	A	46 29	H = 16 42 51 h = 44 km (ISC)
	e	A	46 32	D = 9.7

Day	Phase		h m s	Remarks
28.	eP	A	18 00 57	Unimak Island Region 53.71 N 163.65 W H = 17 49 10.2 h = 23 km MB = 4.5 (USCGS) D = 75.8 PV(A):1.2s 12.7nm MPV(A)=4.9
28.	-eP1	ABC	21 06 21.5	Turkey 39.18 N 29.49 E
	iP2	AB	06 25	H = 21 02 23.4 h = 20 km MB = 6.0 (USCGS)
	eS	B	09 24	D = 17.0
	LmV	B	14.0	P1V(A):2.4s 525.0nm MP1V(A)=5.2
	LmH	B	14.1	P2V(A):1.6s 1485.0nm MP2V(A)=5.9 LmV(B):14s 924.0 µm MLV(B)=7.3 LmH(B):15s 1262.0 µm MLH(B)=7.3
28.	eP	A	21 45 18	Turkey 38.99 N 29.26 E H = 21 41 18.6 h = normal MB = 4.6 (USCGS) D = 17.1 PV(A):1.8s 54.0nm MPV(A)=4.4
28.	eP	A	22 03 08	Turkey 39.23 N 29.29 E H = 21 59 10.9 h = 17 km MB = 4.8 D = 16.89 Az = 318.3 (USCGS)
28.	eP	A	22 11 50	Turkey H = 22 07 48 (UPP) PV(A):1.8s 47.3nm
28.	eP	A	22 09 34	Turkey 38.78 N 29.79 E H = 22 05 27.8 h = 7 km MB = 4.6 (USCGS) D = 17.5 PV(A):1.6s 35.7nm MPV(A)=5.3
28.	eP	A	22 44 14	Turkey 39.13 N 29.29 E
	e	A	44 21.5	H = 22 40 14.9 h = normal MB = 4.4 D = 16.96 Az = 318.5 (USCGS) PV(A)=1.8s 33.8nm MPV(A)=4.2

Day	Phase		h m s	Remarks
28.	eP1	A	23 15 40.5	Turkey 39.22 N 29.52 E
	iP2	A	15 44	H = 23 11 44.0 h = 37 km MB = 5.2 (USCGS)
	e	A	19 27	D = 17.0
				P2V(A):2.7s 959.0nm MPV(A)=5.5
28.	eP	A	23 32(15)	Turkey 39.35 N 29.34 E
	eX	A	32 27	H = 23 28 26.6 h = 39 km MB = 4.5
				D = 16.83 Az = 318.0 (USCGS)
				XV(A):2.0s 85.5nm
28.	eP	A	23 48 02.5	Turkey 39.09 N 29.78 E
	e	A	48 08.5	H = 23 43 58.6 h = 22 km MB = 5.1 (USCGS)
	e	A	48 28	D = 17.2
				PV(A):2.2s 284.0nm MPV(A)=5.0
29.	eP	A	01 32 57	Turkey 39.02 N 29.79 E
				H = 01 28 52.6 h = normal MB = 4.3 (USCGS)
				D = 17.2
29.	eP	A	01 51 25	Turkey
				H = 01 47 17 (UPP)
29.	e(P)	A	02 09 26	Turkey 39.26 N 29.17 E
				H = 02 05 25.0 h = 22 km MB = 4.7 (USCGS)
				D = 16.9
29.	eP	A	02 30 53.5	Mindoro, Philippine Islands
				13.90 N 120.68 E
				H = 02 18 01.2 h = 118 km MB = 5.3
				D = 91.02 Az = 323.0 (USCGS)
				PV(A):1.5s 35.2nm MPV(A)=5.3
29.	e	A	02 44 46	Turkey 38.95 N 29.46 E
				H = 02 40 34.0 h = normal MB = 4.5 (USCGS)
				D = 17.1
29.	eP	A	02 49 47	Turkey 39.29 N 29.34 E
				H = 02 45 50.3 h = normal MB = 4.5 (USCGS)
				D = 16.9

Day	Phase		h m s	Remarks
29.	eP	A	02 58 52	Turkey 39.14 N 29.76 E H = 02 54 50.8 h = 25 km MB = 4.3 (USCGS) D = 17.2 PV(A):1.8s 33.8nm MPV(A)=4.2
29.	eP	A	03 14 48	Turkey 39.09 N 30.09 E H = 03 10 45.0 h = normal MB = 4.4 D = 17.41 Az = 317.6 (USCGS) PV(A):1.8s 33.8nm MPV(A)=4.2
29.	eP	A	03 57 01	Southern Sinkiang, Prov. China
	LmH	B	04 14.3	39.56 N 75.37 E
	LmV	B	17.0	H = 03 48 47.4 h = normal MB = 5.1 D = 45.03 Az = 306.2 (USCGS) PV(A):1.2s 20.3nm MPV(A)=4.9 LmH(B):15s 2.1 μm MLH(B)=5.2 LmV(B):11s 1.4 μm MLV(B)=5.2
29.	eP1	AB	07 00 26	Turkey 39.02 N 29.73 E
	iP2	AB	00 30	H = 06 56 21.9 h = 15 km MB = 5.3 (USCGS)
	iS	BC	03 48	D = 17.2
	LmH	B	06.4	P1V(A):1.5s 100.0nm MP1V(A)=4.7
	LmV	B	08.3	P2V(A):2.0s 299.0nm MP2V(A)=5.1 P2V(B):8s 2.2 μm MP2V(B)=5.3 LmH(B):16s 9.6 μm MLH(B)=5.1 LmV(B):12s 9.6 μm MLV(B)=5.3
29.	eP	A	09 56 14	Turkey 39.19 N 29.20 E H = 09 52 14.9 h = 23 km MB = 4.4 D = 16.88 Az = 318.5 (USCGS)
29.	ePKP	A	10 27 21.5	New Hebrides Islands 17.07 S 168.51 E
	iX	AB	27 26	H = 10 08 20.3 h = 232 km MB = 6.0
	epPKP	BC	28 18	D = 141.73 Az = 336.2 (USCGS)
	e	BC	29 47	PKPV(A):1.2s 28.5nm
	iPP	ABC	30 42	XV(A):1.5s 120.5nm
	iSKP	A	30 57	PPV(A):2.0s 535.0nm MPPV(A)=6.2
	eSKKS	BC	37 04	SKPV(A):1.8s 311.0nm
	eSS	C	48 50	

Day	Phase		h m s	Remarks
29.	e	A	12 18 53	Flores Sea 7.24 S 120.59 E H = 12 00 32.6 h = 480 km MB = 5.3 (USCGS) D = 107.5
29.	+iP	A	14 41 11.8	Turkey 38.87 N 27.92 E
	e	A	41 16	H = 14 37 16.6 h = 27 km MB = 4.6
	LmH	B	46.9	D = 16.47 Az = 320.9 (USCGS)
	LmV	B	48.3	PV(A):1.4s 69.7nm MPV(A)=4.6
29.	eP1	A	14 44 20	Turkey 28.73 N 28.00 E
	eP2	A	44 23	H = 14 40 26.6 h = 47 km MB = 4.5
	LmH	B	50.2	D = 16.62 Az = 321 (ISC)
	LmV	B	50.5	P1V(A):1.4s 11.6nm MP1V(A)=3.8 P2V(A):1.7s 60.6nm MP2V(A)=4.5 LmH(B):12s 1.6/um MLH(B)=4.5 LmV(B):12s 1.8/um MLV(B)=4.6
29.	eP1	A	18 09 51.5	Ionian Sea 37.66 N 20.09 E H = 18 06 23.1 h = 41 km MB = 4.4 D = 14.32 Az = 337.7 (USCGS)
29.	eP	AB	19 15 42.5	Turkey 39.12 N 29.33 E
	PmV	A	15 48	H = 19 11 39.6 h = 6 km MB = 4.8
	LmH	B	23.0	D = 16.99 Az = 318.5 (USCGS) PmV(A):1.7s 121.0nm MPmV(A)=4.8 LmH(B):15s 1.0/um MLH(B)=4.2
29.	eP	A	19 43 21	Ryukyu Islands 28.84 N 129.94 E
	e	A	43 29	H = 19 30 51.4 h = 23 km MB = 5.4
	eS	BC	53 42	D = 83.98 Az = 325.7 (USCGS)
	LmH	B	20 24.8	PV(A):1.8s 47.3nm MPV(A)=5.4
	LmV	B	24.9	LmH(B):17s 2.6/um MLH(B)=5.7 LmV(B):16s 2.3/um MLV(B)=5.7
29.	eP	A	22 16 12	Talaud Islands 4.92 N 125.97 E
	epP	A	16 38.5	H = 22 02 28.0 h = 97 km MB = 5.2
	LmH	B	23 00.0	D = 101.30 Az = 323.7 (USCGS) h = 100 km

Day	Phase		h m s	Remarks
30.	eP	A	00 19 44	Turkey 38.96 N 29.73 E H = 00 15 44.5 h = 55 km D = 17.31 Az = 318 (ISC) PV(A):2.3s 61.0nm MPV(A)=4.3
30.	e(P)	A	01 27 44	Turkey 39.18 N 29.6 E H = 01 23 37.2 h = 0 km MB = 4.1 D = 17.09 Az = 318 (ISC)
30.	eP	A	06 50 21	Turkey 38.97 N 28.80 E H = 06 46 24.4 h = 23 km MB = 4.5 D = 16.83 Az = 319.5 (USCGS) PV(A):1.6s 16.5nm MPV(A)=3.9
30.	eP1	A	06 52 58.5	Turkey 39.30 N 29.26 E
	iP2	A	53 02.5	H = 06 49 04.2 h = normal MB = 4.7
	LmH	B	58.8	D = 16.81 Az = 308.2 (USCGS)
	LmV	B	07 01.2	P1V(A):0.7s 11.5nm MP1V(A)=4.1 P2V(A):2.0s 102.5nm MP2V(A)=4.6 LmH(B):14.5s 1.4/um MLH(B)=4.3 LmV(B):11s 1.3/um MLV(B)=4.6
30.	eP1	ABC	08 03 46	Turkey 39.32 N 29.22 E
	eP2	A	03 51	H = 07 59 50.7 h = 17 km MB = 5.2
	eS	B	06 56	D = 16.79 Az = 318.2 (USCGS)
	eS	C	07 00	P1V(A):1.4s 14.0nm MP1V(A)=3.9
	LmH	B	09.6	P1V(B):9s 1.0/um MP1V(B)=5.0
	LmV	B	11.8	P2V(A):2.8s 300.0nm MP2V(A)=4.9 LmH(B):15.5s 9.3/um MLH(B)=5.1 LmV(B):10s 5.5/um MLV(B)=5.2
30.	eP	A	08 34 53	Hokkaido, Japan 43.23 N 145.70 E H = 08 22 57.4 h = 60 km MB = 4.8 D = 78.31 Az = 332.2 (USCGS)
30.	eP	A	08 39 10	Turkey 39.15 N 28.88 E
	e	A	39 14.5	H = 08 35 16.2 h = 27 km MB = 4.6
	e	A	39 19	D = 16.74 Az = 319.0 (USCGS)

Day	Phase		h m s	Remarks
cont. 30.	LmV	B	08 47.2	LmV(B):13s 1.1 um MLV(B)=4.5
	LmH	B	47.3	LmH(B):13s 1.7 um MLH(B)=4.6
30.	e	A	09 30 24	Turkey 39.01 N 29.39 E
				H = 09 26 09.4 h = 21 km MB = 4.5 (USCGS)
				D = 17.1
30.	e	A	13 52 17	Turkey 39.26 N 29.12 E
				H = 13 48 05.4 h = normal MB = 4.5
				D = 16.78 Az = 318.4 (USCGS)
30.	eP1	A	16 36 36	Turkey 39.09 N 29.62 E
	eiP2	ABC	36 37.5	H = 16 32 33.4 h = 11 km MB = 4.9
	eS	B	40 00	D = 17.16 Az = 318.2 (USCGS)
	LmH	B	42.5	P2V(A):3.2s 645.0nm MP2V(A)=5.2
	LmV	B	44.5	LmH(B):16s 8.4 um MLH(B)=5.1
				LmV(B):13s 5.4 um MLV(B)=5.1
30.	eP1	ABC	17 00 24.5	Mindanao, Philippine Islands
	iP2	A	00 26	6.80 N 126.65 E
	ePP	B	04 44	H = 16 46 45.6 h = 76 km MB = 5.9
	iSKS	BC	10 56	D = 100.19 Az = 324.1 (USCGS)
	eS	B	11 52	P1V(A):2.0s 77.0nm MPV(A)=5.9
	i	C	12 16	P1V(B):12s 4.3 um MP1V(B)=6.9
	-iSP	BC	13 18	P2V(A):1.2s 114.0nm MP2V(A)=6.4
	eSS	C	18 44	PPV(B):14s 3.8 um MPPV(B)=6.8
	+iPSPS	BC	19 24	SH(B):12s 6.1 um MSH(B)=6.9
	iSSS	C	23 09	LmH(B):22s 47.6 um MLH(B)=7.0
	LmH	B	49.9	LmV(B):20s 28.8 um MLV(B)=6.8
	LmV	B	54.3	
30.	eX	A	20 42 09	Turkey 38.98 N 29.62 E
	LmH	B	50.8	H = 20 38 00.9 h = 5 km MB = 4.5
	LmV	B	51.0	D = 17.25 Az = 308.4 (USCGS)
				XV(A):1.7s 42.4nm
				LmH(B):8.5s 1.0 um MLH(B)=4.4
				LmV(B):8s 0.7 um MLV(B)=4.4

Day	Phase		h m s	Remarks
30.	ePKP2	A	21 01 48.5	**Auckland Islands Region** 49.62 S 164.31 E H = 20 40 50.1 h = normal MB = 5.4 (USCGS) D = 162.8 PKP2V(A):1.3s 13.1nm
30.	eP	A	21 03 22	**Turkey** 39.23 N 29.28 E H = 20 59 30.3 h = normal MB = 4.7 (USCGS) D = 16.8
31.	eP	A	00 24 00.5	**Southwestern Ryukyu Islands** 24.35 N 123.08 E
	LmH	B	01 17.6	H = 00 11 36.5 h = 70 km MB = 5.0
	LmV	B	19.9	D = 84.05 Az = 323.4 (USCGS) LmH(B):12.5s 0.8 μm LmV(B):12s 0.6 μm
31.	eP	A	00 55 36	**Turkey** 39.37 N 29.33 E H = 00 51 35.6 h = 14 km MB = 4.7 D = 16.80 Az = 317.9 (USCGS)
31.	eP	AB	03 50 53.5	**Turkey** 39.06 N 29.95 E
	e	B	54 37	H = 03 46 48.3 h = 15 km MB = 4.8
	LmH	B	56.8	D = 17.36 Az = 317.8 (USCGS)
	LmV	B	58.7	PV(A):2.3s 183.0nm MPV(A)=4.8 LmH(B):16s 2.7 μm MLH(B)=4.6 LmV(B):12s 2.4 μm MLV(B)=4.8
31.	eP	A	04 14 04	**Turkey** 39.02 N 29.09 E H = 04 10 07.3 h = normal MB = 4.3 D = 16.94 Az = 319.0 (USCGS) PV(A):1.7s 21.2nm MPV(A)=4.0
31.	eP	A	12 02 02	**Turkey** 38.98 N 29.68 E H = 11 57 58.0 h = 26 km MB = 4.7 D = 17.28 Az = 318.3 (USCGS) PV(A):2.5s 107.6nm MPV(A)=4.5

Day	Phase		h m s	Remarks
31.	ePKIKP	A	15 45 52	Fiji Islands 21.20 S 178.60 W
	ePKHKP	A	45 57.5	H = 15 27 09.5 h = 550 km MB = 5.0
	iPKP2	A	46 05	D = 149.53 Az = 347.1 (USCGS)
				PKIKPV(A):1.6s 19.2nm
				PKHKPV(A):1.4s 116.0nm
				PKP2V(A):1.5s 55.3nm
31.	eP	A	18 29 57	Chagos Archipelago Region
	e	A	30 14.5	3.82 S 69.72 E
	e	A	30 28	H = 18 18 25.2 h = normal MB = 5.5
	eS	C	39 20	D = 73.46 Az = 325.7 (USCGS)
	eSS	C	43 59	PV(A):1.4s 32.6nm MPV(A)=5.2
	LmH	B	55.0	LmH(B):30s 3.1 µm MLH(B)=5.4

Day	Phase		h m s	Remarks
1.	eP	A	03 32 06.5	Gulf of Alaska 59.44 N 144.70 W
				H = 03 20 57.0 h = normal MB = 4.2 (USCGS)
				D = 68.7
				PV(A):1.0s 7.9nm MPV(A)=4.9
1.	LmH	B	11 53.3	Probably Atlantik (USCGS)
	LmV	B	53.9	LmH(B):16s 0.4/um
				LmV(B):18s 0.4/um
1.	eiP	AB	14 35 27	Honshu, Japan 39.77 N 141.79 E
	ipP	A	35 46	H = 14 23 25.1 h = 81 km MB = 5.8
	esP	A	35 57	D = 79.93 Az = 330.4 (USCGS)
	eS	BC	45 24	PV(A):1.4s 111.8nm MPV(A)=5.6
	LmH	B	15 14.7	LmH(B):13.5s 1.6/um MLH(B)=5.5
	LmV	B	17.8	LmV(B):11s 0.9/um MLV(B)=5.7
				pPV(A):2.0s 102.5nm
				sPV(A):1.2s 40.7nm
1.	eP	A	16 00 00	Turkey 39.39 N 29.24 E
	LmH	B	05.7	H = 15 56 02.1 h = 16 km MB = 4.8
	LmV	B	09.3	D = 16.74 Az = 318.0 (USCGS)
				PV(A):2.1s 105.0nm MPV(A)=4.6
				LmH(B):17s 1.3/um MLH(B)=4.2
				LmV(B):8s 0.6/um MLV(B)=4.4
1.	e	A	24 01 39.5	Southern Iran 28.00 N 56.74 E
				H = 23 54 05.6 h = 62 km MB = 5.1
				D = 40.77 Az = 316.3 (USCGS)
2.	eP	A	00 32 31.5	Turkey 39.15 N 29.40 E
	LmH	B	40.2	H = 00 28 32.2 h = 28 km MB = 4.4
	LmV	B	41.8	D = 17.00 Az = 318.3 (USCGS)
				PV(A):1.8s 33.8nm MPV(A)=4.2
2.	eiPKIKP	ABC	11 31 30	Tonga Islands 20.36 S 173.91 W
	LmH	B	12 40.7	H = 11 11 42.0 h = normal MB = 5.7

Day	Phase		h m s	Remarks
cont. 2.	LmV	B	12 46.5	D = 149.47 Az = 353.1 (USCGS) PKIKPV(A):2.0s 184.0nm LmH(B):21s 1.1/um MLH(B)=5.6 LmV(B):18s 0.9/um
2.	eP	A	20 39 10	Turkey 39.05 N 29.12 E
	LmH	B	45.2	H = 20 35 09 h = 35 km MB = 4.4
	LmV	B	47.1	D = 17.24 Az = 318 (ISC) PV(A):1.2s 12.2nm MPV(A)=3.9 LmH(B):16s 1.3/um MLH(B)=4.2 LmV(B):12s 1.0/um MLV(B)=4.4
2.	eP	A	21 29 33	Sicily 38.29 N 14.28 E H = 21 26 39.5 h = 226 km MB = 4.6 (ISC) D = 12.6
3.	ePKP	AB	07 12 20	Tonga Islands 20.51 S 174.05 W
	eX	A	12 23	H = 06 52 33.8 h = 39 km MB = 5.7 (USCGS)
	LmH	B	08 29.3	D = 149.7
	LmV	B	28.0	XV(A):1.7s 69.7nm LmH(B):17s 1.7/um MLH(B)=5.8 LmV(B):19s 1.9/um
3.	eP	A	07 20 33	Mid-Indian Rise 20.54 S 67.89 E H = 07 07 51.1 h = normal MB = 5.1 (USCGS) D = 86.3 PV(A):1.6s 22.0nm MPV(A)=5.1
3.	eP	A	12 20 47.5	Turkey 38.95 N 29.5 E H = 12 16 45.8 h = 0 km MB = 4.2 D = 17.21 Az = 319 (ISC) PV(A):1.5s 20.1nm MPV(A)=4.0
3.	eP	A	14 10 54	Andreanof Islands, Aleutian Is. 51.91 N 175.32 W H = 13 59 02.5 h = 57 km MB = 5.0 D = 77.65 Az = 355 (ISC) PV(A):1.0s 11.8nm MPV(A)=5.0

Day	Phase		h m s	Remarks
3.	eP	A	21 00 32.5	Iran-USSR Border Region 37.07 N 54.58 E
	eipP	A	00 38.5	H = 20 53 54.5 h = 43 km MB = 5.2 (USCGS)
	LmH	B	17.0	D = 33.3
	LmV	B	17.0	PV(A):2.0s 42.7nm MPV(A)=5.0
				pPV(A):1.7s 69.7nm
3.	eP	A	23 23 44	Turkey 38.90 N 29.70 E
				H = 23 19 38 h = 3 km MB = 4.3
				D = 17.35 Az = 318 (ISC)
				PV(A):1.8s 20.3nm MPV(A)=4.0
4.	eP	A	11 05 16.5	Iran-USSR Border Region 37.00 N 59.42 E
	eX	A	05 20.5	H = 10 58 08.9 h = 8 km MB = 4.8 (USCGS)
	LmH	B	23.3	D = 36.4
	LmV	B	26.2	PV(A):1.1s 22.2nm MPV(A)=4.9
				XV(A):1.1s 26.2nm
				LmH(B):16s 0.8μm MLH(B)=4.6
				LmV(B):12s 0.3μm MLV(B)=4.4
4.	eP	A	13 11 30.5	Off W. Coast of Northern Sumatra
				2.33 N 95.12 E
				H = 12 59 00.2 h = 29 km MB = 5.1 (USCGS)
				D = 84.1
				PV(A):1.0s 11.8nm MPV(A)=5.1
4.	-iPKP	ABC	23 05 45	Fiji Islands 16.58 S 177.32 W
				H = 22 46 51.8 h = 394 km MB = 5.2 (USCGS)
				D = 145.27
				PKPV(A):1.4s 112.0nm
5.	eP	A	04 59 56	Crete 34.75 N 25.24 E
	e	A	05 00 14	H = 04 55 40.2 h = 54 km MB = 4.5 (USCGS)
	LmH	C	06.3	D = 18.8
	LmV	C	08.5	LmH(C):20s 0.8μm MLH(C)=5.7
5.	eP	A	05 51 31	Turkey 38.87 N 29.79 E
	LmH	C	57.6	H = 05 47 24.8 h = 16 km MB = 4.5 (USCGS)
				D = 17.4
				PV(A):2.3s 79.1nm MPV(A)=4.4

Day	Phase		h m s	Remarks
5.	e	A	06 53 15.5	Pyrenees 42.5 N 1.6 E
	eSg	A	55 43	H = 06 50 02 (BCIS)
	e	A	55 45	D = 10.7
	LmV	B	56.8	LmV(B):7s 0.6/um
	LmH	B	57.2	LmH(B):11.5s 0.4/um MLH(B)=3.6
5.	ePKP	A	11 29 58	New Hebrides Islands 17.69 S 167.82 E
				H = 11 10 31.2 h = 27 km MB = 5.0
				D = 142.02 Az = 335.3 (USCGS)
6.	e(P)	A	12 33 49	Turkey 39.31 N 29.18 E
				H = 12 29 47.7 h = 0 km MB = 4.3
				D = 16.77 Az = 318 (ISC)
				PV(A):2.2s 65.4nm MPV(A)=4.4
6.	+iP	AB	01 07 27.5	Mindoro, Philippine Islands
	eS	C	18 16	13.86 N 120.18 E
	LmH	C	50.0	H = 00 54 30.8 h = 75 km MB = 5.4
				D = 90.77 Az = 322.9 (USCGS)
				PV(A):1.6s 60.4nm MPV(A)=5.6
6.	ePKP	A	06 27 59.5	Fiji Islands 17.84 S 178.37 W
	e	A	28 02	H = 06 09 18.2 h = 543 km MB = 4.4
				D = 146.31 Az = 348.5 (USCGS)
				PKPV(A):1.4s 16.3nm
6.	ePKP	A	12 42 52.5	New Hebrides Islands 20.09 S 174.03 E
	e	A	42 59	H = 12 23 13.3 h = 31 km MB = 4.5
	e	A	43 03	D = 146.49 Az = 339.6 (USCGS)
6.	eP	A	13 32 14	Sea of Okhotsk 46.93 N 144.79 E
				H = 13 21 12.0 h = 369 km MB = 4.3
				D = 74.75 Az = 331.2 (USCGS)
6.	ePn	A	21 55 52	Yugoslavia 45.43 N 17.29 E
	eSn	A	57 09	H = 21 54 13.8 h = 0 km
	e	A	57 40	D = 6.46 Az = 326 (ISC)

Day	Phase		h m s	Remarks
7.	eP	A	04 16 32	Turkey 39.29 N 29.07 E
	LmH	B	22.2	H = 04 12 33.4 h = normal MB = 4.7 (USCGS)
	LmV	B	24.5	D = 17.0
				PV(A):2.0s 42.7nm MPV(A)=4.2
				LmH(B):12s 0.6 μm MLH(B)=4.1
				LmV(B):12s 0.5 μm MLV(B)=4.1
7.	+eP1	ABC	05 47 03.5	Luzon, Philippine Islands
	+iP2	A	47 05.5	15.76 N 121.71 E
	ePP	BC	50 35	H = 05 34 05.6 h = 37 km MB = 6.4
	eS	B	57 22	D = 90.15 Az = 323.3 (USCGS)
	e	B	57 38	P2V(A):1.7s 370.0nm MP2V(A)=6.3
	i	C	57 44	PPV(B):15s 23.7 μm MPPV(B)=7.4
	eSP	B	58 52	LmH(B):16s 157.0 μm MLH(B)=7.5
	eSS	B	06 04 08	LmV(B):20s 162.0 μm MLV(B)=7.5
	ePKKP	A	04 39	
	eSSSS	B	10 52	
	LmH	B	33.8	
	LmV	B	34.5	
7.	eiP	A	06 06 53.5	Luzon, Philippine Islands
				15.5 N 122.4 E
				H = 05 53 48.6 h = 47 km MB = 5.6
				D = 90.70 Az = 323 (ISC)
				PV(A):1.5s 97.4nm MPV(A)=5.9
7.	eP	A	06 17 37	Luzon, Philippine Islands
				15.40 N 121.84 E
				H = 06 04 39 h = 48 km MB = 5.1
				D = 90.50 Az = 323 (ISC)
7.	eP	A	06 24 54	Luzon, Philippine Islands
				15.68 N 121.85 E
				H = 06 11 52.8 h = 22 km MB = 5.5
				D = 90.29 Az = 323 (ISC)
				PV(A):1.6s 44.0nm MPV(A)=5.4

Day	Phase		h m s	Remarks
7.	eP	A	06 47 19.5	Luzon, Philippine Islands 15.49 N 121.87 E H = 06 34 18.6 h = normal MB = 5.5 D = 90.45 Az = 323.3 (USCGS)
7.	eP	A	08 12 57	Luzon, Philippine Islands 15.44 N 121.79 E H = 07 59 57.1 h = normal MB = 5.2 D = 90.44 Az = 323.3 (USCGS) PV(A):1.2s 20.3nm MPV(A)=5.3
7.	eP e	A A	09 20 39 20 50	Morocco 34.84 N 3.90 W H = 09 16 13.9 h = normal MB = 4.9 D = 19.41 Az = 30.8 (USCGS) PV(A):2.0s 68.4nm MPV(A)=4.5
7.	eP1 +iX	A A	09 23 08 23 15.5	Crete 34.57 N 26.14 E H = 09 18 44 h = 20 km MB = 4.9 D = 19.24 Az = 331 (ISC) P1V(A):1.8s 40.6nm MP1V(A)=4.4 XV(A):0.7s 53.6nm
7.	eP +iX	A A	09 38 32.5 38 39.5	Central Mid-Atlantic Ridge 0.36 S 24.68 W H = 09 28 31.1 h = normal MB = 5.0 D = 59.45 Az = 26.0 (USCGS) XV(A):1.0s 17.7nm
7.	LmH LmV	B B	11 04.4 04.7	Probably Luzon, Philippine Islands (USCGS) LmH(B):13s 8.7 μm LmV(B):14s 0.4 μm
7.	eP	A	14 58 03.5	Sea of Okhotsk 49.22 N 147.21 E H = 14 47 25.9 h = 560 km MB = 4.5 D = 73.49 Az = 332.3 (USCGS) PV(A):1.0s 9.8nm MPV(A)=4.3

Day	Phase		h m s	Remarks
7.	eP1	ABC	15 46 20.5	<u>Central Mid-Atlantic Ridge</u>
	eP2	A	46 27	0.27 S 24.76 W
	eS	BC	55 35	H = 15 36 19.2 h = normal MB = 5.2
	LmH	B	16 13.8	D = 59.41 Az = 26.0 (USCGS)
	LmV	B	20.5	P1V(A):2.0s 59.8nm MP1V(A)=5.3
				P2V(A):2.0s 98.2nm MP2V(A)=5.5
				LmH(B):17s 0.7/um MLH(B)=4.9
				LmV(B):12s 0.4/um MLH(V)=4.8
7.	eP	A	16 25 03	<u>Kodiak Islands Region</u> 57.2 N 154.9 W
				H = 16 13 37 h = 68 km MB = 4.3 (ISC)
				D = 71.8
				PV(A):1.1s 12.1nm MPV(A)=4.9
7.	eP	ABC	17 09 08	<u>Turkey</u> 39.38 N 29.15 E
	eX	A	09 14	H = 17 05 11.9 h = 33 km MB = 5.1
	eS	B	12 26	D = 16.70 Az = 318.1 (USCGS)
	LmH	B	14.9	PV(A):1.7s 54.5nm MPV(A)=4.4
	LmV	B	17.1	XV(A):1.7s 87.8nm
				LmH(B):15s 8.4/um MLH(B)=5.1
				LmV(B):11s 6.8/um MLV(B)=5.3
7.	eP	A	19 24 16	<u>Luzon, Philippine Islands</u>
				15.31 N 121.86 E
				H = 19 11 15.7 h = 33 km MB = 4.7
				D = 90.59 Az = 323 (ISC)
8.	eiP	A	00 53 23	<u>Near East Coast of Kamchatka</u>
				54.06 N 160.77 E
				H = 00 41 59.5 h = 45 km MB = 4.7
				D = 72.50 Az = 340 (ISC)
				PV(A):1.1s 22.2nm MPV(A)=5.2
8.	eP	A	09 02 16.5	<u>Luzon, Philippine Islands</u>
				15.31 N 121.63 E
				H = 08 49 16.6 h = normal MB = 5.2
				D = 90.45 Az = 323.3 (USCGS)
				PV(A):1.5s 25.1nm MPV(A)=5.3

Day	Phase		h m s	Remarks
8.	e	A	10 06 07	Luzon, Philippine Islands
				15.36 N 121.73 E
				H = 09 52 58.0 h = 13 km MB = 4.9
				D = 90.47 Az = 323.3 (USCGS)
8.	+eP	A	11 10 49	Mid-Indian Rise 11.08 S 65.90 E
	eiX	A	10 56.5	H = 10 58 54.6 h = normal MB = 5.1
				D = 77.43 Az = 328.0 (USCGS)
				PV(A):2.2s 54.5nm MPV(A)=5.3
				XV(A):1.6s 32,9nm
8.	eP1	AB	13 53 52	Greece 38.43 N 22.66 E
	eiP2	A	53 57.5	H = 13 50 27.2 h = 17 km MB = 5.8
	Pm	A	54 31	D = 14.51 Az = 330.9 (USCGS)
	eS	B	56 28	P1V(A):1.8s 54.1nm
	LmH	B	14 00.8	P2V(A):1.5s 100.5nm
	LmV	B	00.8	PmV(A):1.5s 502.0nm
				LmH(B):12s 136.0 µm MLH(B)=6.3
				LmV(B):14s 171.5 µm
8.	ePKIKP	A	15 00 27	New Hebrides Islands 13.77 S 166.79 E
	epPKIKP	A	00 41	H = 14 41 08.4 h = 64 km MB = 5.5 (USCGS)
	ePP	A	03 12	D = 138.2
				PKIKPV(A):1.3s 30.6nm
8.	eP	A	18 07 30	Luzon, Philippine Islands
	ePP	A	11 03.5	15.57 N 21.72 E
				H = 17 54 29.5 h = 22 km MB = 5.3
				D = 90.30 Az = 323.3 (USCGS)
				PV(A):1.6s 22.0nm MPV(A)=5.2
8.	eP	A	20 03 01	Alaska Peninsula 56.26 N 156.53 W
				H = 19 51 32 h = 30 km MB = 4.9
				D = 73.01 Az = 8 (ISC)
				PV(A):1.1s 20.2nm MPV(A)=5.2
8.	eP1	AB	21 36 57	Luzon, Philippine Islands
	iP2	A	37 00	15.43 N 121.80 E
	iX	A	37 14	H = 21 23 56.6 h = normal MB = 5.7 (USCGS)

Day	Phase		h m s	Remarks
cont.				
8.	iPP	AB	21 40 37	D = 90.4
	eiS	BC	47 48	P1V(B):7.5s 2.3/um MP1V(B)=6.5
	eSS	C	54 56	P2V(A):1.6s 132.0nm MP2V(A)=6.0
	eSSSS	C	22 00 28	XV(A):1.6s 110.0nm
	LmV	B	23.9	PPV(B):11.5s 1.4/um MPPV(B)=6.3
	LmH	B	24.4	SH(B):15s 4.5/um MSH(B)=6.3
				LmV(B):16s 25.8/um MLV(B)=6.8
				LmH(B):16s 16.7/um MLH(B)=6.6
8.	eP	A	21 59 40	<u>Luzon, Philippine Islands</u>
				15.41 N 121.93 E
				H = 21 46 40.0 h = 36 km MB = 5.1
				D = 90.55 Az = 323.3 (USCGS)
8.	eP	A	23 46 45.5	<u>Luzon, Philippine Islands</u>
				15.54 N 121.74 E
				H = 23 33 46.2 h = normal MB = 5.2
				D = 90.33 Az = 323.3 (USCGS)
				PV(A):1.8s 33.8nm MPV(A)=5.3
9.	iP	A	00 12 02.2	<u>Kurile Islands</u> 45.30 N 149.02 E
	-i	AB	12 02.5	H = 00 00 14.7 h = 100 km MB = 5.4
				D = 77.56 Az = 333.8 (USCGS)
				PV(A):1.1s 129.0nm MPV(A)=5.7
9.	e	A	08 27 35	<u>Austria</u> 47.8 N 16.2 E
	iSg	A	27 43	H = 08 25 31 (BCIS)
				D = 4.1
9.	eP1	A	10 16 27	<u>Turkey</u> 39.20 N 29.53 E
	eiP2	A	16 30	H = 10 12 29.8 h = 33 km MB = 4.8
	LmH	C	24.0	D = 17.04 Az = 318.1 (USCGS)
				P2V(A):1.8s 57.4nm MP2V(A)=4.4
9.	eP	A	16 37 22	<u>Off Coast of Chiapas, Mexico</u>
	e	A	37 30.5	13.21 N 92.26 W

Day	Phase		h m s	Remarks
cont. 9.	LmH	C	17 17.3	H = 16 24 31.0 h = 41 km MB = 5.3 D = 88.49 Az = 38.2 (USCGS) PV(A):1.2s 16.3nm MPV(A)=5.2 LmH(C):19s 1.1/um MLH(C)=5.3
9.	eP LmH	A C	21 55 21.5 22 41.8	Atlantic-Indian Rise 40.89 S 43.26 E H = 21 41 52.3 h = normal MB = 5.3 D = 95.27 Az = 340.4 (USCGS) LmH(C):20s 0.4/um MLH(C)=4.8
10.	eP eX	A A	01 18 37.5 18 42	Turkey 39.4 N 29.3 E H = 01 14 43 D = 16.8 (ANUSSR) PV(A):1.5s 17.6nm MPV(A)=4.0 XV(A):2.6s 130.0nm
10.	epP	A	08 31 41	Afghanistan-USSR Border Region 36.36 N 71.10 E H = 08 22 59.6 h = 125 km MB = 4.7 (USCGS) D = 44.2 h = 123 km pPV(A):1.3s 8.7nm
10.	e	A	09 10 56	Kurile Islands 50.02 N 154.74 E H = 09 10 00.1 h = 152 km MB = 4.4 D = 74.89 Az = 336.7 (USCGS)
10.	eP eX1 eX2 LmH	A A A C	10 32 43 32 49.5 32 56 56.0	West Pakistan 25.29 N 66.66 E H = 10 23 58.2 h = normal MB = 5.1 D = 48.87 Az = 316.1 (USCGS) PV(A):1.5s 37.7nm MPV(A)=5.2 X1V(A):1.2s 24.4nm X2V(A):1.2s 18.4nm LmH(C):21s 0.4/um MLH(C)=4.4
10.	ePKIKP ePKHKP iPKP2 epPKP2	AB A AB A	14 28 52 29 02 29 19 29. 59	Kermadec Islands 27.46 S 177.91 W H = 14 09 16.0 h = 158 km MB = 5.5 D = 155.73 Az = 345.1 (USCGS) PKHKPV(A):1.6s 27.5nm

Day	Phase		h m s	Remarks
cont. 10.	ePP	AB	14 32 54	PKP2V(A):1.6s 71.5nm
	epPP	B	33 42	LmH(C):37s 0.4 µm
	e(SKSP)	C	43 15	
	eSS	C	52 25	
	esSS	C	53 28	
	LmH	C	15 20.0	
10.	iPn	A	20 19 50	<u>Federal Rep. Germany</u> 48.55 N 8.94 E
	iPg	A	19 57	H = 20 19 07.6 h = 17 km
	iSn	A	20 25	D = 2.73 Az = 39 (ISC)
	iSg	A	20 36	LmH(B):6.5s 2.1 µm MLH(B)=3.6
	LmH	B	20.8	LmV(B):4.5s 0.8 µm
	LmV	B	20.9	
11.	eP	A	01 06 38	<u>Greece</u> 38.24 N 23.06 E
	LmH	C	11.6	H = 01 03 11.2 h = 70 km MB = 4.5
				D = 14.84 Az = 330.4 (USCGS)
11.	eP1	AB	04 16 43	<u>Gulf of Alaska</u> 59.71 N 142.74 W
	iP2	A	16 46	H = 04 05 41.1 h = 7 km MB = 5.2
	ePP	BC	19 18	D = 68.05 Az = 17.3 (USCGS)
	eS	BC	25 44	P1V(B):14s 1.5 µm MP1V(B)=6.0
	eSKS	BC	26 45	P2V(A):1.2s 85.4nm MP2V(A)=5.9
	eSS	BC	30 00	SH(B):15.5s 2.3 µm MSH(B)=6.1
	ei	BC	30 41	LmH(B):20s 10.3 µm MLH(B)=6.0
	LmH	B	49.3	LmV(B):16s 6.6 µm MLV(B)=6.0
	LmV	B	51.5	
11.	ePKHKP	ABC	06 41 02	<u>Tonga Islands</u> 19.33 S 173.57 W
	e	A	41 20	H = 06 21 16.3 h = normal MB = 5.3
	LmV	B	07 48.3	D = 148.48 Az = 353.1 (USCGS)
	LmH	B	51.5	PKHKPV(A):2.0s 59.9nm
				LmV(B):20s 1.0 µm
				LmH(B):20s 1.4 µm MLH(B)=5.7
11.	eP	A	10 10 46.5	<u>Gulf of Alaska</u> 59.51 N 142.75 W
				H = 09 59 46.3 h = normal MB = 4.6
				D = 68.24 Az = 17.3 (USCGS)
				PV(A):1.4s 23.3nm MPV(A)=5.2

Day	Phase		h m s	Remarks
11.	eP	A	13 06 36	<u>Gulf of Alaska</u> 59.77 N 142.66 W
				H = 12 55 37.9 h = normal MB = 4.5
				D = 67.98 Az = 17.3 (USCGS)
				PV(A):0.6s 9.6nm MPV(A)=5.2
11.	eP	A	17 28 30	<u>Turkey</u> 39.09 N 29.76 E
	LmH	B	34.4	H = 17 24 25 h = 22 km (ISC)
	LmV	B	36.4	D = 17.2
				LmH(B):16s 0.6 µm MLH(B)=3.9
				LmV(B):12s 0.5 µm MLV(B)=4.0
11.	e	A	19 51 24	Probably <u>Greece</u> (ISC)
12.	eP	A	02 22 30.5	<u>Andreanof Islands, Aleutian Is.</u>
				51.52 N 178.50 W
				H = 02 10 36.0 h = 47 km MB = 5.2
				D = 77.85 Az = 353.4 (USCGS)
12.	eP1	ABC	04 14 46.5	<u>Philippine Islands</u> 15.06 N 122.05 E
	eiP2	A	14 53.5	H = 04 01 44.0 h = 24 km MB = 5.9
	eX	A	15 10.5	D = 90.89 Az = 323.4 (USCGS)
	ePP	A	18 25	P1V(A):2.0s 111.0nm MP1V(A)=5.8
	e	A	18 30	P1V(B):12s 5.4 µm MP1V(B)=6.7
	e	A	18 34	P2V(A):1.9s 170.0nm MP2V(A)=6.0
	iSKS	BC	25 20	XV(A):1.9s 288.0nm MPmV(A)=6.3
	iS	BC	25 43	SH(B):12s 18.2 µm MSH(B)=7.1
	eiSS	C	31 35	LmH(B):19s 89.9 µm MLH(B)=7.2
	LmH	B	54.0	LmV(B):16s 126.0 µm MLV(B)=7.5
	LmV	B	05 02.3	
12.	eiP	A	04 29 59	<u>Philippine Islands</u> 15.17 N 122.03 E
	eX	A	30 06.5	H = 04 16 57.2 h = normal MB = 5.5
				D = 90.80 Az = 323.4 (USCGS)
				PV(A):1.2s 24.4nm MPV(A)=5.4
				XV(A):2.0s 51.3nm
12.	eP	A	04 39 48.5	<u>Philippine Islands</u> 15.11 N 122.08 E
				H = 04 26 38.5 h = normal MB = 5.2 (USCGS)
				D = 90.9
				PV(A):1.0s 9.8nm MPV(A)=5.1

Day	Phase		h m s	Remarks
12.	eP	A	05 31 11.5	Philippine Islands Region
	ePP	A	34 47	15.17 N 122.51 E
	e	A	34 52	H = 05 18 09.1 h = 32 km MB = 5.4
				D = 91.07 Az = 323 (ISC)
				PV(A):1.7s 39.4nm MPV(A)=5.5
12.	e	A	06 43 22.5	Probably Luzon, Philippine Islands (USCGS)
12.	eP	A	11 01 46.5	Philippine Islands 15.08 N 122.04 E
	LmH	B	49.0	H = 10 48 42.9 h = 22 km MB = 5.2
	LmV	B	49.0	D = 90.87 Az = 323.4 (USCGS)
				PV(A):1.8s 27.0nm MPV(A)=5.3
				LmH(B):16s 0.5/um MLH(B)=5.1
				LmV(B):16s 0.3/um MLV(B)=4.8
12.	eP	A	14 35 42	Philippine Islands 15.06 N 122.54 E
	e	A	35 46	H = 14 22 38.9 h = normal MB = 5.4
	e	A	39 16	D = 91.18 Az = 323.5 (USCGS)
	ePP	A	39 20	PV(A):1.3s 21.8nm MPV(A)=5.3
	LmH	B	15 23.3	PPV(A):1.6s 22.0nm MPPV(A)=5.3
	LmV	B	23.5	LmH(B):15.5s 1.0/um MLH(B)=5.4
				LmV(B):16s 1.0/um MLV(B)=5.4
12.	eX	A	19 59 35	Turkey 38.90 N 29.41 E
				H = 19 55 25.0 h = 15 km MB = 4.3
				D = 17.20 Az = 318.8 (USCGS)
				XV(A):2.2s 43.6nm
12.	eX	A	24 05 17	Philippine Islands 15.07 N 122.41 E
	LmV	(B)	52.4	H = 23 52 07.6 h = normal MB = 5.0
				D = 91.09 Az = 323.5 (USCGS)
				XV(A):1.2s 16.3nm
				LmV(B):13s 0.5/um MLV(B)=5.1
13.	e	A	05 20 00	Turkey 39.29 N 29.11 E
				H = 05 15 58.3 h = 8 km MB = 4.5
				D = 16.75 Az = 318.4 (USCGS)

Day	Phase		h m s	Remarks
13.	eX1	A	08 41 31	Philippine Islands 15.19 N 122.17 E
	eX2	A	41 35	H = 08 28 21.8 h = 5 km MB = 5.2
	LmH	B	09 23.0	D = 90.86 Az = 323.4 (USCGS)
	LmV	B	28.8	X2V(A):1.1s 16.1nm
				LmH(B):20s 1.2 /um MLH(B)=5.3
				LmV(B):15s 1.0 /um MLV(B)=5.4
14.	iSg	A	03 37 18	Federal Rep. Germany 48.3 N 9.1 E
				H = 03 35 45 (BCIS)
				D = 2.8
				SgV(A):0.6s 11.5nm
14.	eP	A	06 41 01	Off East Coast of Honshu, Japan
				40.26 N 143.51 E
				H = 06 28 55.1 h = 62 km MB = 4.6 (USCGS)
				D = 80.0
14.	ePKIKP	A	14 06 16.5	Tonga Islands 21.08 S 174.41 W
	iPKHKP	AB	06 19.3	H = 13 46 34 h = 64 km MB = 5.4
	ePKP2	A	06 25	D = 150.11 Az = 352 (ISC)
	epPKP	A	06 30	PKHKPV(A):1.4s 93.0nm
14.	eP	A	19 20 49	Republic of South Africa 33.30 S 19.22 E
	eS	C	31 16	H = 19 08 21.3 h = normal MB = 5.7 (USCGS)
	ePS	C	32 05	D = 83.8
	eSS	C	37 20	PV(A):1.4s 20.9nm MPV(A)=5.2
	eSSS	C	40 10	LmH(B):21s 2.9 /um MLH(B)=5.6
	LmH	B	55.2	LmV(B):14s 1.1 /um MLV(B)=5.4
	LmV	B	20 01.7	
14.	eP	A	24 02 29.5	Southwestern Ryukyu Islands
				24.33 N 125.18 E
				H = 23 49 59.5 h = 79 km MB = 4.9
				D = 85.19 Az = 324.2 (USCGS)
				PV(A):1.2s 12.2nm MPV(A)=4.8
15.	ePKP	A	05 20 09	Tonga Islands 19.05 S 175.77 W
				H = 05 00 43.4 h = 215 km MB = 4.4
				D = 147.94 Az = 351.1 (USCGS)

Day	Phase		h m s	Remarks
15.	+iP	ABC	13 27 28.2	Philippine Islands 15.07 N 122.67 E
	ePP	A	31 01	H = 13 14 21.4 h = 12 km MB = 5.7
	eSKS	BC	37 57	D = 91.24 Az = 323.5 (USCGS)
	eiS	BC	38 22	PV(A):1.5s 65.4nm MPV(A)=5.7
	eSS	C	44 34	PV(B):13s 0.9 µm MPV(B)=6.0
	eSSS	C	48 05	SH(B):13s 2.1 µm MSH(B)=6.1
	LmH	B	14 16.4	LmH(B):16.5s 11.0 µm MLH(B)=6.4
	LmV	B	16.4	LmV(B):16s 12.5 µm MLV(B)=6.5
15.	eP	A	16 33 55	Turkey 39.26 N 29.27 E
	eS	C	37 15	H = 16 29 56.0 h = 18 km MB = 4.6
	LmH	B	39.6	D = 16.85 Az = 318.2 (USCGS)
	LmV	B	41.9	PV(A):2.0s 68.4nm MPV(A)=4.5
				LmH(B):15s 2.8 µm MLH(B)=4.6
				LmV(B):12s 1.8 µm MLV(B)=4.6
16.	eP	A	01 11 (02)	Turkey 39.25 N 29.36 E
				H = 01 07 00.8 h = 15 km MB = 4.4
				D = 16.91 Az = 318.2 (USCGS)
16.	eP	AC	02 08 26.5	Off East Coast of Honshu, Japan
	epP	A	08 37	34.47 N 141.57 E
	eS	BC	18 13	H = 01 55 56.4 h = 35 km MB = 5.2
	LmH	B	49.7	D = 84.45 Az = 330.6 (USCGS)
	LmV	B	53.2	PV(A):1.4s 37.2nm MPV(A)=5.4
				LmH(B):15s 2.6 µm MLH(B)=5.8
				LmV(B):14s 1.7 µm MLV(B)=5.7
16.	eP	A	05 17 20.5	Costa Rica 9.60 N 84.24 W
				H = 05 04 41.7 h = 50 km MB = 4.3
				D = 86.35 Az = 39.4 (USCGS)
16.	eP	AB	05 44 19	Gulf of Alaska 59.77 N 142.60 W
	ePP	B	46 58	H = 05 33 17.5 h = 7 km MB = 5.5
	e	B	47 00	D = 67.97 Az = 17.4 (USCGS)
	eS	B	53 24	PV(A):1.6s 96.2nm MPV(A)=5.8
	+i	B	53 26	PV(B):15s 6.6 µm MPV(B)=6.6
	+i	B	53 38	PPV(B):16s 3.6 µm MPPV(B)=6.4
	eSS	B	57 50	SH(B):16s 16.9 µm MSH(B)=6.9

Day	Phase		h m s	Remarks
cont. 16.	e(PKPPKP)	A	06 12 33	LmH(B):17s 46.7/um MLH(B)=6.8
	LmH	B	17.1	LmV(B):14s 33.9/um MLV(B)=6.8
	LmV	B	18.8	
16.	eP	A	06 18 47	<u>Kodiak Island Region</u> 57.6 N 151.2 W
	e	A	18 55.5	H = 06 07 25 h = 0 km MB = 4.6
				D = 71.23 Az = 11 (ISC)
16.	ePKP2	A	08 09 14	<u>South of Kermadec Islands</u>
				33.42 S 179.92 E
				H = 07 48 41.8 h = 76 km MB = 4.6 (USCGS)
				D = 160.3
				PKP2V(A):1.1s 10.1nm
16.	eP1	ABC	10 46 22.5	<u>Turkey</u> 39.03 N 30.60 E
	+iP2	A	46 25.5	H = 10 42 18.8 h = 9 km MB = 5.5
	eS	BC	49 36	D = 17.40 Az = 317.8 (USCGS)
	LmH	C	52.5	P1V(A):1.7s 30.3nm MP1V(A)=4.2
	LmV	C	54.2	P2V(A):2.1s 460.0nm MP2V(A)=5.2
				SH(B):8s 1.7/um MSH(B)=5.9
				LmH(C):15s 11.9/um MLH(C)=5.2
				LmV(C):12s 10.6/um MLV(C)=5.5
16.	eP	A	11 47 23	<u>Turkey</u> 39.02 N 30.10 E
				H = 11 43 19.2 h = 20 km MB = 4.6
				D = 17.47 Az = 317.7 (USCGS)
				PV(A):1.8s 60.8nm MPV(A)=4.4
16.	eX	A	13 03 18	<u>Turkey</u> 38.99 N 29.5 E
				H = 12 59 18 h = 67 km MB = 4.3 (ISC)
				D = 17.2
				XV(A):1.5s 15.1nm
16.	eP	A	18 30 39	<u>Gulf of Alaska</u> 59.85 N 142.56 W
				H = 18 19 37.5 h = 16 km MB = 4.1
				D = 67.89 Az = 17.4 (USCGS)

Day	Phase		h m s	Remarks
16.	eP	AB	22 42 35.5	<u>Greece</u> 40.70 N 23.48 E
	eS	BC	45 05	H = 22 39 28.3 h = 4 km MB = 5.1
	LmH	B	47.0	D = 12.93 Az = 324.2 (USCGS)
	LmV	B	47.9	PV(A):1.4s 20.9nm
				LmH(B):9.5s 11.6 μm MLH(B)=5.3
				LmV(B):10.5s 7.9 μm
16.	LmH	B	23 19.3	<u>Greece</u> 40.7 N 23.9 E
	LmV	B	20.0	H = 23 11 47.7 h = normal MB = 4.2 (USCGS)
				D = 13.2
				LmH(B):10s 1.0 μm MLH(B)=4.2
17.	eP	A	01 35 42	<u>Turkey</u> 37.06 N 27.00 E
	LmH	(B)	42.6	H = 01 31 38.1 h = normal
	LmV	B	43.1	D = 17.49 Az = 325.8 (USCGS)
				LmH(B):10s 0.6 μm MLH(B)=4.1
				LmV(B):8s 0.6 μm MLV(B)=4.4
17.	eP	A	15 42 49	<u>Gulf of Alaska</u> 59.55 N 142.70 W
				H = 15 31 47.3 h = 15 km MB = 4.3
				D = 68.20 Az = 17.3 (USCGS)
				PV(A):1.0s 9.8nm MPV(A)=5.0
17.	e	A	20 31 25	<u>Turkey</u> 39.06 N 29.72 E
				H = 20 27 20.3 h = normal MB = 4.5
				D = 17.23 Az = 318.1 (USCGS)
18.	eP	A	05 41 23	<u>Turkey</u> 39.38 N 29.28 E
	LmH	C	47.0	H = 05 37 26.9 h = normal MB = 4.5
				D = 16.77 Az = 310.0 (USCGS)
				LmH(C):18s 0.5 μm MLH(C)=3.8
18.	LmH	C	06 02.5	Probably <u>Turkey</u> (ISC)
				LmH(C):16.5s 0.8 μm
18.	-iP	AB	09 01 38.3	<u>Southern Alaska</u> 59.86 N 152.82 W
	eipP	AB	02 03.5	H = 08 50 40.8 h = 94 km MB = 5.7
	isP	AB	02 12	D = 69.14 Az = 10.5 (USCGS)
	iS	BC	10 35	PV(A):1.0s 130.0nm MPV(A)=5.8

Day	Phase		h m s	Remarks
cont. 18.	ePS	BC	09 11 20	pPV(A):1.3s 157.3nm
	eSS	C	15 00	sPV(A):1.3s 153.0nm
18.	eP	A	19 34 45.5	Fox Islands, Aleutian Is. 53.61 N 166.07 W H = 19 23 01.2 h = 50 km MB = 4.7 D = 76.10 Az = 1.5 (USCGS) PV(A):1.3s 13.1nm MPV(A)=4.9
18.	eP	A	20 33 07	Off Coast of Hokkaido, Japan 42.84 N 147.27 E
	LmH	B	21 09.5	H = 20 21 03.9 h = 39 km MB = 4.5
	LmV	B	13.0	D = 79.19 Az = 333.1 (USCGS) PV(A):1.4s 9.3nm MPV(A)=4.6 LmH(B):14s 0.5/um MLH(B)=5.0 LmV(B):18s 0.5/um MLV(B)=5.0
18.	eP	A	23 10 36	Turkey 39.01 N 29.95 E H = 23 06 31.0 h = normal MB = 4.1 D = 17.40 Az = 317.9 (USCGS)
18.	eP	AB	23 37 37.5	Kurile Islands 43.25 N 147.23 E
	ipP	A	37 49.6	H = 23 25 35.5 h = 38 km MB = 5.2
	ePP	B	40 28	D = 78.81 Az = 333.0 (USCGS)
	eS	BC	47 24	PV(A):1.8s 54.0nm MPV(A)=5.3
	eSS	C	52 50	PV(B):10s 1.0/um MPV(B)=5.8
	eSSS	B	56 20	pPV(A):1.1s 36.3nm
	LmV	B	24 17.5	PPV(B):8s 0.4/um MPPV(B)=5.6
	LmH	B	17.8	SH(B):13.5s 1.1/um MSH(B)=5.9 LmV(B):18s 6.6/um MLV(B)=6.1 LmH(B):15s 5.6/um MLH(B)=6.0
18.	eP	A	23 47 32	Kurile Islands 43.39 N 147.18 E H = 23 35 30.6 h = normal MB = 4.4 D = 78.67 Az = 333.0 (USCGS) PV(A):1.6s 16.5nm MPV(A)=4.8

Day	Phase		h m s	Remarks
19.	iP1	AB	01 26 47.1	Gulf of Alaska 59.64 N 142.83 W
	+iP2	A	26 49.4	H = 01 15 46.8 h = 20 km MB = 5.8
	eiX	A	26 52	D = 68.13 Az = 17.2 (USCGS)
	e	B	27 24	P1V(A):0.9s 27.3nm MP1V(A)=5.5
	e	B	27 54	P1V(B):8s 0.7/um MP1V(B)=5.9
	e	B	29 54	P2V(A):1.6s 126.0nm MP2V(A)=5.9
	eS	B	35 48	XV(A):1.5s 120.8nm
	ePKPPKP	A	55 05	PKPPKPV(A):1.4s 18.6nm
	LmH	B	59.6	LmH(B):16s 7.7/um MLH(B)=6.0
	LmV	B	02 02.9	LmV(B):16s 5.8/um MLV(B)=6.0
19.	eP	A	04 53 02.5	Near East Coast of Kamchatka
	ePP	A	56 06	51.81 N 157.11 E
				H = 04 41 37.8 h = 105 km MB = 5.1
				D = 73.82 Az = 337.9 (USCGS)
				PV(A):1.4s 25.6nm MPV(A)=4.9
				PPV(A):2.0s 25.6nm MPPV(A)=5.2
19.	eP	A	05 41 40	Tadzhik SSR 38.99 N 70.76 E
				H = 05 33 46.2 h = normal MB = 4.7
				D = 42.46 Az = 305.9 (USCGS)
19.	eP	A	10 23 25	Kurile Islands 43.41 N 147.71 E
				H = 10 11 24.2 h = 40 km MB = 4.7
				D = 78.83 Az = 333.3 (USCGS)
				PV(A):1.3s 10.9nm MPV(A)=4.7
19.	eiP	A	11 38 46.5	North Atlantic Ridge 12.80 N 44.60 W
	e	A	38 50	H = 11 28 48.0 h = normal MB = 4.8
	e	A	38 54.5	D = 59.00 Az = 38.1 (USCGS)
				PV(A):1.2s 24.4nm MPV(A)=5.1
19.	eP1	ABC	13 33 37.5	Turkey 39.07 N 29.79 E
	iP2	A	33 41.5	H = 13 29 36.4 h = 20 km MB = 5.4
	iP3	A	33 46	D = 17.27 Az = 318.0 (USCGS)
	eS	B	36 54	P1V(A):1.5s 17.6nm MP1V(A)=4.0

Day	Phase		h m s	Remarks
cont. 19.	LmH	B	13 39.6	P1V(B):10s 6.9/um MP1V(B)=5.7
	LmV	B	41.5	P2V(A):1.6s 264.0nm MP2V(A)=5.1
				P3V(A):2.2s 1550.0nm MP3V(A)=5.8
				SH(B):12s 6.7/um
				LmH(B):16.5s 38.2/um MLH(B)=5.7
				LmV(B):13s 39.6/um MLV(B)=6.0
19.	eP1	ABC	13 51 36.5	<u>Turkey</u> 39.06 N 29.83 E
	eP2	A	51 39	H = 13 47 35.2 h = 26 km MB = 5.5
	iP3	A	51 46	D = 17.29 Az = 318.0 (USCGS)
	eiS	B	55 00	P1V(A):1.4s 27.9nm MP1V(A)=4.2
	LmH	B	57.7	P2V(A):2.5s 815.0nm MP2V(A)=5.4
	LmV	B	59.5	P3V(A):2.2s 1090.0nm MP3V(A)=5.6
	ePKKP	A	14 19 35	LmH(B):16s 34.2/um MLH(B)=5.7
				LmV(B):13s 27.9/um MLV(B)=5.9
				PKKPV(A):0.8s 7.7nm
19.	ePKIKP	A	17 21 34.5	<u>Fiji Islands</u> 19.59 S 177.68 W
	+iPKHKP	A	21 39	H = 17 02 59.4 h = 609 km MB = 5.0
	ePKP2	A	21 43	D = 148.15 Az = 348.8 (USCGS)
	epPKIKP	A	23 55.5	PKHKPV(A):0.7s 28.3nm
	epPKHKP	A	23 58	pPKHKPV(A):1.5s 35.2nm
19.	ePn	A	18 17 47.5	<u>Northern Italy</u> 45.55 N 10.39 E
	iPg	A	18 09	H = 18 16 30.8 h = 0 km
	iSn	A	18 46	D = 5.17 Az = 9 (ISC)
	iSg	A	19 16.5	LmV(B):7s 1.3/um
	LmV	B	19.9	LmH(B):7.5s 1.0/um MLH(B)=3.6
	LmH	B	20.0	
20.	eP	AB	02 27 30.5	<u>Turkey</u> 38.96 N 30.10 E
	LmH	B	33.5	H = 02 23 26.7 h = normal MB=4.5
				D = 17.51 Az = 317.9 (USCGS)
				LmH(B):16s 1.0/um MLH(B)=4.2
20.	ePKIKP	A	02 28 15	<u>South of Kermadec Islands</u>
	e(PKP2)	A	28 56	32.05 S 179.32 W

Day	Phase	h m s	Remarks
cont. 20.	e(pPKP2) A	02 29 38	H = 02 08 33.3 h = 137 km MB = 5.4 D = 159.73 Az = 340 (ISC) (PKP2)V(A):1.4s 32.6nm (pPKP2)V(A):1.7s 42.4nm
20.	eX A	04 01 00.5	Tonga Islands 15.05 S 173.86 W H = 03 41 14.0 h = normal MB = 4.8 D = 144.23 Az = 354.0 (USCGS) XV(A):1.5s 15.1nm
20.	ePKP A	07 04 31	New Hebrides Islands 20.80 S 172.15 E H = 06 44 45.8 h = 20 km MB = 4.2 D = 146.50 Az = 337.4 (USCGS)
20.	eP A	07 23 07	Turkey 39.06 N 28.86 E H = 07 19 05.3 h = normal MB = 4.4 D = 16.79 Az = 319.2 (USCGS)
20.	+iPKP ABC	10 58 17	New Hebrides Islands 18.83 S 169.30 E
	-iX A	58 20.7	H = 10 39 12.5 h = 246 km MB = 6.3
	epPKP B	59 18	D = 143.63 Az = 335.9 (USCGS)
	esPKP B	11 00 00	PKPV(A):1.2s 447.0nm
	ePP BC	01 32	XV(A):1.2s 711.0nm
	epPP B	02 32	ScPPKPV(A):1.9s 45.5nm
	eScPPKP A	09 38.5	
	e BC	09 50	
	e B	11 48	
	e B	16 35	
	eSS B	19 50	
	esSS B	21 32	
	eSSS B	24 50	
	eScPPKP A	11 09 38.5	
20.	eP1 AB	15 43 01	Greece 38.41 N 22.79 E
	eiP2 A	45 05	H = 15 39 29.2 h = 20 km MB = 5.2
	eiP3 A	45 07.5	D = 14.58 Az = 330.7 (USCGS)
	eS B	46 00	P2V(A):0.8s 38.5nm

Day	Phase		h m s	Remarks
cont. 20.	LmH	B	15 49.3	P3V(A):1.5s 95.4nm
	LmV	B	49.3	LmH(B):13.5s 12.2/um MLH(B)=5.2
				LmV(B):12s 11.5nm MLV(B)=5.3
20.	LmH	B	18 12.5	Probably <u>Turkey</u> (USCGS)
	LmV	B	12.5	LmH(B):10s 0.6/um
				LmV(B):11s 0.7/um
20.	eP	A	18 37 23	<u>Turkey</u> 38.90 N 30.66 E
	LmH	B	45.2	H = 18 33 13.3 h = normal MB = 4.5
	LmV	B	45.2	D = 17.85 Az = 317.3 (USCGS)
				LmH(B):10s 0.5/um MLH(B)=4.0
				LmV(B):11s 0.6/um MLV(B)=4.2
20.	e	A	22 01 41	<u>Sumba Island</u> 9.82 S 119.34 E
	ePP	A	02 00.5	H = 21 43 00.5 h = normal MB = 5.6
	LmH	B	52.0	D = 108.77 Az = 320.2 (USCGS)
	LmV	B	52.6	PPV(A):1.3s 21.8nm MPPV(A)=5.6
				LmH(B):20s 1.0/um MLH(B)=5.4
				LmV(B):21s 0.5/um MLV(B)=5.1
21.	+iPKHKP	A	03 39 25	<u>Fiji Islands</u> 20.49 S 178.30 W
	ePKP2	A	39 30.5	H = 03 20 37.5 h = 550 km MB = 4.5
				D = 148.90 Az = 347.7 (USCGS)
				PKHKPV(A):1.0s 27.6nm
21.	eP	A	04 55 39	<u>Kurile Islands</u> 43.41 N 147.49 E
				H = 04 43 37.3 h = normal MB = 5.1
				D = 78.76 Az = 333.1 (USCGS)
				PV(A):1.3s 30.6nm MPV(A)=5.2
21.	eP	A	06 55 28.5	<u>Gulf of Alaska</u> 59.62 N 142.65 W
				H = 06 44 24.8 h = 1 km MB = 4.9
				D = 68.13 Az = 17.3 (USCGS)
				PV(A):1.7s 42.4nm MPV(A)=5.4

Day	Phase		h m s	Remarks
21.	ePKP2	A	07 24 00	South of Fiji Islands 26.06 S 177.21 W H = 07 03 55.3 h = 90 km MB = 4.6 D = 154.54 Az = 346.9 (USCGS) PKP2V(A):1.3s 10.9nm
21.	+eP	A	07 41 02	Turkey 39.19 N 29.75 E H = 07 36 58.7 h = 25 km MB = 4.7 D = 17.15 Az = 317.8 (USCGS)
21.	ePKP	A	10 56 25	New Hebrides Islands 20.70 S 172.45 E H = 10 36 45.9 h = 33 km MB = 4.7 D = 146.51 Az = 337.7 (USCGS) PKPV(A):1.2s 16.2nm
21.	eP	A	14 24 45	North of Ascension Island
	eX	A	24 50.5	3.33 S 12.14 W
	LmH	B	50.0	H = 14 14 57.4 h = normal MB = 5.3
	LmV	B	50.6	D = 57.51 Az = 17.7 (USCGS) PV(A):2.2s 87.2nm MPV(A)=5.4 XV(A):1.4s 55.8nm LmV(B):15s 0.6/um MLV(B)=4.9
22.	eP	A	14 57 04	Turkey 39.22 N 41.4 E H = 14 51 53 h = 28 km MB = 4.4 D = 23.81 Az = 308 (ISC)
22.	+iP	AB	05 28 08	Turkey 39.08 N 29.83 E
	e	A	28 17.5	H = 05 24 05.7 h = normal MB = 5.1
	e	B	31 36	D = 17.28 Az = 317.9 (USCGS)
	LmH	B	34.0	PV(A):2.0s 154.0nm MPV(A)=4.8
	LmV	B	35.9	LmH(B):16s 3.1/um MLH(B)=4.6 LmV(B):12s 3.2/um MLV(B)=5.0
22.	ePP	A	11 57 10.5	South of Sumbia Island
	e	A	57 20	11.22 S 120.01 E H = 11 38 03.7 h = normal MB = 5.6 (USCGS) D = 110.2

Day	Phase		h m s	Remarks
22.	ePKIKP	A	12 24 35	Banda Sea 6.29 S 128.72 E H = 12 06 31.5 h = 295 km MB = 5.1 (USCGS) D = 111.8
22.	eP	A	14 01 54	Luzon, Philippine Islands 15.31 N 121.82 E H = 13 48 53.5 h = 28 km MB = 5.0 D = 90.56 Az = 323.3 (USCGS) PV(A):1.8s 23.7nm MPV(A)=5.2
22.	ePKP	A	15 59 32	Tonga Islands 16.85 S 174.11 W
	e(pPKP)	A	16 00 07	H = 15 40 02.1 h = 98 km MB = 4.1 D = 145.97 Az = 353.5 (USCGS)
22.	eP1	A	18 42 47.5	Turkey 39.15 N 29.78 E
	eP2	A	42 52	H = 18 38 48.6 h = 41 km MB = 4.6
	LmH	B	48.7	D = 17.20 Az = 317.8 (USCGS)
	LmV	B	51.0	P1V(A):1.3s 21.8nm MPV1(A)=4.1 P2V(A):1.5s 40.2nm MPV2(A)=4.3
23.	-eiP	AB	01 03 50.5	East of Severnaya Zemlya
	eipP	A	03 58.5	80.66 N 121.99 E
	esP	A	04 02.5	H = 00 55 48.2 h = normal MB = 5.1
	eS	C	10 20	D = 43.56 Az = 300.0 (USCGS)
	eSS	C	13.4	h = 36 km
	LmH	B	24.9	PV(A):1.4s 102.3nm MPV(A)=5.4
	LmV	B	26.7	pPV(A):1.4s 114.0nm sPV(A):1.2s 48.7nm LmH(B):14s 0.8/um MLH(B)=4.8 LmV(B):14s 1.0/um MLV(B)=5.0
23.	eP	A	04 33 24	Southern Greece 37.58 N 22.87 E
	LmH	A	39	H = 04 29 50.2 h = 92 km MB = 4.8 D = 15.34 Az = 332.0 (USCGS) PV(A):1.0s 33.5nm MPV(A)=4.5
23.	eP1	A	07 22 35	Turkey 38.97 N 30.11 E
	iP2	A	22 37	H = 07 18 29.2 h = 15 km MB = 5.0

Day	Phase		h m s	Remarks
cont. 23.	eX	A	07 22 43.5	D = 17.51 Az = 317.8 (USCGS)
	LmH	B	28.6	P2V(A):2.0s 120.0nm MPV(A)=4.7
	LmV	B	30.5	XV(A):2.2s 120.0nm LmH(B):16s 2.1 /um MLH(B)=4.5 LmV(B):13s 1.4 /um MLV(B)=4.6
23.	eP1	AB	09 05 19	Turkey 39.13 N 28.70 E
	eP2	A	05 23	H = 09 01 24.7 h = 18 km MB = 5.2
	+iP3	A	05 28	D = 16.66 Az = 319.3 (USCGS)
	eS	B	08 28	P1V(A):1.5s 35.2nm MP1V(A)=4.3
	LmH	B	11.2	P1V(B):10s 1.6 /um MP1V(B)=5.1
	LmV	B	12.8	P2V(A):2.4s 415.0nm MP2V(A)=5.1 P3V(A):1.5s 186.0nm MP3V(A)=5.0 LmH(B):12.5s 17.9 /um MLH(B)=5.5 LmV(B):10s 8.4 /um MLV(B)=5.4
23.	eP	A	15 49 29	North Atlantic Ocean 55.56 N 35.04 W
	LmH	B	16 02.5	H = 15 43 38.6 h = normal MB = 4.5
	LmV	B	02.5	D = 28.01 Az = 80.4 (USCGS) LmH(B):13s 0.4 /um MLH(B)=4.2 LmV(B):13s 0.5 /um MLV(B)=4.4
23.	eP	A	18 10 28	Tadzhik SSR 37.46 N 72.63 E H = 18 02 18.8 h = 46 km MB = 5.1 D = 44.56 Az = 307.5 (USCGS) PV(A):1.0s 19.7nm MPV(A)=4.9
24.	eP	A	00 30 31	Leyte, Philippine Islands 10.47 N 125.33 E H = 00 17 09.3 h = 80 km MB = 5.1 D = 96.47 Az = 324.1 (USCGS) PV(A):1.6s 38.4nm MPV(A)=5.7
24.	+eP	AB	00 44 04.5	Turkey 39.06 N 29.91 E
	LmH	B	50.0	H = 00 40 01.7 h = 38 km MB = 4.8
	LmV	B	51.9	D = 17.34 Az = 317.9 (USCGS) PV(A):1.6s 71.4nm MPV(A)=4.6 PV(B):8s 0.5 /um MPV(B)=4.7 LmH(B):16s 2.7 /um MLH(B)=4.6 LmV(B):12s 2.9 /um MLV(B)=4.9

Day	Phase		h m s	Remarks
24.	eP	A	01 27 09	North Atlantic Ocean 55.59 N 35.35 W H = 01 21 16.7 h = normal MB = 4.6 D = 28.18 Az = 80.2 (USCGS)
24.	+eP	ABC	01 29 04.5	North Atlantic Ocean 55.68 N 34.99 W
	eS	BC	33 48	H = 01 23 12.0 h = 10 km MB = 5.4
	LmH	B	40.3	D = 27.96 Az = 80.6 (USCGS)
	LmV	B	41.6	PV(A):1.7s 115.0nm MPV(A)=5.4 SH(B):8s 1.6/um MSH(B)=5.6 LmH(B):15.5s 5.2/um MLH(B)=5.2 LmV(B):14s 6.9/um MLV(B)=5.5
24.	eP	A	01 53 04	North Atlantic Ocean 55.48 N 35.08 W H = 01 47 13.7 h = normal MB = 4.9 D = 28.05 Az = 80.2 (USCGS) PV(A):1.3s 19.7nm MPV(A)=4.8
24.	eP	A	01 57 11	North Atlantic Ocean 55.38 N 35.18 W H = 01 51 18.9 h = normal MB = 4.8 D = 28.12 Az = 79.9 (USCGS) PV(A):1.5s 22.6nm MPV(A)=4.8
24.	eP	A	02 44 09	Turkey 39.06 N 28.64 E
	eX	A	44 14.5	H = 02 40 11.3 h = 5 km MB = 4.5 D = 16.69 Az = 319.5 (USCGS) XV(A):1.2s 16.3nm
24.	ePKHKP	A	03 17 08	Tonga Islands 18.56 S 174.55 W H = 02 57 37.6 h = 133 km MB = 4.6 D = 147.61 Az = 352.7 (USCGS) PKHKPV(A):1.4s 14.0nm
24.	eP	A	11 20 03	Greece-Albania Border Region 39.87 N 19.47 E H = 11 17 10.3 h = 23 km MB = 4.5 D = 12.10 Az = 335.5 (USCGS)

Day	Phase		h m s	Remarks
24.	eP	A	14 41 36	Dodecanese Islands 36.78 N 28.84 E
	eX	A	41 41	H = 14 37 19.2 h = 33 km MB = 4.6
				D = 18.56 Az = 323.7 (USCGS)
				PV(A):0.9s 17.5nm MPV(A)=4.2
				XV(A):1.2s 24.4nm
25.	eP	A	03 55 13.5	Chagos Archipelago Region
	eX1	A	55 17	6.41 S 69.81 E
	eX2	A	55 22.5	H = 03 43 30.3 h = normal MB = 5.1
				D = 75.64 Az = 326.0 (USCGS)
				PV(A):1.2s 16.3nm MPV(A)=5.0
				X1V(A):1.6s 38.5nm
				X2V(A):0.8s 29.9nm
25.	e	A	12 12 53	Hokkaido, Japan 41.54 N 141.87 E
				H = 12 00 54.2 h = 79 km MB = 4.6
				D = 78.43 Az = 330.3 (USCGS)
25.	eP	A	19 38 18	Off Coast of Ecuador 2.68 S 81.98 W
				H = 19 24 57.0 h = 20 km MB = 5.1
				D = 94.34 Az = 39.6 (USCGS)
				PV(A):1.6s 27.5nm MPV(A)=5.4
25.	ePKHKP	A	20 49 34	Fiji Islands 20.54 S 178.20 W
				H = 20 30 50.7 h = 595 km MB = 4.7
				D = 148.97 Az = 347.8 (USCGS)
				PKHKPV(A):1.3s 48.0nm
26.	eP	A	06 45 41.5	North Atlantic Ocean 55.52 N 35.12 W
	eS	BC	50 25	H = 06 39 50.9 h = normal MB = 5.0
	LmV	B	58.3	D = 28.06 Az = 80.3 (USCGS)
	LmH	B	58.7	PV(A):2.8s 193.0nm MPV(A)=5.4
				LmV(B):13s 2.6/um MLV(B)=5.1
				LmH(B):13s 2.4/um MLH(B)=5.0
26.	+iP	ABC	14 32 11.6	Near Islands, Aleutian Islands
	eipP	A	32 25	52.97 N 171.46 E
	ePP	BC	35 00	H = 14 20 30.6 h = 41 km MB = 5.8

Day	Phase		h m s	Remarks
cont. 26.	eS	BC	14 41 50	D = 75.36 Az = 346.9 (USCGS)
	ePS	BC	42 22	h = 52 km
	eSS	B	46 45	PV(A):1.5s 176.0nm MPV(A)=5.9
	LmV	B	15 11.8	PV(B):14s 1.2/um MPV(B)=5.8
	LmH	B	12.0	pPV(A):2.0s 188.0nm
				PPV(B):14s 0.8/um MPPV(B)=5.6
				SH(B):16s 0.7/um MSH(B) =5.5
				LmV(B):16s 1.6/um MLV(B)=5.5
				LmH(B):17.5s 3.0/um MLH(B)=5.6
26.	eiPKHKP	A	15 59 44.5	<u>Fiji Islands</u> 21.69 S 176.19 W
	epPKHKP	A	16 00 22	H = 15 40 06.9 h = 131 km MB = 5.2 (USCGS)
				D = 150.4 h = 142 km
				PKHKPV(A):0.8s 34.6nm
				pPKHKPV(A):0.9s 33.1nm
26.	eP	A	23 19 08	<u>Turkey</u> 38.85 N 29.91 E
	LmH	B	25.2	H = 23 15 02.7 h = normal MB = 4.4
				D = 17.49 Az = 318.3 (USCGS)
				PV(A):1.8s 20.3nm MPV(A)=4.0
27.	+iP	A	01 58 15.8	<u>Turkey</u> 38.99 N 30.13 E
	LmH	B	02 06.0	H = 01 54 10.5 h = 25 km MB = 4.6
	LmV	B	06.0	D = 17.50 Az = 317.8 (USCGS)
				PV(A):1.1s 20.2nm MPV(A)=4.2
				LmH(B):12s 0.4/um MLH(B)=3.9
				LmV(B):10s 0.5/um MLV(B)=4.2
27.	+iP	A	04 08 20.5	<u>Near East Coast of Kamchatka</u>
	epP	A	08 30	54.68 N 161.61 E
	esP	A	08 34	H = 03 56 59.6 h = 51 km MB = 5.2
				D = 72.09 Az = 340.5 (USCGS)
				h = 37 km
				PV(A):1.0s 37.4nm MPV(A)=5.5
27.	eP	A	08 16 52	<u>Kurile Islands</u> 43.59 N 146.51 E
	epP	A	17 08.5	H = 08 04 55.7 h = 52 km MB = 4.7
				D = 78.27 Az = 332.6 (USCGS)
				h = 63 km

Day	Phase		h m s	Remarks
27.	eP	AB	09 38 38	<u>Turkey</u> 38.99 N 30.02 E
				H = 09 34 34.3 h = normal MB = 4.5
				D = 17.45 Az = 317.9 (USCGS)
				PV(A):1.7s 30.3nm MPV(A)=4.2
27.	eP1	AB	09 39 14.5	<u>Turkey</u> 39.04 N 29.51 E
	eP2	A	39 17.5	H = 09 35 12.9 h = normal MB = 4.9 (USCGS)
	eS	C	42 32	D = 17.2
	e	B	42 37	P1V(A):1.6s 38.5nm MP1V(A)=4.3
	e	B	44 27	P2V(A):1.8s 81.1nm MP2V(A)=4.6
	e	A	45 19.5	LmH(B):10.5 2.0/um MLH(B)=4.7
	LmH	B	46.8	LmV(B):9s 1.6/um MLV(B)=4.8
	LmV	B	47.2	
27.	ePKP	A	12 38 22	<u>New Hebrides Islands</u> 19.12 S 168.46 E
				H = 12 18 43.2 h = 22 km MB = 4.4
				D = 143.56 Az = 335.1 (USCGS)
				PKPV(A):0.9s 9.7nm
27.	ePKP	A	22 11 07.5	<u>Samoa Islands</u> 15.67 S 172.65 W
	eX	A	11 30	H = 21 51 32.3 h = normal MB = 5.0
				D = 144.94 Az = 355.3 (USCGS)
				PKPV(A):1.1s 16.1nm
				XV(A):1.6s 22.0nm
27.	eP	AB	22 28 47	<u>Turkey</u> 39.04 N 29.48 E
	e	B	34 10	H = 22 24 45.0 h = normal MB = 4.7
	LmH	B	36.7	D = 17.13 Az = 318.5 (USCGS)
	LmV	B	36.7	PV(A):1.5s 42.8nm MPV(A)=4.4
				LmH(B):9s 1.1/um MLH(B)=4.4
				LmV(B):9s 0.9/um MLV(B)=4.5
28.	ePKIKP	A	00 48 33	<u>Salomon Islands</u> 8.10 S 156.41 E
	ePP	B	50(30)	H = 00 29 21.0 h = 3 km MB = 5.2
	ePKS	BC	50 55	D = 128.54 Az = 332.0 (USCGS)
	eSS	C	01 07 50	PKIKPV(A):1.5s 30.2nm
	LmH	B	39.6	LmH(B):17s 0.6/um MLH(B)=5.4

Day	Phase		h m s	Remarks
28.	ePKIKP	A	01 31 25	Salomon Islands 8.12 S 156.39 E
				H = 01 12 14.4 h = 5 km MB = 5.6
				D = 128.54 Az = 332.0 (USCGS)
				PKIKPV(A):1.2s 18.3nm
28.	eP	A	03 26 33	United Arab Republic 27.67 N 33.63 E
	e	A	26 37	H = 03 20 34.7 h = 5 km MB = 4.9
	eX	A	26 43	D = 28.40 Az = 329.9 (USCGS)
	eS	BC	31 24	PV(A):1.9s 60.5nm MPV(A)=5.1
	e	BC	31 46	XV(A):1.5s 45.2nm
28.	LmH	B	21 09.4	North of Halmahera 4.0 N 128.6 E
	LmV	B	09.4	H = 20 11 12.5 h = 46 km MB = 5.3 (USCGS)
				D = 103.7
				LmH(B):18s 0.5/um MLH(B)=5.1
				LmV(B):18s 0.4/um MLV(B)=5.0
29.	+iP	A	06 06 59.3	Kurile Islands 43.47 N 146.44 E
	ipP	A	07 10	H = 05 55 02.4 h = 50 km MB = 5.3
	isP	A	07 15	D = 78.35 Az = 332.5 (USCGS)
	LmH	B	45.0	pPV(A):1.5s 55.2nm
	LmV	B	48.0	sPV(A):1.5s 75.4nm
				LmH(B):17s 0.6/um MLH(B)=5.0
				LmV(B):19s 0.6/um MLV(B)=5.0
29.	eP	B	11 35 21	Near Coast of Chiapas, Mexico
	e	A	35 27	14.62 N 92.69 W
	eS	BC	45 55	H = 11 22 36.4 h = 41 km MB = 5.6
	eSS	BC	52 08	D = 87.64 Az = 38.1 (USCGS)
	eSSS	B	56 00	LmH(B):17s 24.6/um MLH(B)=6.7
	LmH	B	12 16.6	LmV(B):17s 24.5/um MLV(B)=6.7
	LmV	B	16.6	
29.	eP	A	12 12(30)	Rat Islands, Aleutian Is.
	epP	A	12 47	51.76 N 177.00 E
				H = 12 00 39.5 h = 52 km MB = 5.0
				D = 77.22 Az = 350.5 (USCGS)

Day	Phase		h m s	Remarks
29.	eP	A	14 14 05.5	<u>Near Coast of Chiapas, Mexico</u>
				14.58 N 92.75 W
				H = 14 01 19.1 h = normal MB = 5.6
				D = 87.71 Az = 38.1 (USCGS)
				PV(A):1.7s 30.3nm MPV(A)=5.3
29.	eiP1	AB	14 14 20	<u>Near Coast of Chiapas, Mexico</u>
	eP2	A	14 23	14.52 N 92.60 W
	eiP3	AB	14 27	H = 14 01 32.8 h = normal MB = 5.8
	e	B	17 08	D = 87.67 Az = 38.1 (USCGS)
	iSKS	B	24 44	P1V(A):2.0s 145.2nm MP1V(A)=5.9
	iS	B	25 08	P2V(A):2.0s 222.0nm MP2V(A)=6.1
	LmH	B	55.5	P3V(A):3.5s 2210.0nm MP3V(A)=6.9
	LmV	B	55.5	P3V(B):10s 26.6 µm MP3V(B)=7.5
				SH(B):17s 35.1 µm MSH(B)=7.1
				LmH(B):19s 144.0 µm MLH(B)=7.4
				LmV(B):19s 138.0 µm MLV(B)=7.4
29.	e	A	14 52 22	<u>Dodekanese Islands</u> 36.3 N 26.6 E
	e	A	52 28	H = 14 48 15
				D = 17.95 Az = 328
29.	ePKIKP	A	18 21 17	<u>Easter Island Cordillera</u>
	iPKHKP	A	21 25.5	55.48 S 124.37 W
	iPKP2	A	21 37	H = 18 01 29.6 h = normal MB = 5.6
				D = 153.46 Az = 82.0 (USCGS)
				PKIKPV(A):1.8s 40.5nm
				PKHKPV(A):1.6s 121.0nm
				PKP2V(A):1.3s 48.0nm
29.	eP	A	19 42 42.5	<u>Near Coast of Chiapas, Mexico</u>
	LmV	B	20 23.0	14.67 N 93.50 W
	LmH	B	23.5	H = 19 29 52.3 h = 29 km MB = 5.2
				D = 88.09 Az = 38.0 (USCGS)
				PV(A):2.0s 68.4nm MPV(A)=5.6
				LmV(B):19s 1.1 µm MLV(B)=5.3
				LmH(B):19s 1.2 µm MLH(B)=5.3

Day	Phase		h m s	Remarks
29.	eP	AB	21 33 14	Near Coast of Chiapas, Mexico
	eX	A	33 21	14.57 N 93.57 W
	eSKS	B	43 53	H = 21 20 24.1 h = 35 km MB = 5.3
	LmV	B	22 13.3	D = 88.21 Az = 37.9 (USCGS)
	LmH	B	13.6	PV(A):2.8s 150.0nm MPV(A)=5.8
				XV(A):2.8s 129.0nm
				LmV(B):18s 2.3/um MLV(B)=5.6
				LmH(B):18s 2.0/um MLH(B)=5.6
29.	eP	A	22 01 51	Near Coast of Chiapas, Mexico
				14.19 N 93.38 W
				H = 21 49 00.7 h = normal MB = 5.1
				D = 88.40 Az = 38.0 (USCGS)
30.	eP	AB	08 45 50	Near Coast of Chiapas, Mexico
	Pm	A	45 58	14.70 N 93.19 W
	ePP	C	49 08	H = 08 32 59.1 h = 19 km MB = 5.6
	eiS	BC	56 28	D = 87.88 Az = 38.0 (USCGS)
	ePS	C	57 38	PV(B):14s 5.3/um MPV(B)=6.7
	eSSS	C	09 06 00	PmV(A):3.2s 710.0nm MPmV(A)=6.3
	LmH	B	28.2	SH(B):12s 5.7/um MSH(B)=6.5
	LmV	B	28.5	LmH(B):16s 21.5/um MLH(B)=6.7
				LmV(B):17s 23.8/um MLV(B)=6.7
30.	eP	AB	13 04 28	Near Coast of Chiapas, Mexico
	eSKS	B	15 00	14.45 N 93.43 W
	eS	B	15 20	H = 12 51 36.3 h = 24 km MB = 5.3
	LmH	B	45.0	D = 88.22 Az = 38.0 (USCGS)
	LmV	B	47.4	PV(A):1.8s 37.2nm MPV(A)=5.4
				LmH(B):17s 2.8/um MLH(B)=5.7
				LmV(B):16s 2.5/um MLV(B)=5.7
30.	eP	A	13 33 55	Near Coast of Chiapas, Mexico
	e	A	34 01	14.48 N 93.50 W
	LmH	B	14 14.8	H = 13 21 03.2 h = 22 km MB = 5.4
	LmV	B	17.6	D = 88.24 Az = 37.9 (USCGS)
				PV(A):2.0s 59.8nm MPV(A)=5.6
				LmH(B):16.5s 1.7/um MLH(B)=5.5
				LmV(B):15s 1.6/um MLV(B)=5.6

Day	Phase		h m s	Remarks
30.	eP1	AB	14 52 35	<u>Northeast of Taiwan</u> 27.17 N 125.20 E
	iP2	A	52 36.5	H = 14 40 33.4 h = 220 km MB = 5.1
				D = 82.93 Az = 324.0 (USCGS)
				P2V(A):1.6s 82.5nm MPV(A)=5.7
30.	eP	A	15 02 19.5	<u>Turkey</u> 39.34 N 29.34 E
				H = 14 58 21.8 h = 25 km MB = 4.6
				D = 16.84 Az = 318.0 (USCGS)
30.	eP	A	16 48 45	<u>Turkey</u> 39.39 N 29.07 E
	LmV	B	56.6	H = 16 14 45.9 h = 23 km MB = 4.8
	LmH	B	56.8	D = 16.66 Az = 318.2 (USCGS)
				PV(A):1.8s 47.3nm MPV(A)=4.3
				LmV(B):10s 1.5 μm MLV(B)=4.7
				LmH(B):10s 1.8 μm MLH(B)=4.6
30.	eP	A	18 22 33	<u>Taiwan</u> 24.10 N 121.74 E
	eX	A	22 51	H = 18 10 08.7 h = 59 km MB = 5.1
				D = 83.51 Az = 323.0 (USCGS)
				XV(A):2.0s 25.6nm
30.	eP	A	19 43 20	<u>Near Coast of Oaxaca, Mexico</u>
	eX	A	43 30	15.05 N 93.99 W
				H = 19 30 28.7 h = 23 km MB = 5.1
				D = 88.08 Az = 37.8 (USCGS)
				PV(A):1.8s 27.0nm MPV(A)=5.3
				XV(A):2.0s 34.2nm
30.	+eP	A	24 03 11	<u>Turkey</u> 39.17 N 29.38 E
	LmH	B	10.9	H = 23 59 10.7 h = normal MB = 4.6
	LmV	B	10.9	D = 16.98 Az = 318.3 (USCGS)
				PV(A):2.3s 91.4nm MPV(A)=4.5

Day	Phase		h m s	Remarks
1.	eP	AB	03 35 13	Luzon, Philippine Islands
	e	A	37 50	15.67 N 121.83 E
	e	A	38 22	H = 03 22 13.4 h = 29 km MB = 5.5 (USCGS)
	eSKS	BC	45 40	D = 90.2
	eS	C	46 00	PV(A):2.0s 55.5nm MPV(A)=5.5
	eSS	C	52 00	LmH(B):18s 2.9/um MLH(B)=5.7
	LmH	B	04 19.4	LmV(B):18s 2.2/um MLV(B)=5.7
	LmV	B	19.4	
1.	eP	A	03 55 29.5	Kazakh-Sinkiang Border Region
				47.62 N 82.68 E
				H = 03 47 15.7 h = normal MB = 4.9
				D = 44.96 Az = 301.5 (USCGS)
1.	eP	A	04 20 56.5	Near Coast of Chiapas, Mexico
				14.61 N 93.55 W
				H = 04 08 07.2 h = 35 km MB = 5.1
				D = 88.17 Az = 37.9 (USCGS)
1.	eP1	A	08 48 10.5	Near Coast of Chiapas, Mexico
	eP2	A	48 15	14.64 N 93.16 W
				H = 08 35 24.2 h = 44 km MB = 5.4
				D = 87.91 Az = 38.0 (USCGS)
				P2V(A):1.6s 22.0nm MPV(A)=5.2
1.	eP	AB	20 16 16	Near Coast of Chiapas, Mexico
	eSKS	B	26 52	14.62 N 93.61 W
	eS	B	27 05	H = 20 03 27.9 h = 38 km MB = 5.0
	ePS	B	28 04	D = 88.19 Az = 37.9 (USCGS)
	eSSS	C	36 40	PV(A):2.8s 150.0nm MPV(A)=5.8
	LmH	B	56.5	PV(B):8s 0.7/um MPV(B)=6.0
	LmV	B	56.5	SH(B):11s 0.7/um MSH(B)=5.6
				LmH(B):20s 1.8/um MLH(B)=5.5
				LmV(B):20s 1.2/um MLV(B)=5.3
1.	eP	A	20 47 44	Off Coast of Hokkaido, Japan
	epP	A	47 55.5	42.90 N 147.34 E
				H = 20 35 38.4 h = 15 km MB = 5.1
				D = 79.16 Az = 333.1 (USCGS)
				PV(A):1.6s 38.5nm MPV(A)=5.2

Day	Phase		h m s	Remarks
1.	eP	A	22 08 49	USSR-Mongolia Border Region
				51.3 N 93.2 E
				H = 22 00 06 D = 48.8 (ANUSSR)
2.	eiP	ABC	02 19 45.5	Near Coast of Chiapas, Mexico
	eS	BC	30 25	14.72 N 93.71 W
	ePS	BC	31 33	H = 02 06 56.3 h = 32 km MB = 5.4
	eSSS	C	40 00	D = 88.17 Az = 37.9 (USCGS)
	LmH	B	03 00.2	PV(A):2.6s 211.0nm MPV(A)=6.0
	LmV	B	00.3	PV(B):7.5s 0.8/um MPV(B)=6.1
				SH(B):10s 0.9/um MSH(B)=5.7
				LmH(B):18s 1.7/um LmH(B)=5.5
				LmV(B):18s 1.2/um LmV(B)=5.4
2.	e	A	06 19 03.5	Probably Adriatic Sea (ISC)
	e	A	19 45	
3.	ePn	A	04 19 05	Italy 44.9 N 10.4 E
	ePg	A	19 33.5	H = 04 17 41 (BCIS)
	iSn	A	20 13.5	D = 5.85
	eSg	AB	20 59.5	
3.	eP	A	04 31 36.5	Kurile Islands 43.61 N 146.59 E
	epP	A	31 54	H = 04 19 40.8 h = 57 km MB = 4.8
				D = 78.28 Az = 332.6 (USCGS)
				PV(A):1.2s 16.3nm MPV(A)=4.8
3.	e	A	11 04 25	Greenland Sea 79.68 N 3.54 E
				H = 10 58 03.5 h = normal MB = 4.1
				D = 29.29 Az = 169.5 (USCGS)
3.	eP	A	12 24 55	Northeast of Taiwan 26.6 N 125.9 E
				H = 12 12 36 h = 112 km (ISC)
				D = 83.7
4.	eP	A	07 19 03	Andreanof Islands, Aleutian Is.
				51.40 N 179.22 W
				H = 07 07 08.0 h = 44 km MB = 5.2
				D = 77.91 Az = 353.0 (USCGS)
				PV(A):1.3s 21.8nm MPV(A)=5.1

Day	Phase		h m s	Remarks
4.	ePKP	B	08 00 35	New Hebrides Islands 20.69 S 173.49 E
	ePKP	A	00 39	H = 07 40 52.3 h = 14 km MB = 5.2
	e	B	01 35	D = 146.86 Az = 338.8 (USCGS)
	ePP	A	04 18	PKPV(A):2.5s 46.1nm
	LmH	B	09 04.5	LmV(B):22s 0.9/um
	LmV	B	04.5	
4.	ePKIKP	A	11 44 07.5	Loyalty Islands 21.53 S 170.51 E
	-iPKHKP	A	44 10	H = 11 24 42.6 h = 135 km MB = 5.2
	epPKHKP	A	44 45	D = 146.55 Az = 335.4 (USCGS)
				PKHKPV(A):1.0s 65.0nm
				pPKHKPV(A):1.9s 53.0nm
4.	eP	A	12 22 06.5	Luzon, Philippine Islands
				14.62 N 124.23 E
				H = 12 08 57.0 h = normal MB = 4.9
				D = 92.50 Az = 324.0 (USCGS)
				PV(A):1.2s 10.2nm
4.	ePP	ABC	19 12 24	Mid-Indian Rise 41.57 S 80.09 E
	e	B	19 55	H = 18 53 19.7 h = normal MB = 5.3 (USCGS)
	eSS	B	27 50	D = 109.6
	LmH	B	20 06.5	PPV(A):2.0s 25.6nm MPPV(A)=6.2
	LmV	B	07.7	LmV(B):16s 1.2/um MLV(B)=5.6
4.	+iPKP	A	20 40 34	Fiji Islands 20.29 S 177.90 W
				H = 20 21 35.3 h = 438 km MB = 4.2
				D = 148.79 Az = 348.3 (USCGS)
				PKPV(A):1.3s 26.2nm
5.	eP	A	01 34 41	South of Honshu, Japan 32.33 N 141.58 E
	e	A	34 54.5	H = 01 22 00.3 h = normal MB = 4.9
				D = 86.32 Az = 330.7 (USCGS)
				PV(A):1.4s 9.3nm MPV(A)=4.8
5.	ePn	A	12 51 25	Italy 44.8 N 10.2 E
	e	A	51 43.5	H = 12 49 56 (BCIS)
	eiSn	A	52 32	D = 5.95
	ei	A	52 44	
	eSg	A	53 16	

Day	Phase		h m s	Remarks
5.	eP	A	15 42 19	Southern Nevada 37.22 N 116.18 W
				H = 15 30 00.2 h = 0 km MB = 5.2
				D = 81.21 Az = 30.6 (USCGS)
				Nevada test site "Mint Leaf"
				37°12'59.52" N 116°11'02.64" W (USAEC)
				PV(A):1.2s 12.2nm MPV(A)=4.8
5.	eiPKP	AB	20 25 44	New Hebrides Islands 20.16 S 170.23 E
	eiX	A	25 47.5	H = 20 06 08.5 h = 36 km MB = 5.3
				D = 145.20 Az = 336.0 (USCGS)
				PKPV(A):1.1s 64.5nm
				XV(A):1.5s 45.2nm
6.	eP	A	02 48 15.5	Luzon, Philippine Islands
				15.70 N 121.75 E
				H = 02 35 17.2 h = 39 km MB = 5.2
				D = 90.22 Az = 323.3 (USCGS)
6.	eP	A	12 33 46	North Atlantik Ridge 49.95 N 29.02 W
	LmH	B	43.3	H = 12 28 16.0 h = normal MB = 4.8
	LmV	B	43.7	D = 25.74 Az = 72.7 (USCGS)
				PV(A):1.5s 30.1nm MPV(A)=4.7
				LmH(B):18s 2.2/um MLH(B)=4.7
				LmV(B):18s 1.1/um MLV(B)=4.6
6.	eP	A	15 33 47	Nicobar Islands 9.84 N 92.95 E
	e	A	33 58	H = 15 21 55.1 h = normal MB = 5.3
	eX1	A	34 00	D = 76.97 Az = 319.8 (USCGS)
	eX2	A	34 09.5	PV(A):1.4s 34.9nm MPV(A)=5.2
	LmH	B	16 12.3	X1V(A):1.5s 40.2nm
	LmV	B	12.3	X2V(A):1.9s 72.0nm
				LmH(B):20s 0.8/um MLH(B)=5.0
				LmV(B):17s 0.7/um MLV(B)=5.1
6.	eP	A	21 06 08	Turkey 39.03 N 30.04 E
	LmH	B	12.2	H = 21 02 05.7 h = 42 km MB = 4.5
	LmV	B	15.0	D = 17.42 Az = 317.8 (USCGS)

Day	Phase		h m s	Remarks
7.	ePKP	A	06 58 56.5	<u>Loyalty Islands</u> 22.37 S 170.25 E H = 06 39 07.6 h = 37 km MB = - D = 147.20 Az = 334.6 (USCGS) PKPV(A):1.6s 27.5nm
7.	e	A	19 35 15	<u>Halmahera</u> 1.65 N 127.05 E H = 19 17 00.8 h = 106 km MB = 5.5 (USCGS) D = 104.6
8.	eP	A	02 53 20	<u>Turkey</u> 38.90 N 29.95 E
	LmH	B	59.3	H = 02 49 15.6 h = 30 km MB = 4.7
	LmV	B	03 01.2	D = 17.47 Az = 318.2 (USCGS) PV(A):2.0s 64.0nm MPV(A)=4.4 LmH(B):16s 1.3 μm MLH(B)=4.3 LmV(B):12s 1.0 μm MLV(B)=4.4
8.	eP	A	07 03 04	<u>Turkey</u> 38.88 N 29.94 E H = 06 58 59.1 h = 34 km MB = 4.4 (USCGS) D = 17.5
8.	eP	A	15 55 13.5	<u>Kurile Islands</u> 44.70 N 148.29 E H = 15 43 25.2 h = 102 km MB = 4.6 D = 77.87 Az = 333.5 (USCGS)
8.	eP	A	18 33 59	<u>Greece</u> 38.77 N 22.41 E
	e	A	34 16	H = 18 30 41.4 h = 47 km MB = 4.3
	e	A	34 23	D = 14.12 Az = 330.7 (USCGS)
	LmH	B	40.0	LmH(B):12s 0.5 μm MLH(B)=3.8
	LmV	B	40.0	LmV(B):10s 0.4 μm
8.	eP	A	19 34 38	<u>Kurile Islands</u> 43.96 N 146.68 E
	e	A	34 49	H = 19 22 37.6 h = 13 km MB = 4.9 D = 78.00 Az = 332.6 (USCGS) PV(A):1.1s 16.1nm MPV(A)=5.1
9.	e	A	13 36 01	Probably <u>West New Guinea Region</u> (USCGS)
	LmH	B	14 17.0	
	LmV	B	22.0	

Day	Phase		h m s	Remarks
9.	ePKIKP	A	18 19 25	New Britain Region 4.36 S 151.71 E
	epPKIKP	A	21 09	H = 18 00 50.0 h = 203 km MB = 5.9
	ePKKP	A	29 13	D = 123.05 Az = 330.8 (USCGS)
	eSS	B	37 35	PKIKPV(A):1.0s 27.6nm
	esSS	B	39 53	LmH(B):22s 1.4/um
	LmH	B	19 04.0	
	LmV	B	19.0	
10.	ePn	A	01 49 58.5	Austria 47.2 N 9.6 E
	ePg	A	50 12.5	H = 01 49 03 (BCIS)
	eSn	A	50 41.5	D = 3.6
	iSg	A	50 57.5	
	e	A	51 02	
10.	eP1	A	04 06 10	Ionian Sea 37.37 N 20.69 E
	eP2	A	06 16.5	H = 04 02 43.1 h = 53 km MB = 4.1 (USCGS)
	e(pP)	A	06 29	D = 14.8
	LmH	B	13.0	P2V(A):0.8s 21.2nm MP2V(A)=4.3
	LmV	B	13.0	LmH(B):12s 0.4/um
				LmV(B):12s 0.5/um
10.	ePKP	A	10 58 08	Fiji Islands 20.53 S 178.24 W
				H = 10 39 18.2 h = 521 km MB = 4.5 (USCGS)
				D = 149.0
				PKPV(A):1.1s 16.1nm
10.	eP	A	20 17 58	Mariana Islands 18.58 N 145.24 E
	e	A	22 03	H = 20 05 15.9 h = 602 km MB = 5.6 (USCGS)
	eiPP1	AB	22 12	D = 99.9
	iPP2	A	22 14	PV(A):1.5s 25.1nm MPV(A)=5.4
	eS	B	28 36	PP2V(A):1.4s 58.1nm MPPV(A)=5.7
	e	B	34 12	
	eSS	B	35 40	
	LmH	B	21 07.0	
	LmV	B	07.0	
10.	eP	A	21 58 55.5	Off East Coast of Honshu, Japan
				34.47 N 141.60 E
				H = 21 46 23.6 h = 43 km MB = 4.7
				D = 84.47 Az = 330.7 (USCGS)

Day	Phase		h m s	Remarks
11.	eP	A	03 19 35.5	Southern Iran 28.54 N 52.27 E
	e	A	19 40	H = 03 12 19.7 h = 22 km MB = 5.1
	LmH	B	39.1	D = 37.67 Az = 317.3 (USCGS)
	LmV	B	39.1	PV(A):1.2s 28.5nm MPV(A)=4.9
11.	eP	A	03 55 32	Mexico (UPP)
				PV(A):2.3s 48.7nm
11.	ePn	A	05 53 46.5	Yugoslavia 45.0 N 17.3 E
	eSn	A	55 02.5	H = 05 52 07 (BCIS)
	e	A	55 20	D = 6.80
	e	A	55 33	PnV(A):0.8s 17.3nm
11.	e(P)	A	09 19 19	Turkey 39.61 N 29.37 E
				H = 09 15 26 h = 56 km
				D = 16.65 Az = 317 (ISC)
11.	eP	A	10 21 45	Near Islands, Aleutian Is.
				52.50 N 173.37 E
				H = 10 10 01.2 h = 59 km MB = 4.8
				D = 76.08 Az = 348.1 (USCGS)
11.	ePKHKP	A	15 25 05.5	Fiji Islands 19.57 S 178.04 W
	ePKP2	A	25 09.5	H = 15 06 13.6 h = 480 km MB = 4.9 (USCGS)
	epPKHKP	A	26 57	D = 148.1
				PKHKPV(A):1.6s 44.0nm
				pPKHKPV(A):1.8s 40.6nm
11.	eP	A	16 32 38	Turkey 38.92 N 30.11 E
	LmH	B	38.6	H = 16 28 30.8 h = normal MB = 4.3
	LmV	B	40.4	D = 17.54 Az = 317.9 (USCGS)
				PV(A):2.0s 30.0nm MPV(A)=4.1
12.	ePKIKP	A	01 21 27.5	South of Fiji Islands
	ePKHKP	A	21 36	23.47 S 179.99 W
	ePKP2	A	21 47	H = 01 02 43.2 h = 550 km MB = 4.9
	epPKP	A	23 44.5	D = 151.41 Az = 344.5 (USCGS)
	e	A	23 51.5	PKHKPV(A):1.1s 20.2nm
				pPKPV(A):1.1s 20.2nm

Day	Phase		h m s	Remarks
12.	ePKIKP	A	17 17 37.5	Tonga Islands 20.60 S 174.42 W
	ePKHKP	A	17 42	H = 16 58 01.4 h = 95 km MB = 5.3
				D = 149.64 Az = 352.4 (USCGS)
				PKHKPV(A):1.5s 65.3nm
12	eP	A	22 16 17	West Pakistan 27.49 N 67.47 E
				H = 22 07 39.4 h = 23 km MB = 4.7
				D = 47.81 Az = 314.7 (USCGS)
12.	eP	A	22 52 28.5	Greece 38.24 N 22.69 E
	eX1	A	52 34	H = 22 59 02.2 h = 35 km MB = 4.9
	eX2	A	52 38.5	D = 14.69 Az = 331.2 (USCGS)
	eX3	A	52 41.5	X3V(A):1.3s 30.6nm
	LmH	B	59.2	LmH(B):12s 1.1 µm MLH(B)=5.2
	LmV	B	59.4	LmV(B):15s 1.4 µm
13.	ePKP2	A	16 03 59	South of Kermadec Islands
				33.73 S 178.79 W
				H = 15 43 21.2 h = 60 km MB = -
				D = 161.46 Az = 338.8 (USCGS)
13.	iPn	A	17 52 40.5	Federal Rep. Germany 47.8 N 11.1 E
	iPg	A	52 45.5	H = 17 51 57 (BCIS)
	eiSn	A	53 15	D = 2.85
	eiSg	A	53 26	
	i	A	53 29	
14.	eP	A	03 48 04	Off Coast of Oregon 42.52 N 126.39 W
				H = 03 35 54.3 h = normal MB = 4.7
				D = 80.25 Az = 25.6 (USCGS)
14.	ePKIKP1	A	08 51 29.5	Near North Coast of New Guinea
	ePKIKP2	A	51 33	3.38 S 145.15 E
	ePP	AB	52 55	H = 08 32 42.2 h = 29 km MB = 5.6
	eSS	B	09 05	D = 118.88 Az = 328.2 (USCGS)
	eSSS	B	13 40	PKIKP2V(A):2.0s 29.9nm

Day	Phase		h m s	Remarks
14.	+eP	AB	09 25 47.5	<u>Eastern Caucasus</u> 43.02 N 47.08 E
	Pm	A	25 52	H = 09 20 22.0 h = 17 km MB = 5.8
	e	A	26 09	D = 25.24 Az = 300.0 (USCGS)
	i	B	26 16	PV(B):7s 2.1 /um MPV(B)=6.0
	e	A	26 19	PmV(A):1.6s 121.0nm MPmV(A)=5.4
	eS	B	30 19	LmH(B):14s 9.4 /um MLH(B)=5.5
	LmH	B	36.5	LmV(B):14s 11.4 /um MLV(B)=5.7
	LmV	B	37.2	
14.	ePn	A	10 01 04.5	<u>Czechoslovakia</u> 50.76 N 14.43 E
	iSg	A	01 29	H = 10 00 00
				explosion 21.6 to
				D = 1.79 Az = 267 (PRU)
14.	+iP1	AB	18 17 50	<u>Eastern Caucasus</u> 43.03 N 47.09 E
	Pm	A	18 13	H = 18 12 28.0 h = 44 km MB = 5.6
	iS	B	22 16	D = 25.24 Az = 299.9 (USCGS)
	LmH	B	28.5	PV(A):1.2s 69.1nm MPV(A)=5.2
	LmV	B	29.3	PV(B):6s 4.0 /um MPV(B)=6.3
				PmV(A):1.6s 867.0nm MPmV(A)=6.2
				LmH(B):16s 113.0 /um MLH(B)=6.5
				LmV(B):16s 136.0 /um MLV(B)=6.7
14.	e	A	19 26 02	<u>Eastern Caucasus</u> 42.98 N 46.79 E
				H = 19 20 26.1 h = normal MB = 4.8
				D = 25.08 Az = 300.1 (USCGS)
14.	eP	A	19 37 06	<u>Eastern Caucasus</u> 42.87 N 47.00 E
	e	A	37 19	H = 19 31 38.1 h = normal MB = 4.6
				D = 25.27 Az = 300.3 (USCGS)
14.	eP	A	21 21 14	<u>Eastern Caucasus</u> 43.05 N 47.15 E
				H = 21 15 50.5 h = normal MB = 4.9
				D = 25.27 Az = 299.9 (USCGS)
14.	eP	A	21 38 10	<u>Turkey</u> 38.86 N 30.04 E
				H = 21 34 04.8 h = 19 km MB = 4.2
				D = 17.56 Az = 318.2 (USCGS)
				PV(A):2.2s 32.7nm MPV(A)=4.1

Day	Phase		h m s	Remarks
15.	e	A	02 02 23	Eastern Caucasus 42.92 N 46.76 E H = 01 56 47.3 h = normal MB = 4.6 D = 25.08 Az = 300.2 (USCGS)
15.	eP e	A A	02 12 50 13 30	Eastern Caucasus 43.00 N 46.81 E H = 02 08 04.1 h = normal MB = 4.8 D = 25.08 Az = 300.0 (USCGS)
15.	eP	A	06 24 47	Eastern Caucasus 43.06 N 47.28 E H = 06 19 16.6 h = normal MB = 4.5 D = 25.35 Az = 299.9 (USCGS)
15.	eP LmH LmV	AB B B	09 57 33 10 38.5 38.9	Near Coast of Chiapas, Mexico 14.51 N 92.81 W H = 09 44 45.2 h = normal MB = 5.2 D = 87.80 Az = 38.1 (USCGS) PV(A):0.5s 11.5nm MPV(A)=5.4 PV(B):7s 0.4/um MPV(B)=5.8 LmH(B):17s 1.3/um MLH(B)=5.4 LmV(B):17s 1.0/um MLV(B)=5.3
15.	+iPKIKP -iPKHKP epPKP	AB AB A	10 59 31.5 59 36.8 11 00 37	Fiji Islands 21.47 S 176.72 W H = 10 40 14.1 h = 251 km MB = 5.3 D = 150.16 Az = 349.3 (USCGS) PKIKPV(A):1.2s 30.5nm PKHKPV(A):1.1s 186.0nm
15.	eP	A	13 42 18.5	Southern Nevada 37.16 N 116.04 W H = 13 30 00.0 h = 0 km MB = 5.3 D = 81.19 Az = 30.7 (USCGS) Nevada test site "Cornice" 37°09'50" N 116°02'13" W (USAEC) PV(A):1.4s 18.6nm MPV(A)=4.9
15.	eP	A	14 21 55	South Atlantic Ridge 22.49 S 12.71 W H = 14 10 10.1 h = normal MB = 4.9 D = 75.92 Az = 15.7 (USCGS) PV(A):1.4s 18.6nm MPV(A)=5.0

Day	Phase		h m s	Remarks
15.	iPKP	A	15 37 30	Loyalty Islands 20.26 S 168.80 E H = 15 18 00.7 h = 64 km MB = 4.4 D = 144.73 Az = 334.7 (USCGS) PKPV(A):1.1s 24.2nm
15.	+iP1	AB	17 21 54.5	USSR-Mongolia Border Region
	iP2	A	22 00.5	50.17 N 91.25 E
	+iPP	B	23 46	H = 17 13 15.1 h = normal MB = 5.9
	-i	A	23 50	D = 48.38 Az = 303.1 (USCGS)
	eS	B	28 55	P1V(A):0.8s 50.0nm MP1V(A)=5.6
	eSS	B	32 20	P1V(B):5.5s 5.2 um MP1V(B)=6.8
	LmH	B	41.8	P2V(A):1.9s 985.0nm MP2V(A)=6.5
	LmV	B	43.4	PPV(B):8s 7.1 um MPPV(B)=6.6 SH(B):11s 7.5 um MSH (B)=6.5 LmH(B):13s 218.0 um MLH (B)=7.3 LmV(B):14s 133.5 um MLV(B) =7.2
15.	eP	A	18 07 09	USSR-Mongolia Border Region 50.18 N 91.34 E H = 17 58 28.3 h = normal MB = 5.1 D = 48.42 Az = 303.1 (USCGS) PV(A):0.8s 15.4nm MPV(A)=5.2
15.	eP	A	18 44 46.5	USSR-Mongolia Border Region 50.25 N 91.34 E H = 18 36 04.0 h = normal MB = 4.5 D = 48.38 Az = 303.1 (USCGS)
15.	eP	A	18 58 47	USSR-Mongolia Border Region 50.29 N 91.19 E H = 18 50 07.4 h = normal MB = 4.6 D = 48.28 Az = 303.0 (USCGS)
15.	eP1	A	20 20 57.5	USSR-Mongolia Border Region
	eP2	A	21 00.5	50.23 N 91.26 E
	LmH	B	40.7	H = 20 12 16.9 h = normal MB = 5.0
	LmV	B	43.5	D = 48.36 Az = 303.1 (USCGS) P1V(A):1.6s 22.0nm MP1V(A)=5.0 P2V(A):1.5s 40.2nm MP2V(A)=5.3

Day	Phase		h m s	Remarks
cont. 15.				LmH(B):13s 1.4 µm MLH(B)=5.1
				LmV(B):16s 0.9 µm MLV(B)=4.9
15.	eP1	A	20 59 56	<u>East of Lake Baikal</u> 56.85 N 117.83 E
	eP2	AB	21 00 00	H = 20 50 12.7 h = normal MB = 4.9
	eSS	B	12 28	D = 56.86 Az = 313.1 (USCGS)
	LmH	B	26.6	P1V(A):1.0s 13.6nm MP1V(A)=4.9
	LmV	B	26.6	P2V(A):2.0s 55.6nm MP2V(A)=5.2
				LmH(B):13s 2.8 µm MLH(B)=5.5
				LmV(B):14s 2.9 µm MLV(B)=5.6
16.	e	A	10 49 11	<u>Eastern Caucasus</u> 42.88 N 47.11 E
				H = 10 43 23.0 h = normal MB = 4.4
				D = 25.33 Az = 300.2 (USCGS)
16.	eP	A	21 32 24	<u>Eastern Caucasus</u> 42.99 N 46.96 E
	e	A	32 31.5	H = 21 26 55.3 h = normal MB = 4.8
	LmH	B	44.0	D = 25.18 Az = 300.0 (USCGS)
	LmV	B	45.4	LmH(B):13s 0.5 µm MLH(B)=4.2
				LmV(B):11s 0.5 µm MLV(B)=4.4
17.	eP	A	01 05 32	<u>USSR-Mongolia Border Region</u>
	e	A	07 32.5	50.19 N 91.26 E
	e	A	08 20	H = 00 56 48.0 h = normal MB = 4.5
	LmH	B	25.2	D = 48.37 Az = 303.1 (USCGS)
	LmV	B	28.0	LmH(B):13s 0.5 µm MLH(B)=4.7
				LmV(B):13s 0.4 µm MLV(B)=4.7
17.	ePKP	A	03 53 56	<u>Fiji Islands</u> 17.55 S 178.89 W
				H = 03 35 22.8 h = 610 km MB = 4.4
				D = 145.93 Az = 348.0 (USCGS)
				PKPV(A):1.0s 9.8nm
17.	eP1	A	05 07 41.5	<u>Eastern Caucasus</u> 43.00 N 46.88 E
	eP2	A	07 43.5	H = 05 02 17.7 h = normal MB = 4.7
	LmH	B	19.0	D = 25.13 Az = 300.0 (USCGS)
	LmV	B	19.7	P2V(A):2.0s 21.4nm MP2V(A)=4.5
				LmH(B):14s 0.6 µm MLH(B)=4.3
				LmV(B):12s 0.7 µm MLV(B)=4.5

Day	Phase		h m s	Remarks
17.	eP	AB	06 54 31	<u>Eastern Caucasus</u> 42.97 N 46.88 E
	i	A	54 38.5	H = 06 49 06.1 h = normal MB = 5.0
	i	A	54 43.5	D = 25.14 Az = 300.1 (USCGS)
	Pm	A	55 04	PV(A):1.5s 27.6nm MPV(A)=4.8
	eS	B	59 06	PmV(A):1.9s 91.0nm MPmV(A)=5.2
	e	B	07 00 32	LmH(B):16s 4.5/um MLH(B)=5.1
	ei	B	01 00	LmV(B):14s 4.1/um MLV(B)=5.2
	LmH	B	05.0	
	LmV	B	05.8	
17.	eP	A	15 04 18.5	<u>Near East Coast of Honshu, Japan</u>
				34.84 N 140.89 E
				H = 14 51 55.6 h = 72 km MB = 5.0
				D = 83.85 Az = 330.3 (USCGS)
				PV(A):1.7s 18.2nm MPV(A)=4.8
17.	eP	A	23 07 03	<u>Near Islands, Aleutian Is.</u>
				53.14 N 171.04 E
				H = 22 55 23.2 h = 38 km MB = 4.6
				D = 75.14 Az = 346.6 (USCGS)
18.	eiP	AB	01 35 35.5	<u>North Atlantic Ridge</u> 52.08 N 30.20 W
	ei	A	35 37.5	H = 01 30 02.6 h = normal MB = 4.9
	i	A	35 47.5	D = 25.34 Az = 54.2 (USCGS)
	eS	B	40 12	PV(A):1.0s 35.5nm MPV(A)=5.0
	LmH	B	45.3	LmH(B):14s 2.5/um MLH(B)=4.9
	LmV	B	46.0	LmV(B):13s 1.6/um MLV(B)=4.9
18.	ePKIKP	A	03 26 31	<u>Santa Cruz Islands</u> 12.38 S 166.59 E
				H = 03 07 15.5 h = 76 km MB = 5.1
				D = 136.72 Az = 336.9 (USCGS)
18.	eP	A	07 40 56.5	<u>USSR-Mongolia Border Region</u>
	LmV	B	08 03.3	50.26 N 91.30 E
				H = 07 32 16.2 h = normal MB = 4.8
				D = 48.36 Az = 303.0 (USCGS)
				PV(A):1.0s 15.8nm MPV(A)=5.1
				LmV(B):14s 0.6/um MLV(B)=4.8

Day	Phase		h m s	Remarks
18.	eP1	A	14 46 27	East of Lake Baikal 56.89 N 117.62 E
	eP2	A	46 31	H = 14 36 40.7 h = normal MB = 4.6
	LmV	B	15 12.9	D = 56.75 Az = 313.0 (USCGS)
	LmH	B	15 13.0	P2V(A):1.0s 15.7nm MP2V(A)=5.0
				LmV(B):14s 0.6 µm MLV(B)=4.8
				LmH(B):14s 1.1 µm MLH(B)=5.0
18.	eP	A	19 22 52	Fox Islands, Aleutian Is.
	e	A	23 05	51.32 N 170.83 W
				H = 19 10 52.1 h = normal MB = 4.3
				D = 78.39 Az = 358.4 (USCGS)
19.	LmH	B	01 52.5	Probably Near Coast of Chiapas, Mexico
	LmV	B	52.8	(USCGS)
				LmH(B):16s 0.4 µm
				LmV(B):16s 0.5 µm
19.	eP1	A	02 13 40	Greenland Sea 79.24 N 2.50 E
	eP2	A	13 48	H = 02 07 41.5 h = normal MB = 4.8
	e	A	13 56	D = 28.90 Az = 167.9 (USCGS)
				P2V(A):1.5s 30.2nm MP2V(A)=4.9
19.	ePKP	A	07 35 10	New Hebrides Islands 18.78 S 16[illegible].23 E
				H = 07 16 03.0 h = 212 km MB = 4.6
				D = 143.56 Az = 335.9 (USCGS)
				PKPV(A):1.1s 12.1nm
19.	eP	A	10 34 56.5	Near Coast of Venezuela 10.91 N 69.94 W
				H = 10 22 57.6 h = 16 km MB = 5.1
				D = 75.66 Az = 40.4 (USCGS)
				PV(A):1.0s 7.9nm MPV(A)=4.8
19.	ePKIKP	A	15 07 29	New Britain Region 5.19 S 152.25 E
				H = 14 48 35.3 h = 65 km MB = 5.6
				D = 124.02 Az = 330.8 (USCGS)
19.	e(PKP2)	A	20 38 56	Kermadec Islands Region 31.4 S 178.1 W
				H = 20 18 10.6 h = 41 km MB = 4.6 (USCGS)
				D = 159.7

Day	Phase		h m s	Remarks
20.	ePKP2	A	00 39 27	<u>Kermadec Islands Region</u> 31.81 S 179.94 E H = 00 19 39.6 h = 396 km MB = 4.4 (USCGS) D = 159.3
20.	ePKIKP	A	20 22 08	<u>South Sandwich Islands</u> 55.89 S 28.33 W
	ePP	A	22 44	H = 20 03 42.3 h = 70 km MB = 6.0
	e	B	23 09	D = 111.23 Az = 26.0 (USCGS)
	e	B	23 42	PPV(A):1.5s 20.1nm MPPV(A)=5.5
	e	B	30 28	PKKP1V(A):1.2s 14.2nm
	eSP	B	32 12	PKKP2V(A):1.3s 26.2nm
	ePKKP1	A	33 05	LmH(B):22s 3.5 µm MLH(B)=5.9
	ePKKP2	A	33 13.5	LmV(B):18s 2.7 µm MLV(B)=5.9
	eSS	B	38 00	
	LmH	B	21 02.6	
	LmV	B	07.0	
20.	eP	AB	20 42 49	<u>Andreanof Islands, Aleutian Is.</u> 51.47 N 178.52 W H = 20 30 54.7 h = 48 km MB = 5.7 D = 77.90 Az = 353.4 (USCGS) PV(A):1.2s 61.0nm MPV(A)=5.0
20.	eP	A	22 14 44.5	<u>Near West Coast of Columbia</u> 5.66 N 77.50 W H = 22 02 09.0 h = normal MB = 4.9 D = 85.10 Az = 39.7 (USCGS) PV(A):1.5s 20.1nm MPV(A)=5.1
21.	eiPKHKP	A	00 55 30.5	<u>Fiji Islands</u> 20.29 S 178.00 W
	ePKP2	A	55 36	H = 00 36 43.8 h = 549 km MB = 5.0 D = 148.76 Az = 348.2 (USCGS) PKHKPV(A):1.0s 31.5nm
21.	eP	A	14 27 19	<u>Southern Nevada</u> 37.07 N 116.01 W H = 14 15 00.0 h = 0 km MB = 5.1 D = 81.26 Az = 30.7 (USCGS) Nevada test site "Morrones" 37°04'15" N 116°00'47" W (USAEC) PV(A):1.1s 6.0nm MPV(A)=4.6

Day	Phase		h m s	Remarks
22.	iPn	A	14 45 29.5	Adelebsen/Solling, Fed. Rep. Germany
	iSn	A	45 50	$51^{\circ}36.51'$ N $9^{\circ}44.78'$ E
	iSg	A	45 53	explosion, H = 14 45 00.84 yield 7.5 to
				(Hannover)
				D = 1.5
23.	LmV	B	08 44.8	Probably Southern Pacific Ocean (USCGS)
	LmH	B	46.0	LmV(B):18s 0.5µm
				LmH(B):20s 0.5µm
23.	eP	A	15 00 16.5	USSR-Mongolia Border Region
	LmH	B	20.0	50.06 N 91.63 E
	LmV	B	23.4	H = 14 51 34.7 h = 43 km MB = 4.5
				D = 48.64 Az = 303.3 (USCGS)
				PV(A):0.8s 7.7nm MPV(A)=4.8
				LmH(B):12s 0.4µm MLH(B)=4.7
				LmV(B):12s 0.4µm MLV(B)=4.7
23.	eP	A	16 29 19	Near Coast of Chiapas, Mexico
				14.91 N 92.23 W
				H = 16 16 43.8 h = 109 km MB = 4.9
				D = 87.14 Az = 38.2 (USCGS)
				PV(A):0.6s 11.5nm MPV(A)=5.0
23.	eP	AB	23 21 54	Kurile Islands 43.55 N 147.91 E
	epP	A	22 03.5	H = 23 09 52.8 h = 35 km MB = 5.2
	LmV	B	24 00.3	D = 78.77 Az = 333.4 (USCGS)
	LmH	B	01.7	PV(A):1.0s 19.7nm MPV(A)=5.1
				LmV(B):12s 3.7µm MLV(B)=6.0
				LmH(B):16s 1.5µm MLH(B)=5.4
24.	eP	A	00 53 21	Kurile Islands 43.82 N 147.63 E
	epP	A	53 31.5	H = 00 41 19.5 h = normal MB = 4.5
				D = 78.44 Az = 333.2 (USCGS)
24.	e	A	11 07 16	Dodecanese Islands 36.17 N 25.61 E
	LmH	B	15.6	H = 11 02 56.3 h = 13 km MB = 4.5
	LmV	B	15.8	D = 17.64 Az = 329.5 (USCGS)

Day	Phase		h m s	Remarks
24.	eP	A	16 44 41	Near East Coast of Kamchatka
	LmH	B	17 18.5	55.00 N 162.59 E
	LmV	B	19.0	H = 16 33 17.7 h = normal MB = 4.7
				D = 71.97 Az = 341.0 (USCGS)
25.	ePn	A	12 57 19.5	Italy 42.6 N 12.5 E
	ePg	A	58 02	H = 12 55 24 (BCIS)
	eSn	A	58 49	D = 8.10
				PnV(A):1.1s 14.1nm
25.	eP	A	16 27 49.5	Near East Coast of Honshu, Japan
				37.26 N 140.99 E
				H = 16 15 33.8 h = 50 km MB = 4.5
				D = 81.80 Az = 330.2 (USCGS)
25.	ePKIKP	A	17 07 25	Kermadec Islands 29.43 S 177.80 W
	ePKHKP	A	07 37	H = 16 47 36.0 h = 63 km MB = 5.5
	ePKP2	A	07 58	D = 157.65 Az = 344.1 (USCGS)
				PKIKPV(A):1.4s 16.3nm
				PKHKPV(A):1.3s 15.3nm
				PKP2V(A):1.5s 40.0nm
25.	ePn	A	17 45 50	Swabian Jura, Fed.Rep.Germany 48.3 N 9.1 E
	ePg	A	46 04	H = 17 45 06 (BCIS)
	iSg	A	46 37	D = 2.84
25.	eP	A	18 17 12	Dodecanese Islands 36.23 N 25.68 E
				H = 18 13 07.8 h = 29 km MB = 4.4
				D = 17.61 Az = 329.3 (USCGS)
26.	eSg	A	03 22 09	Swabian Jura, Fed.Rep.Germany 48.3 N 9.1 E
				H = 03 20 37 (BCIS)
				D = 2.84
26.	+iP	A	10 05 13	Unimak Island 54.19 N 164.69 W
				H = 09 53 31.0 h = 36 km MB = 5.2
				D = 75.50 Az = 2.4 (USCGS)
				PV(A):1.1s 42.3nm MPV(A)=5.4

Day	. Phase		h m s	Remarks
26.	ePn	A	12 06 29.5	Italy 42.6 N 12.5 E
	eSn	A	08 00	H = 12 04 34 (BCIS)
	LmH	B	10.4	D = 8.1
	LmV	B	10.4	PnV(A):1.0s 19.7nm
26.	LmH	B	13 43.0	Probably New Britain Region (USCGS)
	LmV	B	43.0	
26.	eP	A	15 12 18	Southern Nevada 37.11 N 116.11 W
				H = 15 00 00.0 h = 0 km MB = 5.6
				D = 81.27 Az = 30.6 (USCGS)
				Nevada test site "Flask"
				37°06'50" N 116°03'48" W (USAEC)
				PV(A):1.3s 41.5nm MPV(A)=5.3
27.	-eP1	A	12 17 23.5	Bonin Islands 27.22 N 140.12 E
	-iP2	ABC	17 24.2	H = 12 05 06.0 h = 382 km MB = 6.2
	epP	A	18 54	D = 90.13 Az = 330.1 (USCGS)
	-isP	B	19 35	P2V(A):2.4s 1800.0nm MP2V(A)=6.5
	ePP	B	21 05	P2V(B):10s 15.1/um MP2V(B)=6.8
	epPP	A	22 18	LmH(B):16s 32.6/um
	isPP	B	23 00	LmV(B):15s 27.2/um
	eisPPPP	B	26 30	(P'P'P')V(A):1.8s 20.3nm
	eiSP	B	28 44	
	e	A	28 50	
	iPS	B	29 36	
	isS	B	30 15	
	iSS	B	33 46	
	ePKKP	A	34 52	
	esSS	B	36 18	
	e	B	42 50	
	eP'P'	A	42 55	
	eSKPPKP	A	45 42	
	LmH	B	57.3	
	LmV	B	13 01.5	
	e(P'P'P')	A	03 47	

Day	Phase		h m s	Remarks
27.	eP	AB	19 17 46	Off East Coast of Honshu, Japan
	ePP	B	20 52	40.30 N 142.99 E
	eS	B	27 50	H = 19 05 39.1 h = 33 km MB = 5.7
	eSS	B	34 10	D = 79.92 Az = 331.0 (USCGS)
	LmH	B	52.7	PV(A):2.0s 342.0nm MPV(A)=5.9
	LmV	B	58.7	PV(B):12s 3.1 /um MPV(B)=6.1
				PPV(B):(16s) 1.8 /um MPPV(B)=5.9
				SH(B):20s 5.2 /um MSH(B)=6.1
				LmH(B):18s 47.3 /um MLH(B)=6.9
				LmV(B):16s 41.7 /um MLV(B)=6.9
27.	eP	A	20 14 54	Near Coast of Chiapas, Mexico
				15.37 N 93.74 W
				H = 20 02 13.7 h = 89 km MB = 4.6
				D = 87.68 Az = 37.9 (USCGS)
				PV(A):1.4s 11.6nm MPV(A)=4.8
27.	eP	AB	22 47 56.5	Off East Coast of Honshu, Japan
	eS	B	58 00	40.21 N 143.16 E
	LmH	B	23 23.0	H = 22 35 46.4 h = 16 km MB = 5.5
	LmV	B	27.3	D = 80.07 Az = 331.1 (USCGS)
				PV(A):2.2s 262.0nm MPV(A)=5.8
				LmH(B):18s 9.8 /um MLH(B)=6.2
				LmV(B):15.5s 5.5 /um MLV(B)=6.0
27.	+iP	AB	24 08 46.5	Off East Coast of Honshu, Japan
	ePP	B	11 44	40.30 N 142.99 E
	ePP	A	11 50	H = 23 50 40.0 h = 38 km MB = 5.4
	eS	B	18 50	D = 79.92 Az = 331.0 (USCGS)
	LmH	B	43.7	PV(A):2.0s 171.0nm MPV(A)=5.6
	LmV	B	49.6	LmH(B):18s 5.7 /um MLH(B)=6.0
				LmV(B):17s 4.3 /um MLV(B)=5.9
28.	eP	A	06 41 22	Kurile Islands 46.47 N 153.41 E
				H = 06 29 25.7 h = 30 km MB = 4.7
				D = 77.79 Az = 336.2 (USCGS)

Day	Phase		h m s	Remarks
28.	ePKP	A	11 53 03.5	New Hebrides Islands 20.49 S 169.68 E
	epPKP	A	53 35.5	H = 11 33 41.3 h = 144 km MB = 4.9
				D = 145.28 Az = 335.3 (USCGS)
				PKPV(A):1.4s 48.8nm
28.	ePKP	A	14 32 23	New Hebrides Islands 18.84 S 169.49 E
				H = 14 13 18.0 h = 232 km MB = 4.1
				D = 143.72 Az = 336.1 (USCGS)
				PKPV(A):1.2s 12.2nm
28.	eP	A	15 27 06	Bonin Islands 27.25 N 140.14 E
				H = 15 14 46.8 h = 369 km MB = 4.9
				D = 90.10 Az = 330.1 (USCGS)
				PV(A):1.5s 15.1nm MPV(A)=4.7
28.	eiP	A	17 50 16.5	Vancouver Island 48.45 N 126.66 W
				H = 17 38 32.1 h = 3 km MB = 4.9
				D = 74.99 Az = 26.0 (USCGS)
				PV(A):1.0s 21.6nm MPV(A)=5.1
29.	-eiP	AB	04 42 55	Kurile Islands 44.19 N 146.73 E
				H = 04 31 03.7 h = 13 km MB = 5.3
				D = 77.83 Az = 332.6 (USCGS)
				PV(A):1.3s 96.0nm MPV(A)=5.8
29.	ePKP	AB	05 34 11.5	Tonga Islands 14.99 S 173.52 W
	e(PSPS)	B	52 08	H = 05 14 38.0 h = normal MB = 5.5
	LmH	B	06 39.0	D = 144.19 Az = 354.4 (USCGS)
	LmV	B	39.8	PKPV(A):3.0s 131.5nm
				LmH(B):20s 1.2/um MLH(B)=5.6
				LmV(B):20s 1.6/um
29.	iPn	A	07 29 08	Swabian Jura, Fed.Rep.Germany 48.3 N 9.1 E
	iPg	A	29 17	H = 07 28 23 (BCIS)
	iSn	A	29 42.5	D = 2.84
	eiSg	A	29 52	
	i	A	29 55	
	i	A	29 57	

Day	Phase		h m s	Remarks
29.	e(P)	A	10 44 49.5	Burma-India Border Region
	epP	A	45 05	24.01 N 94.09 E
				H = 10 33 58.6 h = 47 km MB = 5.0
				D = 67.16 Az = 316.8 (USCGS)
				PV(A):1.0s 15.7nm MPV(A)=5.2
				pPV(A):1.5s 30.1nm
29.	ePKIKP	AB	19 21 36	Santa Cruz Islands 11.63 S 166.34 E
	epPKIKP	A	21 50	H = 19 02 19.0 h = 50 km MB = 5.9
	ePP	B	24 12	D = 135.92 Az = 337.0 (USCGS)
	LmH	B	20 19.6	PKIKPV(A):3.0s 131.5nm
	LmV	B	23.9	LmH(B):21s 2.8/um MLH(B)=5.9
				LmV(B):21s 2.6/um
29.	eP	A	19 45 18	Kurile Islands 45.44 N 149.99 E
				H = 19 33 24.4 h = 45 km MB = 5.0
				D = 77.73 Az = 334.4 (USCGS)
				PV(A):1.2s 18.3nm MPV(A)=5.1
29.	ePKIKP	AB	20 49 22	Fiji Islands 20.57 S 178.74 W
	eiPKHKP	AB	49 26.8	H = 20 30 45.1 h = 610 km MB = 5.1
	iPKP2	A	49 35	D = 148.89 Az = 347.2 (USCGS)
	epPKP	A	51 46	PKIKPV(A):0.8s 13.5nm
	e	A	51 51	PKHKPV(A):1.1s 117.0nm
				PKP2V(A):1.2s 75.5nm
30.	ePKP	A	03 44 49.5	Loyalty Islands 21.85 S 170.09 E
				H = 03 25 12.1 h = 58 km MB = 4.7
				D = 146.68 Az = 334.8 (USCGS)
				PKPV(A):1.3s 21.8nm
30.	eP1	A	04 28 04.5	Taiwan Region 21.37 N 121.86 E
	eP2	A	28 07	H = 04 15 20.8 h = normal MB = 4.5
	LmH	B	05 11.5	D = 85.76 Az = 323.2 (USCGS)
	LmV	B	11.5	P2V(A):1.0s 11.8nm MP2V(A)=5.0
				LmH(B):12s 0.4/um MLH(B)=5.1
				LmV(B):14s 0.6/um MLV(B)=5.1

Day	Phase		h m s	Remarks
30.	ePKIKP	A	11 12 37	**Fiji Islands** 20.06 S 178.52 W
	eiPKHKP	A	12 42.5	H = 10 54 03.1 h = 615 km MB = 4.8
	ePKP2	A	12 48.5	D = 148.44 Az = 347.6 (USCGS)
				PKHKPV(A):1.3s 43.6nm
30.	eP	AB	13 29 39	**Samar, Philippine Islands**
	e	A	30 08	12.16 N 124.48 E
	e	AB	30 18	H = 13 16 27.1 h = 93 km MB = 5.8
	ePP	AB	33 27	D = 94.62 Az = 323.9 (USCGS)
	eSKS	B	40 00	PV(A):1.6s 55.0nm MPV(A)=5.8
	eS	B	40 48	PPV(A):1.5s 45.2nm MPPV(A)=5.8
	ePS	B	42 22	LmH(B):17.5s 0.9 µm
	ePSPS	B	47 44	LmV(B):16s 1.1 µm
	eSSS	B	51 05	
	LmH	B	14 17.0	
	LmV	B	19.0	
30.	iPn	A	16 38 47.5	**Swabian Jura, Fed.Rep.Germany** 48.3 N 9.1 E
	ePg	A	38 56	H = 16 38 03 (BCIS)
	iSn	A	39 23	D = 2.84
	iSg	A	39 34	LmH(B):4.5s 0.5 µm
	i	A	39 35	LmV(B):4.5s 0.2 µm
	LmH	B	39.9	
	LmV	B	39.9	
30.	iPn	A	16 56 11.5	**Yugoslavia** 45.3 N 15.1 E
	e	A	56 33	H = 16 54 45 (BCIS)
	ePg	A	56 37	D = 5.82
	eSn	A	57 14	PnV(A):0.4s 11.6nm
	eSg	A	57 48	
30.	e(PKHKP)	A	18 25 47	**Tonga Islands** 22.53 S 174.51 W
	eX	A	25 50	H = 18 05 54.4 h = 55 km MB = 4.6
	e(PKP2)	A	25 59	D = 151.53 Az = 351.8 (USCGS)
	e	A	26 05	XV(A):1.6s 19.2nm
	e	A	26 18	

Day	Phase		h m s	Remarks
30.	eP	A	23 31 23	Unimak Island 53.67 N 164.12 W
	e	A	31 32	H = 23 19 37.4 h = normal MB = 4.9
	eX	A	31 35	D = 76.00 Az = 2.8 (USCGS)
	LmV	B	24 11.0	XV(A):1.8s
	LmH	B	11.5	
31.	eP	A	03 02 45	Unimak Island 53.62 N 164.51 W
				H = 02 50 58.5 h = normal MB = 4.4
				D = 76.06 Az = 2.5 (USCGS)
31.	+iPn	A	08 12 12.8	Federal Rep. Germany 48.36 N 9.18 E
	iPg	A	12 19.2	H = 08 11 29.3 h = 16 km MB = 3.9
	iSg	A	12 59	D = 2.78 Az = 33.9 (USCGS)
	LmH	B	13.2	LmH(B):7.5s 2.3 /um
	LmV	B	13.4	LmV(B):5s 1.3 /um
31.	e	A	10 31 26.5	Eastern Caucasus 42.98 N 47.02 E
	LmH	B	41.8	H = 10 25 50.4 h = 25 km MB = 4.6
	LmV	B	42.4	D = 25.22 Az = 300.1 (USCGS)
31.	-iP1	AB	20 36 57.3	Near Coast of Northern Peru
	+iP2	B	37 20	9.18 S 78.82 W
	eiP3	B	37 32	H = 20 23 27.3 h = 43 km MB = 6.6 (USCGS)
	ePP1	B	41 00	D = 97.3
	e	B	41 08	Multiple shock
	ePP2	B	41 20	P1V(A):1.8s 203.0nm MP1V(A)=6.5
	ePP3	B	41 35	P3V(A):2.2s 1450.0nm MP2V(A)=7.2
	eSKS1	B	47 25	P3V(B):15s 40.0 /um MP2V(B)=7.9
	eSKS2	B	47 46	PP3V(B):16s 32.4 /um MPP3V(B)=7.5
	eiSKS3	B	47 59	LmV(B):19s 405.0 /um MLV(B)=8.0
	eS	B	48 13	LmH(B):20s 325.0 /um MLH(B)=7.9
	esS	B	48 45	
	eSPP	B	50 25	
	eSS	B	55.0	
	eP'P'	A	21 02 06	
	LmV	B	20.0	
	LmH	B	20.1	
	e(P'P'P')	A	23 20	

Day	Phase		h m s	Remarks
31.	e	A	22 01 51	<u>Near Coast of Peru</u> 10.22 S 78.60 W H = 21 48 04.4 h = 51 km MB = 5.6 D = 97.95 Az = 40.0 (USCGS)

Day	Phase		h m s	Remarks
1.	eP1	A	01 18 59	Greece-Albania Border Region
	eP2	A	19 02	39.06 N 20.79 E
				H = 01 15 49.0 h = 65 km MB = 4.2
				D = 13.28 Az = 333.8 (USCGS)
				P2V(A):0.8s 9.6nm
1.	eP	A	01 49 40	Off Coast of Northern Peru
	epP	AB	49 53	9.30 S 79.00 W
	e	A	50 09	H = 01 36 10.2 h = 49 km MB = 6.0
	eSKS	B	02 00 12	D = 97.50 Az = 40.0 (USCGS)
	eS	B	00 59	PV(A):1.8s 33.8nm MPV(A)=5.7
	eSS	B	07 40	LmH(B):19s 0.7/um MLH(B)=5.1
	LmH	B	33.1	LmV(B):19s 1.1/um MLV(B)=5.4
	LmV	B	33.2	
1.	+eP	A	02 58 53.3	Near Coast of Peru 10.15 S 78.73 W
	epP	AB	59 07	H = 02 45 21.5 h = 66 km MB = 5.8
	iSKS	B	03 09 28	D = 97.98 Az = 40.0 (USCGS)
	eS	B	10 14	PV(A):1.8s 27.0nm MPV(A)=5.5
	eSP	B	11 32	SKSH(B):5.5s 1.2/um
	eSPP	B	12 32	SH(B):7s 0.5/um MSH(B)=6.1
	LmV	B	41.2	LmV(B):21s 2.6/um
	LmH	B	41.3	LmH(B):21s 2.3/um
1.	eP	A	07 59 10.5	Crete 34.3 N 24.2 E
	LmH	B	08 08.0	H = 07 54 52 h = 51 km (ISC)
	LmV	B	08.0	D = 18.8
1.	eP	AB	17 57 08	South of Panama 5.92 N 82.52 W
	e	A	57 12.5	H = 17 44 15.0 h = 9 km MB = 5.6
	eSKS	B	18 07 40	D = 88.09 Az = 39.4 (USCGS)
	eS	B	07 55	PV(A):1.4s 32.6nm MPV(A)=5.5
	eSS	B	14 55	LmH(B):20.5s 2.5/um MLH(B)=5.6
	LmH	B	29.4	LmV(B):20s 1.7/um MLV(B)=5.5
	LmV	B	31.0	
2.	eP	A	01 50 53.5	Near Coast of Northern Peru
	epP	AB	51 08.5	9.76 S 78.80 W

Day	Phase		h m s	Remarks
cont. 2.	ePP	B	01 54 48	H = 01 37 22.7 h = 58 km MB = 5.7
	eSKS	B	02 01 28	D = 97.73 Az = 40.0 (USCGS)
	ePS	B	04 00	PV(A):0.8s 7.7nm MPV(A)=5.3
	LmH	B	35.0	pPV(A):1.9s 106.0nm
	LmV	B	35.0	LmH(B):17s 0.9/um MLH(B)=5.3
				LmV(B):17s 1.2/um MLV(B)=5.5
2.	eX	A	03 01 48.5	Tonga Islands 17.81 S 174.66 W
				H = 02 41 42.9 h = 137 km MB = 4.7 (USCGS)
				D = 146.8
				XV(A):1.8s 33.8nm
2.	+iP	AB	03 10 17.5	Southern Alaska 61.60 N 151.73 W
	ei	A	10 22.5	H = 02 59 31.3 h = 95 km MB = 5.5
	e(PcP)	B	10 47	D = 67.31 Az = 11.4 (USCGS)
	eS	B	19 04	PV(A):1.0s 68.9nm MPV(A)=5.5
	ePS	B	19 40	PV(B):4.5s 0.4/um MPV(B)=5.6
	eP'P'	A	38 34	SH(B):6.5s 1.0/um MSH(B)=5.9
				P'P'V(A):1.5s 17.6nm
2.	eiPKP	A	21 20 34.5	New Hebrides Islands 18.74 S 168.88 E
	ePP	A	24 42	H = 21 01 21.3 h = 163 km MB = 4.9
				D = 143.39 Az = 335.6 (USCGS)
				PKPV(A):1.2s 34.6nm
2.	+iPKHKP	A	21 49 36.1	Fiji Islands 20.33 S 177.45 W
	ePKP2	A	49 43	H = 21 30 32.2 h = 388 km MB = 5.2
	epPKHKP	A	51 12	D = 148.91 Az = 348.8 (USCGS)
				PKHKPV(A):0.8s 38.5nm
2.	eiP	A	23 45 27	Kurile Islands 45.71 N 150.92 E
	e	A	45 31.5	H = 23 33 30.2 h = 20 km MB = 5.4
	LmH	B	24 22.9	D = 77.77 Az = 334.9 (USCGS)
	LmV	B	23.2	PV(A):1.1s 28.2nm MPV(A)=5.3
				LmH(B):20s 0.5/um MLH(B)=4.8
				LmV(B):20s 0.5/um MLV(B)=4.9

Day	Phase		h m s	Remarks
3.	LmH	B	21 54.3	<u>Off Coast of Michoacan, Mexico</u>
	LmV	B	54.3	15.4 N 105.0 W
				H = 20 57 08.3 h = normal MB = 4.2 (USCGS)
				D = 94.2
				LmH(B):16s 0.3 µm MLH(B)=4.8
				LmV(B):16s 0.4 µm MLV(B)=4.9
3.	ePKIKP	A	22 56 47	<u>New Hebrides Islands</u> 13.13 S 167.20 E
				H = 22 37 46.5 h = 220 km MB = 4.8
				D = 137.63 Az = 337.0 (USCGS)
				PKIKPV(A):1.3s 10.9nm
4.	eP	A	04 22 56	<u>Near Coast of Northern Peru</u>
	epP	AB	23 11	9.80 S 78.63 W
	esP	A	23 20	H = 04 09 26.3 h = 57 km MB = 5.8
	ePP	BC	26 48	D = 97.65 Az = 40.0 (USCGS)
	eSKS	BC	33 28	PV(A):1.4s 21.0nm MPV(A)=5.5
	iSKS	BC	33 32	pPV(A):3.0s 237.0nm
	eSP	BC	35 32	pPV(B):10s 1.2 µm
	eSPP	BC	36 38	LmH(B):19s 4.9 µm MLH(B)=6.0
	e	B	40 24	LmV(B):16s 4.9 µm MLV(B)=6.1
	eSS	BC	40 50	
	eSSS	BC	44 48	
	LmH	B	05 05.4	
	LmV	B	07.5	
4.	ePg	A	23 06 02	Explosion
	eSg	A	06 20	
5.	eP	A	05 01 25	<u>Alma-Ata Region</u> 42.48 N 78.76 E
	iP	A	01 26	H = 04 53 06.4 h = 20 km MB = 6.0
	ei	A	03 16.5	D = 45.43 Az = 304.6 (USCGS)
	ePP	A	03 20	PV(A):1.3s 197.0nm MPV(A)=5.9
	i	B	03 27	PV(B):12s 6.8 µm MPV(B)=6.5
	eiPcS	B	05 48	PPV(B):12s 10.0 µm MPPV(B)=6.6
	LmH	B	22.7	LmH(B):11s 97.0 µm MLH(B)=7.0
	LmV	B	24.0	LmV(B):15s 124.0 µm MLV(B)=7.0

Day	Phase		h m s	Remarks
5.	eP1	A	09 23 39.5	Southern Italy 39.19 N 15.44 E
	eP2	A	23 41.5	H = 09 20 55.6 h = 262 km MB = 4.4
				D = 11.77 Az = 348.0 (USCGS)
				P1V(A):1.5s 12.6nm
				P2V(A):1.3s 48.0nm
5.	+iP	A	10 10 05.8	Greece 38.05 N 22.41 E
				H = 10 06 36.2 h = 100 km MB = 4.2
				D = 14.76 Az = 332.1 (USCGS)
				PV(A):0.8s 19.2nm
5.	-iP	AB	10 42 05.5	Eastern Siberia 63.37 N 146.23 E
	e	B	11 06 22	H = 10 31 34.3 h = normal MB = 5.5 (USCGS)
	LmH	B	08.3	D = 60.8
	LmV	B	12.6	PV(A):1.3s 107.0nm MPV(A)=5.8
				LmH(B):16s 3.4/um MLH(B)=5.6
				LmV(B):14s 1.7/um MLV(B)=5.4
5.	eP	A	12 03 11.5	Rumania 45.66 N 26.62 E
				H = 12 00 32.9 h = 129 km MB = 4.4
				D = 11.19 Az = 301.9 (USCGS)
				PV(A):1.5s 25.1nm
5.	eP	A	14 06 37.5	Fox Islands, Aleutian Is.
				52.20 N 170.52 W
				H = 13 54 44.8 h = 45 km MB = 4.4
				D = 77.51 Az = 358.6 (USCGS)
				PV(A):1.8s 27.0nm MPV(A)=5.1
5.	+iP	AB	22 51 57.5	Off East Coast of Kamchatka
	epP	A	52 12	52.18 N 159.58 E
	eS	BC	23 01 26	H = 22 40 23.1 h = normal MB = 5.5
	esS	BC	01 45	D = 74.02 Az = 339.4 (USCGS)
	LmH	B	22.8	PV(A):1.4s 97.6nm MPV(A)=5.6
	LmV	B	30.7	pPV(A):1.4s 114.0nm
				LmH(B):23s 2.0/um MLH(B)=5.4
				LmV(B):14s 1.3/um MLV(B)=5.4

Day	Phase		h m s	Remarks
6.	eP	A	10 39 00	Mindanao, Philippine Islands 9.12 N 125.31 E H = 10 25 14.1 h = 49 km MB = - (USCGS) D = 97.5 PV(A):0.8s 7.7nm MPV(A)=5.4
6.	eP ePP	A A	13 12 35 16 50	Mariana Islands 18.63 N 145.21 E H = 12 59 53.8 h = 599 km MB = 5.1 D = 99.87 Az = 332.1 (USCGS) PPV(A):1.4s 46.5nm MPPV(A)=5.6
6.	ePKHKP ePKP2	A A	18 02 38 02 44	Fiji Islands 20.39 S 178.99 W H = 17 43 57.7 h = 619 km MB = 4.4 D = 148.67 Az = 347.0 (USCGS) PKHKPV(A):1.4s 18.6nm
7.	eP e	A A	07 55 48 56 05	Eastern Gulf of Aden 12.9 N 51.3 E H = 07 46 59 M = 5 D = 49.6 (ANUSSR) PV(A):1.7s 24.2nm MPV(A)=4.9
7.	eiP ePcP	A A	13 40 52 41 09.5	Sea of Okhotsk 52.99 N 153.53 E H = 13 30 14.9 h = 448 km MB = 4.8 (USCGS) D = 71.6 PV(A):1.0s 27.6nm MPV(A)=4.8 PcPV(A):1.1s 12.1nm
7.	eP LmH LmV	A B B	17 55 05 18 11.5 13.2	Western Iran 33.01 N 49.37 E H = 17 48 33.8 h = 50 km MB = 4.7 (USCGS) D = 32.8
7.	eP	A	20 37 04	Kurile Islands 43.74 N 146.67 E H = 20 25 18.6 h = 132 km MB = 4.2 D = 78.19 Az = 332.6 (USCGS) PV(A):1.0s 7.9nm MPV(A)=4.4
8.	eP e e(Sg)	A A A	06 53 43 56 35 57 12	Albania 41.23 N 20.01 E H = 06 51 04.9 h = 38 km MB = 4.6 D = 11.08 Az = 331.0 (USCGS)

Day	Phase		h m s	Remarks
8.	eP	A	09 31 33	Atlantic-Indian Rise 29.35 S 61.03 E
				H = 09 18 31.7 h = normal MB = 5.1
				D = 90.83 Az = 331.1 (USCGS)
8.	eP	A	12 38 28	Eastern Caucasus 43.18 N 47.33 E
	e	A	46 45	H = 12 33 02.9 h = normal MB = 4.7
				D = 25.32 Az = 299.7 (USCGS)
9.	eP	A	10 41 48	Eastern Gulf of Aden 13.27 N 51.37 E
	ePP	A	43 43	H = 10 33 01.0 h = normal MB = 5.0
				D = 49.28 Az = 327.5 (USCGS)
				PV(A):1.8s 20.3nm MPV(A)=4.8
9.	ePKIKP	AB	11 14 35.5	Samoa Islands Region 15.73 S 172.89 W
	LmH	B	12 20.0	H = 10 55 01.1 h = normal MB = 5.3
	LmV	B	29.5	D = 144.98 Az = 355.0 (USCGS)
				PKIKPV(A):1.5s 30.2nm
				PKIKPV(B):10s 1.0/um
				LmH(B):20s 0.7/um MLH(B)=5.4
				LmV(B):16s 0.6/um
9.	eP	A	14 44 28.5	Turkey 38.8 N 30.0 E
	LmH	B	52.2	H = 14 40 19.6 h = 0 km (ISC)
	LmV	B	52.3	D = 17.6
9.	eP	A	20 47 34	Dodecanese Islands 36.09 N 25.59 E
	LmH	B	53.6	H = 20 43 28.9 h = 54 km MB = 4.4
	LmV	B	56.5	D = 17.70 Az = 329.6 (USCGS)
10.	e	A	04 26 23.5	Southern Alaska 61.31 N 151.09 W
				H = 04 15 16.8 h = 64 km MB = 4.0
				D = 67.54 Az = 11.8 (USCGS)
10.	eP	AB	05 21 14	Turkey 39.16 N 29.46 E
	LmV	B	29.3	H = 05 17 14.0 h = 27 km MB = 4.2
	LmH	B	29.5	D = 17.03 Az = 318.2 (USCGS)
				PV(A):2.0s 34.2nm MPV(A)=4.1
				LmV(B):12s 0.7/um MLV(B)=4.2
				LmH(B):9.5s 0.7/um MLH(B)=4.2

Day	Phase		h m s	Remarks
10.	ePKP2	A	06 17 43	<u>Kermadec Islands</u> 30.28 S 177.74 W
				H = 05 57 13.4 h = 26 km MB = 4.4
				D = 158.48 Az = 343.6 (USCGS)
10.	LmH	B	09 13.6	Probably <u>Kyushu, Japan</u> (USCGS)
	LmV	B	13.6	LmH(B):14.5s 0.6/um
				LmV(B):16s 0.7/um
10.	eiP	AB	16 29 43	<u>Kurile Islands</u> 44.89 N 149.55 E
	eS	B	39 24	H = 16 17 48.7 h = 57 km MB = 5.7 (USCGS)
	eSKS	B	39 50	D = 78.0
	LmH	B	17 08.5	PV(A):2.0s 385.0nm MPV(A)=6.0
	LmV	B	08.6	LmH(B):17s 5.2/um MLH(B)=5.9
				LmV(B):17s 5.8/um MLV(B)=6.0
10.	eP	A	17 48 59.5	<u>Turkey</u> 39.20 N 30.67 E
				H = 17 44 53.5 h = 20 km MB = 4.3
				D = 17.63 Az = 316.7 (USCGS)
				PV(A):2.2s 32.7nm MPV(A)=4.1
11.	eP	AB	06 16 40.5	<u>Chile-Argentinia Border Region</u>
	i	A	16 54	24.53 S 68.50 W
	eX	A	17 10	H = 06 02 54.9 h = 112 km MB = 6.3 (USCGS)
	eiPP	B	20 51	D = 102.6
	ePP	A	20 55	PV(A):1.8s 43.6nm MPV(A)=5.9
	iSKS	B	27 14	XV(A):2.5s 169.0nm
	eSP	B	29 48	PKKPV(A):2.2s 120.0nm
	e	A	32 40	(SKKS)V(A):2.5s 84.5nm
	ePKKP	A	33 02	LmH(B):20s 6.7/um
	eSS	B	35 05	LmV(B):21s 8.3/um
	e(SKKS)	A	40 57	
	LmH	B	07 00.7	
	LmV	B	00.7	
11.	ePKIKP	B	17 06 28	<u>Maquarie Islands Region</u>
	eX	A	06 42	59.11 S 157.77 E
	e(PKIKP2)	AB	06 55	H = 16 46 38.3 h = normal MB = 5.8

Day	Phase		h m s	Remarks
cont. 11.	ePKP21	B	17 07 12	D = 158.97 Az = 261.2
	i	A	07 36.2	Appears to be a complex multiple event.
	i(PKP22)	A	07 40	Many incompatable data (USCGS)
	eSKKP	B	18 10	PKIKPV(B):5s 0.5/um
	eSS	B	31 16	X1V(A): 1.3s 10.9nm
	e	B	34 20	(PKIKP2)V(A):3.3s 965.0nm
	eSSS	B	37 45	LmH(B):18s 48.3/um MLH(B)=7.3
	e	B	46 08	LmV(B):16s 42.7/um
	LmH	B	18 26.5	
	LmV	B	36.2	
11.	eiP	A	17 48 42	<u>Afghanistan-USSR Border Region</u>
	epP	A	49 24	36.45 N 71.10 E
				H = 17 40 50.4 h = 184 km MB = 5.2 (USCGS)
12.	eP	A	03 17 16.5	<u>Kurile Islands</u> 44.87 N 148.91 E
				H = 03 05 20.7 h = normal MB = 4.9
				D = 77.91 Az = 333.8 (USCGS)
				PV(A):1.0s 11.8nm MPV(A)=5.0
12.	+iP	AB	05 05 55	<u>Kodiak Island Region</u> 56.64 N 152.05 W
	e	A	06 02	H = 04 54 31.4 h = normal MB = 5.2 (USCGS
	eS	BC	15 22	D = 68.0
	eSS	BC	20 25	PV(A):1.7s 81.9nm MPV(A)=5.7
	LmH	B	40.6	PV(B):6s 0.6/um MPV(B)=6.0
	LmV	B	47	SH(B):12s 0.7/um MSH(B)=5.7
				LmH(B):17s 1.3/um MLH(B)=5.2
				LmV(B):14s 1.1/um MLV(B)=5.3
12.	ePKIKP	A	08 24 56	<u>Near N. Coast of West New Guinea</u>
	e	B	25 37	2.89 S 139.10 E
	ePP	A	26 01	H = 08 06 16.6 h = 32 km MB = 5.7
	e	B	26 12	D = 115.18 Az = 326.1 (USCGS)
	eSKP	B	28 38	LmH(B):19s 4.6/um MLH(B)=6.0
	e	B	33 40	LmV(B):18s 5.0/um MLV(B)=6.1
	e	B	34 08	
	ePKKP	A	35 38	
	eSS	C	41 50	
	LmH	B	09 17.8	
	LmV	B	19.7	

Day	Phase		h m s	Remarks
12.	eP	A	13 06 42.5	Luzon, Philippine Islands 18.68 N 122.31 E H = 12 53 55.5 h = 48 km MB = 5.0 D = 88.15 Az = 323.4 (USCGS) PV(A):1.2s 21.2nm MPV(A)=5.3
12.	iPn	A	15 05 25.0	Bransrode, Fed.Rep.Germany, explosion
	iPg	A	05 25.5	$51^{o}13.9'N$ $09^{o}51.51'E$
	i	A	05 41	H = 15 05 01.16, yield 23 to (Hannover) D = 1.2
12.	eP	A	16 08 26	Southern Sinkiang Prov., China
	e	A	08 33	40.74 N 78.37 E
	LmH	B	29.4	H = 16 00 01.4 h = normal MB = 5.0 D = 46.20 Az = 305.8 (USCGS) PV(A):1.4s 23.3nm MPV(A)=5.0
12.	ePKP	A	21 56 08	New Hebrides Islands 18.85 S 169.35 E H = 21 37 04.1 h = 245 km MB = 4.7 D = 143.67 Az = 336.0 (USCGS)
13.	eP	A	05 39 47.5	Andreanof Islands, Aleutian Is. 51.59 N 178.34 W H = 05 27 54.4 h = 55 km MB = 5.5 D = 77.79 Az = 353.5 (USCGS) PV(A):0.8s 19.2nm MPV(A)=5.1
13.	eP	A	12 00 40	Kurile Islands 44.55 N 148.27 E H = 11 48 46.3 h = 50 km MB = 5.0 D = 77.99 Az = 333.5 (USCGS) PV(A):1.0s 5.9nm MPV(A)=4.7
14.	+iPKIKP1	ABC	00 19 11	Near Coast of Southern Chile
	eiPKIKP2	AB	19 19	51.95 S 73.85 W H = 00 00 11.3 h = normal MB = 6.0
	ePP	ABC	21 00	D = 125.07 Az = 50.8 (USCGS)
	e	A	21 24	PKIKP1V(A):1.8s 142.0nm
	e	A	22 06	PKIKP2V(A):1.6s 192.0nm
	eSKP	BC	22 22	

Day	Phase		h m s	Remarks
cont. 14.	e	A	00 22 32	LmH(B):22s 20.9/um MLH(B)=6.8
	e	A	22 46	LmV(B):24s 25.4/um MLV(B)=6.8
	ePPP	A	23 50	
	e	C	25 00	
	eSKS	BC	26 16	
	eSKKS	BC	28 00	
	eSKSP	BC	30 44	
	eSS	C	37 52	
	eSSSS	BC	46 35	
	LmH	B	01 06.6	
	LmV	B	06.7	
14.	ePKIKP1	A	00 31 24	Near Coast of Southern Chile
	eiPKIKP2	A	31 28	52.05 S 74.17 W
				H = 00 12 24.7 h = normal MB = 5.7
				D = 125.28 Az = 51.1 (USCGS)
				PKIKP1V(A):1.2s 12.4nm
				PKIKP2V(A):1.2s 32.5nm
14.	iP	A	11 59 17	Northern Italy 45.3 N 11.2 E
	iS	A	12 00 26.5	H = 11 57 44 (BCIS)
				D = 5.35
14.	ePKIKP	A	20 21(46)	Near Coast of Southern Chile
				51.96 S 74.22 W
				H = 20 02 45.3 h = normal MB = 5.2
				D = 125.25 Az = 51.0 (USCGS)
14.	ePKIKP	A	21 51 31	Solomon Islands 5.92 S 154.78 E
				H = 21 32 44.2 h = 142 km MB = 5.1
				D = 125.87 Az = 331.9 (USCGS)
14.	ePKHKP	AB	22 01 49	Easter Island Cordillera
	LmH	B	23 11.7	54.79 S 119.66 W
	LmV	B	15.3	H = 21 42 00.0 h = normal MB = 5.1
				D = 150.68 Az = 77.7 (USCGS)
				PKHKPV(A):1.6s 35.8nm
				LmH(B):18s 0.3/um MLH(B)=5.1
				LmV(B):18s 0.4/um

Day	Phase		h m s	Remarks
15.	e	A	06 27 53	Eastern Caucasus 43.22 N 47.07 E
	e	A	28 14	H = 06 22 15 h = 3 km (ISC)
				D = 25.1
15.	ePKIKP	AB	11 33 46	Falkland Islands Region 54.34 S 63.65 W
	ei	A	33 47	H = 11 14 52.4 h = normal MB = 5.6
	ei	A	33 51.5	D = 122.02 Az = 46.6 (USCGS)
	eiX	A	33 55	PKIKPV(A):0.7s 15.3nm
	ei	A	34 02.5	XV(A):1.3s 61.0nm
	ePP	BC	35 22	LmV(B):20s 64.9 μm MLV(B)=7.3
	ePP	A	35 27	LmH(B):17.5s 51.4 μm MLH(B)=7.2
	e	BC	35 44	
	eSKP	BC	37 24	
	eSKS	BC	40 55	
	eSKKS	BC	42 20	
	e(ScSP)	BC	44 55	
	ePS	BC	45 20	
	ePPS	C	46 50	
	eSS	BC	51 55	
	LmV	B	12 27.6	
	LmH	B	35.3	
15.	ePn	A	20 25 44	Albania 41.9 N 20.2 E
	eSn	A	27 40	H = 20 23 11 (BCIS)
				D = 10.50
15.	e	A	20 46 59.5	Albania 41.9 N 20.2 E
	e	A	48 29.5	H = 20 42 22 (BCIS)
				D = 10.50
16.	iPKIKP	AB	02 51 34	South of Fiji Islands 23.12 S 179.14 E
	eiPKHKP	AB	51 39	H = 02 32 50.5 h = 581 km MB = 5.0
	eiPKP2	A	51 49	D = 150.85 Az = 343.6 (USCGS)
	epPKIKP	A	53 50	PKHKPV(A):1.0s 51.2nm
	epPKHKP	A	53 55	
	epPKP2	A	54 05	

Day	Phase		h m s	Remarks
16.	ePn	A	04 20 53	Italy 44.8 N 10.4 E
	ePg	A	21 21	H = 04 19 26 (BCIS)
	eSn	A	22 00	D = 5.90
	eSg	AB	22 43	PgV(A):0.8s 23.1nm
	LmH	B	22.7	LmH(B):5s 0.2 μm
	LmV	B	22.7	LmV(B):4s 0.2 μm
16.	e	A	05 20(16)	Probably Italy (BCIS)
	e	A	21 11	
16.	eP	AB	05 23 28	South of Panama 5.37 N 82.55 W
	eS	B	34 00	H = 05 10 33.0 h = 17 km MB = 5.6
	eSS	B	40 00	D = 88.53 Az = 39.4 (USCGS)
				PV(A):2.0s 64.1nm MPV(A)=5.6
16	eX	A	06 32 44.5	Eastern Caucasus 43.2 N 47.2 E
				H = 06 22 15 (BCIS)
				D = 25.20
				XV(A):0.7s 11.5nm
16.	ePn	A	06 51 40	Yugoslavia 45.4 N 16.0 E
	eSn	A	52 43	H = 06 50 12 (BCIS)
	eSg	A	53 19	D = 6.00
16.	eP	A	17 32 03.5	Southern Iran 29.57 N 51.34 E
	e(PcP)	A	34 42.5	H = 17 25 02.1 h = 38 km MB = 4.3
				D = 36.37 Az = 316.7 (USCGS)
				(PcP)V(A):0.6s 9.6nm
16.	ePKHKP	A	18 14 26	South of Fiji Islands 24.10 S 177.03 W
	ePKP2	A	14 36.5	H = 17 54 41.7 h = 116 km MB = 4.8
				D = 152.66 Az = 348.0 (USCGS)
				PKHKPV(A):0.6s 5.7nm
17.	+eiP	AB	04 57 48.5	Southern Peru 15.75 S 71.84 W
	epP	B	58 19	H = 04 44 20.9 h = 91 km MB = 5.9
	e	A	58 31	D = 97.94 Az = 39.7 (USCGS)
	e	A	58 39	PV(A):1.8s 101.5nm MPV(A)=6.1

Day	Phase		h m s	Remarks
cont. 17.	e	A	05 00 26	LmH(B):21s 2.1 µm
	ePP	A	01 46	LmV(B):20.5s 2.7 µm
	+iS	B	08 20	
	+eisS	B	09 09	
	+ePS	B	10 30	
	LmH	B	39.4	
	LmV	B	39.5	
17.	eP	A	06 03 21	<u>Greece</u> 38.38 N 22.00 E
	e	A	03 26	H = 05 59 58.1 h = 6 km MB = 4.6
	LmH	B	09.5	D = 14.30 Az = 332.3 (USCGS)
	LmV	B	09.6	PV(A):0.7s 5.8nm
				LmH(B):12s 0.8 µm MLH(B)=3.9
				LmV(B):12s 0.7 µm
17.	eP	A	12 01 32	<u>Yunnan Province, China</u> 24.18 N 102.37 E
				H = 11 50 06.9 h = normal MB = 4.8
				D = 72.15 Az = 318.0 (USCGS)
17.	eP	A	18 56 15	<u>Kyushu, Japan</u> 30.20 N 131.11 E
	LmH	B	19 38.5	H = 18 43 48.2 h = 24 km MB = 5.1
	LmV	B	38.6	D = 83.43 Az = 326.1 (USCGS)
				PV(A):2.4s 48.3nm MPV(A)=5.3
				LmH(B):15.5s 2.5 µm MLH(B)=5.7
				LmV(B):15s 2.9 µm MLV(B)=5.8
17.	ePKIKP	A	21 49 22	<u>West Chile Rise</u> 36.30 S 97.59 W
	eX	A	49 34	H = 21 30 14.6 h = normal MB = 5.2
	LmH	B	22 48.2	D = 128.59 Az = 50.3 (USCGS)
	LmV	B	48.3	XV(A):1.8s 27.0nm
18.	ePKP2	AB	06 59 41	<u>Balleny Islands Region</u>
	ePP	B	07 03 10	61.33 S 160.02 E
	LmH	B	08 35.8	H = 06 39 03.3 h = normal MB = 5.0 (USCGS)
	LmV	B	40.0	D = 159.7
				PKP2V(A):2.0s 59.8nm
				LmH(B):17.5s 0.9 µm MLH(B)=5.6
				LmV(B):16s 0.8 µm

Day	Phase		h m s	Remarks
18.	ePn	A	09 05 30	Western Mediterranean Sea
	e	A	05 34	40.95 N 7.42 E
				H = 09 03 00.2 h = 0 km MB = 4.7
				D = 10.13 Az = 15 (ISC)
18.	LmV	B	11 49.8	Probably Kyushu, Japan (USCGS)
	LmH	B	49.9	LmV(B):16s 0.9/um
				LmH(B):14s 0.8/um
19.	eP	A	11 10 06	Near Coast of Northern Chile
	ePP	A	14 15	22.19 S 70.52 W
	e	A	14 29	H = 10 56 14.8 h = 52 km MB = 6.2
	e	A	14 36	D = 102.05 Az = 40.2 (USCGS)
	e	B	15 00	PV(A):1.5s 30.2nm MPV(A)=5.7
	iSKS	B	20 43	PPV(B):14s 0.8/um MPPV(B)=5.9
	eS	B	21 43	SH(B):13s 1.9/um MSH(B)=6.6
	ePKKP	AB	26 32	PKKPV(A):1.7s 24.2nm
	eiSS	AB	28 40	LmH(B):25s 5.3/um MLH(B)=6.0
	LmH	B	52.0	LmV(B):23s 5.8/um MLV(B)=6.1
	LmV	B	52.2	
19.	+iP1	AB	14 35 08.8	North Atlantic Ridge 15.35 N 45.95 W
	+eiP2	A	35 11	H = 14 25 18.4 h = normal MB = 5.5
	eiS	B	43 12	D = 57.84 Az = 39.4 (USCGS)
	eSS	C	46 50	P1V(A):1.8s 108.0nm MP1V(A)=5.6
	LmH	B	15 01.4	P1V(B):10s 1.1/um MP1V(B)=5.8
	LmV	B	03.3	P2V(A):1.7s 188.0nm MP2V(A)=5.8
				SH(B):16s 3.4/um MSH(B) =5.9
				LmH(B):16.5s 5.9/um MLH(B) =5.8
				LmV(B):16s 8.6/um MLV(B) =6.0
19.	e	A	15 31 50	Birresborn/Eifel/Fed.Rep.Germany
				$52^{o}11.69'$ N $06^{o}37.81'$ E
				H = 15 30 00.55
				explosion, yield 4.2 to (Hannover)
				D = 1.6
				traces

Day	Phase		h m s	Remarks
19.	ePKP	AB	18 58 05	Fiji Islands Region 15.40 S 176.29 W H = 18 38 24.9 h = normal MB = 5.3 D = 144.27 Az = 351.4 (USCGS) PKPV(A):1.5s 10.1nm
19.	eP	AB	19 03 43	Near East Coast of Kamchatka 57.34 N 163.27 E H = 18 52 33.6 h = normal MB = 5.2 D = 69.87 Az = 341.2 (USCGS) PV(A):2.0s 59.8nm MPV(A)=5.4
19.	ePKP	AB	19 09 19	Fiji Islands Region 15.20 S 176.34 W
	LmH	B	20 13.0	H = 18 49 46.6 h = normal MB = 5.4
	LmV	B	14.0	D = 144.07 Az = 350.4 (USCGS) PKPV(A):2.0s 59.8nm LmH(B):19.5s 2.0/um MLH(B)=5.9 LmV(B):20s 3.2/um
19.	iP	A	22 30 00	Greece-Albania-Border Region 39.4 N 20.4 E H = 22 27 01 (BCIS) D = 12.85
20.	eP	A	02 36 34	Off East Coast of Honshu, Japan
	LmH	B	03 11.6	40.14 N 143.24 E
	LmV	B	17.9	H = 02 24 26.5 h = 42 km MB = 4.9 D = 80.16 Az = 331.1 (USCGS) PV(A):1.6s 22.0nm MPV(A)=4.9 LmH(B):18s 0.9/um MLH(B)=5.2 LmV(B):15s 0.5/um MLV(B)=5.0
20.	ePKP	A	04 22 35	Fiji Islands Region 17.93 S 178.63 W H = 04 04 01.9 h = 633 km MB = 4.4 D = 146.35 Az = 348.2 (USCGS) PKPV(A):1.0s 13.8nm
20.	eP	A	06 08 32	Turkey 38.95 N 29.92 E
	LmH	B	16.3	H = 06 04 24.8 h = 26 km MB = 4.5

Day	Phase		h m s	Remarks
cont. 20.	LmV	B	06 16.3	D = 17.42 Az = 318.1 (USCGS) LmH(B):12s 0.9/um MLH(B)=4.2 LmV(B):12.5s 1.2/um MLV(B)=4.5
20.	LmH	B	14 03.7	Probably <u>Philippine Islands Region</u> (USCGS)
	LmV	B	03.8	
20.	eP	A	18 16 58	<u>South Atlantic Ridge</u> 26.53 S 13.76 W H = 18 04 50.4 h = normal MB = 4.3 D = 80.05 Az = 16.1 (USCGS)
21.	eP	A	16 21 09	<u>Kirgiziya</u> 40.83 N 72.93 E H = 16 13 08.0 h = 40 km MB = 4.7 D = 42.76 Az = 305 (ISC)
21.	eP	A	20 00 22	<u>South Atlantic Ridge</u> 26.55 S 13.73 W
	e	BC	04 44	H = 19 48 14.3 h = normal MB = 5.2
	eS	C	10 30	D = 80.07 Az = 16.1 (USCGS)
	eSS	C	15 34	PV(A):1.6s 30.2nm MPV(A)=5.0
	e	C	22 00	LmH(B):16s 0.4/um MLH(B)=4.9
	LmH	B	35.0	LmV(B):16s 0.4/um MLV(B)=4.9
	LmV	B	35.0	
22.	ePKHKP	A	06 07 32	<u>Loyalty Islands Region</u> 22.00 S 170.54 E
	ePKP2	A	07 34	H = 05 47 50.1 h = normal MB = - D = 146.99 Az = 335.2 (USCGS) PKHKPV(A):1.2s 10.2nm PKP2V(A):1.2s 20.3nm
22.	ePKP2	A	09 41 24	<u>South of Fiji Islands</u> 26.07 S 177.36 W H = 09 21 18.0 h = 85 km MB = 4.7 D = 154.51 Az = 346.7 (USCGS)
22.	ePKIKP	A	11 58 02	<u>New Hebrides Islands</u> 15.24 S 167.49 E H = 11 38 45.5 h = 111 km MB = 4.8 D = 139.66 Az = 336.3 (USCGS)

Day	Phase		h m s	Remarks
22.	+eP	ABC	14 51 04.5	<u>South of Alaska</u> 55.19 N 156.50 W
	eS	BC	15 00 38	H = 14 39 30.6 h = normal MB = 5.5
	e	B	00 49	D = 74.07 Az = 7.8 (USCGS)
	e	C	00 50	PV(A):1.7s 161.0nm MPV(A)=5.8
	LmH	B	29.4	LmH(B):16s 1.1 µm MLH(B)=5.3
	LmV	B	37.6	LmV(B):13s 1.1 µm MLV(B)=5.4
22.	eP	A	14 55 47	<u>South of Alaska</u> 55.22 N 156.69 W
				H = 14 44 07.7 h = 0 km MB = 5.1
				D = 74.05 Az = 8 (ISC)
				PV(A):1.0s 15.8nm MPV(A)=5.0
22.	eP	A	18 15 04	<u>Near East Coast of Kamchatka</u>
				52.60 N 158.85 E
				H = 18 03 37.2 h = 65 km MB = 4.9
				D = 73.47 Az = 338.9 (USCGS)
				PV(A):1.0s 13.8nm MPV(A)=4.8
22.	eP	ABC	21 45 33.5	<u>Kurile Islands</u> 43.47 N 147.62 E
	+ipP	A	45 45	H = 21 33 32.7 h = normal MB = 5.6
	ePP	B	48 30	D = 78.75 Az = 333.2 (USCGS)
	eS	BC	55 28	PV(A):1.5s 85.4nm MPV(A)=5.6
	iSKS	BC	55 47	PV(B):5.s 0.8 µm MPV(B)=6.0
	LmH	B	22 25.1	pPV(A):1.5s 211.0nm
	LmV	B	32.5	SH(B):6.5s 0.6 µm MSH(B)=5.8
				LmH(B):14.5s 4.2 µm MLH(B)=5.9
				LmV(B):13.5s 2.9 µm MLV(B)=5.8
23.	ePP	A	03 58 14	<u>South Sandwich Islands Region</u>
				60.69 S 25.35 W
				H = 03 38 35.1 h = normal MB = 5.3 (USCGS)
				D = 114.8
23.	e	A	04 27(44)	<u>Macquarie Islands Region</u> 59.61 S 157.90 E
	ePKP2	AB	28 16.5	H = 04 07 46.0 h = 30 km MB = 5.3
	ePP	B	31 55	D = 158.95 Az = 259.7 (USCGS)
				PKP2V(A):2.0s 59.8nm

Day	Phase		h m s	Remarks
23.	ePKHKP	A	08 33 42.5	South of Fiji Islands 21.99 S 179.74 W
	ePKP2	A	33 51	H = 08 15 02.0 h = 649 km MB = 4.8
				D = 150.04 Az = 345.5 (USCGS)
				PKHKPV(A):1.2s 26.4nm
23.	ePKIKP	A	16 28 54	Kermadec Islands Region 31.86 S 178.65 W
	e	A	29 12	H = 16 08 54.9 h = 42 km MB = 4.8
	e	A	29 28	D = 159.74 Az = 340.9 (USCGS)
23.	ePKP	A	18 02 01	Tonga Islands 16.33 S 173.11 W
	eX1	A	02 09.5	H = 17 42 24.9 h = normal MB = 4.7
	eX2	A	02 20	D = 145.55 Az = 354.7 (USCGS)
				PKPV(A):1.5s 45.2nm
				X1V(A): 1.5s 50.2nm
				X2V(A): 1.4s 25.6nm
23.	ePP	A	20 24 20	Northern Chile 19.40 S 69.08 W
				H = 20 06 57.6 h = 118 km MB = 5.3 (USCGS)
				D = 99.0
23.	ePn	A	22 14 18.5	Peissenberg, Fed.Rep.Germany 47.8 N 11.1 E
	e(Pg)	A	14 24.5	H = 22 13 32 (BCIS)
	eSn	A	14 41.5	D = 2.85
	eSg	A	15 05.5	
24.	e(P)	A	00 53 45	India-China Border Region
				28.93 N 95.57 E
				H = 00 43 01.9 h = normal MB = 4.8
				D = 64.54 Az = 315.5 (USCGS)
				(P)V(A):0.7s 11.5nm MPV(A)=5.2
24.	eP	AB	07 42 02	Queen Charlotte Islands Region
	eS	BC	51 32	51.84 N 130.76 W
	e	B	53 15	H = 07 30 30.8 h = normal MB = 4.9 (USCGS)
	eSS	BC	56 17	D = 73.0
	LmH	B	08 14.3	PV(A):1.2s 16.3nm MPV(A)=5.0
	LmV	B	17.4	LmH(B):18s 2.0 μm MLH(B)=5.4
				LmV(B):15s 2.7 μm MLV(B)=5.7

Day	Phase		h m s	Remarks
24.	-eP1	AB	13 20 40.5	Queen Charlotte Islands Region
	+iP2	A	20 49	51.75 N 131.02 W
	ePP	B	23 30	H = 13 09 08.3 h = 12 km MB = 5.6 (USCGS)
	eS	B	30 16	D = 73.0
	eSS	BC	34 58	P1V(A):1.8s 47.3nm MP1V(A)=5.3
	eP'P'	A	48 15.5	P1V(B):10.5s 5.9/um MP1V(B)=6.7
	eP'P'	A	48 25	P2V(A):1.6s 159.2nm MP2V(A)=5.9
	LmH	B	56.2	PPV(B):10s 5.4/um MPPV(B)=6.8
	LmV	B	56.2	SH(B):12s 10.3/um MSH(B)=6.8
				LmH(B):18s 96.4/um MLH(B)=7.1
				LmV(B):17s 147.0/um MLV(B)=7.4
24.	-eiP	A	13 28 30	Queen Charlotte Islands Region
				51.89 N 131.09 W
				H = 13 17 01.2 h = normal MB = 5.5 (USCGS)
				D = 73.0
				PV(A):1.1s 32.3nm MPV(A)=5.4
24.	eP	A	15 24 30	Kurile Islands 50.27 N 156.82 E
				H = 15 12 50.2 h = 36 km MB = 4.8
				D = 75.17 Az = 337.9 (USCGS)
25.	eP	A	05 05 40.5	Ural Mountains Region 52.20 N 55.69 E
				H = 04 59 52.4 h = 0 km MB = 4.9
				D = 27.22 Az = 284.4 (USCGS)
				Underground explosion (UPP)
25.	-ePKIKP	AB	05 33 01	Solomon Islands 7.92 S 158.69 E
	epPKIKP	A	33 15.5	H = 05 13 58.6 h = 69 km MB = 6.1
	ePP	B	35 08	D = 129.42 Az = 333.4 (USCGS)
	e	A	35 11.5	PKIKPV(A):2.0s 162.4nm
	eSKP	B	36 18	pPKIKPV(A):1.6s 126.5nm
	e	BC	36 24	LmV(B):20s 4.6/um
	e	BC	36 46	LmH(B):20s 4.2/um
	e(PS)	BC	45 38	
	e(PPS)	BC	47 10	
	eSS	BC	52 50	
	eSSS	BC	57 30	
	LmV	B	06 34.1	
	LmH	B	34.2	

Day	Phase		h m s	Remarks
25.	ePKIKP	A	13 25 44.5	Solomon Islands 7.81 S 158.66 E H = 13 06 40.3 h = 57 km MB = 5.3 D = 129.31 Az = 333.4 (USCGS)
25.	eP	A	16 18 40.5	Off East Coast of United States 39.60 N 71.00 W H = 16 08 54.6 h = 0 km MB = 5.0 D = 56.44 Az = 49.3 (USCGS) PV(A):1.2s 12.2nm MPV(A)=4.8
26.	+iP	AB	16 02 51.5	North of Ascension Island
	eS	BC	10 40	0.04 S 17.92 W
	LmH	B	26.0	H = 15 53 11.2 h = normal MB = 5.4
	LmV	B	30.5	D = 56.40 Az = 22.1 (USCGS) PV(A):1.6s 60.5nm MPV(A)=5.4 LmH(B):16s 0.7/um MLH(B)=4.8 LmV(B):13s 0.4/um MLV(B)=4.8
27.	eP	A	03 36 40.5	Kurile Islands 48.40 N 155.59 E
	LmH	B	04 15.7	H = 03 24 51.6 h = normal MB = 4.8
	LmV	B	15.7	D = 76.60 Az = 337.4 (USCGS) PV(A):1.2s 26.4nm MPV(A)=5.2
27.	e	A	05 22 22	Caspian Sea 43.25 N 49.1 E H = 05 16 45.1 h = 33 km (ISC) D = 26.2
27.	eP	A	08 04 22	Iran 35.15 N 50.70 E
	e	A	04 31	H = 07 57 53.3 h = 14 km MB = 4.9 D = 32.11 Az = 311.0 (USCGS)
27.	eP	A	09 58 57	Near Coast of Northern Peru
	LmH	B	10 43.0	9.77 S 78.59 W
	LmV	B	43.0	H = 09 45 28.5 h = 62 km MB = 5.6 D = 97.60 Az = 40.0 (USCGS) PV(A):1.1s 16.1nm MPV(A)=5.5 LmH(B):20s 0.2/um MLH(B)=4.6 LmV(B):20s 0.3/um MLV(B)=4.8

Day	Phase		h m s	Remarks
27.	eP	A	13 26 46	<u>Talaud Islands</u> 4.18 N 126.20 E
	e	A	31 24	H = 13 12 59.5 h = 85 km MB = 5.6
	e	A	31 33	D = 102.04 Az = 323.7 (USCGS)
	e	A	31 46	PV(A):1.5s 32.7nm MPV(A)=5.8
	eSKS	BC	37 12	LmH(B):23s 1.7 μm
	eS	BC	38 22	LmV(B):18s 1.1 μm
	e(SPP)	BC	40 45	
	LmH	B	14 07.5	
	LmV	B	18.3	
27.	ePn	A	18 59 46.5	<u>Albania</u> 41.47 N 19.37 E
	e	A	59 56	H = 18 57 12.2 h = 25 km MB = 4.5
	e	A	19 00 16	D = 10.64 Az = 332.3 (USCGS)
	ePg	A	00 47	LmH(B):6s 1.2 μm
	e	A	02 36	LmV(B):5s 0.7 μm
	eSn	B	02 44	
	eSg	A	02 51	
	LmH	B	03.7	
	LmV	B	03.7	
27.	eP	AB	23 06 13	<u>Near Coast of Chiapas, Mexico</u>
	eS	BC	16 48	14.70 N 92.84 W
	ePS	BC	18 00	H = 22 53 27.0 h = 41 km MB = 5.3 (USCGS)
	LmH	B	47.5	D = 87.7
	LmV	B	47.5	LmH(B):18.5s 0.9 μm MLH(B)=5.1
				LmV(B):18s 0.6 μm MLV(B)=5.0
28.	eP diff	AB	01 44 44	<u>Timor</u> 8.66 S 124.16 E
	ePKIKP	A	48(34)	H = 01 30 12.6 h = 41 km MB = 6.0 (USCGS)
	e	AB	49 15	D = 110.9
	iPP	BC	49 25	PdiffV(A):1.9s 41.6nm MPdiffV(A)=6.5
	e	B	49 50	LmH(B):25s 10.3 μm MLH(B)=6.3
	ePPP	BC	51 35	LmV(B):27s 8.5 μm MLV(B)=6.2
	ePS	BC	58 52	
	LmH	B	02 35.2	
	LmV	B	36.9	

Day	Phase		h m s	Remarks
28.	+iP	AB	02 05 46.4	Eastern Kazakh SSR 49.83 N 78.25 E
	ePn	A	07 18.5	H = 01 57 57.7 h = 0 km MB = 5.9
				D = 41.31 Az = 297.7 (USCGS)
				Underground explosion (UPP)
				PV(A):0.7s 318.0nm MPV(A)=6.2
28.	ePKP	A	04 00 29.5	Tonga Islands 16.58 S 173.71 W
	epPKP	A	00 52.5	H = 03 40 59.1 h = 90 km MB = 4.5
				D = 145.75 Az = 354.0 (USCGS)
				PKPV(A):1.4s 32.5nm
28.	e	A	07 59 52	Eastern Caucasus 43.0 N 45.7 E
				H = 07 54 42 h = 148 km MB = 4.5 (ISC)
				D = 24.3
28.	+iP	AB	11 13 23.3	Near East Coast of Kamchatka
	eS	BC	22 45	53.44 N 160.37 E
				H = 11 01 53.5 h = 23 km MB = 5.8
				D = 73.01 Az = 339.8 (USCGS)
				PV(A):1.0s 260.0nm MPV(A)=6.3
28.	iPKIKP	AB	11 28 30.5	Fiji Islands Region 21.63 S 179.53 W
	iPKHKP	AB	28 36.5	H = 11 09 54.2 h = 623 km MB = 5.8
	iPKP2	A	28 43.5	D = 149.74 Az = 345.9 (USCGS)
	epPKP	B	30 50	PKIKPV(A):1.1s 68.5nm
	e	A	30 59.5	PKHKPV(A):1.2s 724.0nm
	esPKP	B	31 52	PKP2V(A):1.2s 488.0nm
	eSKKS	BC	38 05	
	e	BC	42 25	
	eSS	BC	50 40	
	esSS	C	54 24	
28.	eP	A	13 43 40	Tunesia 35.8 N 11.6 E
	e	A	43 46	H = 13 40 14 h = 37 km
	eX	A	44 03	D = 14.82 Az = 360 (ISC)
	eS	BC	46 40	XV(A):1.8s 47.3nm
	LmH	B	49.7	LmH(B):15s 3.5/um
	LmV	B	50.6	LmV(B):13s 3.7/um

Day	Phase		h m s	Remarks
28.	ePKIKP	AB	22 58 26.5	Tonga Islands 21.07 S 174.45 W
	ePKHKP	A	58 32	H = 22 38 37.3 h = 34 km MB = 5.4
	ePKP2	A	58 40	D = 150.10 Az = 352.2 (USCGS)
				PKIKPV(A):1.4s 55.8nm
				PKHKPV(A):1.5s 55.5nm
				PKP2V(A): 1.5s 60.5nm
29.	ePKIKP	A	06 07 41.5	Kermadec Islands Region 31.06 S 179.86 W
	ePKHKP	A	07 54	H = 05 48 23.5 h = 335 km MB = 5.2
	+iPKP2	AB	08 20	D = 158.64 Az = 339.6 (USCGS)
	e	A	08 33	PKP2V(A):1.2s 112.0nm
29.	ePn	A	12 17 50	Yugoslavia 42.1 N 19.6 E
				H = 12 15 23 h = 0 km
				D = 10.20 Az = 330 (ISC)
29.	eP	A	18 12 59	North of Ascension Island 0.14 S 17.94 W
	LmH	B	42.3	H = 18 03 18.6 h = normal MB = 5.0
	LmV	B	42.7	D = 56.49 Az = 22.1 (USCGS)
				PV(A):0.7s 19.2nm MPV(A)=5.2
29.	ePKP2	A	23 33 04	Kermadec Islands Region 30.86 S 179.60 W
				H = 23 13 11.5 h = 361 km MB = 5.3 (USCGS)
				D = 158.7
				PKP2V(A):1.5s 17.6nm
30.	eP	A	03 43 10	Iceland Region 68.05 N 18.66 W
				H = 03 38 09.5 h = normal MB = 4.3
				D = 22.88 Az = 124.4 (USCGS)
30.	eP	A	03 58 01.5	Taiwan Region 22.76 N 121.45 E
				H = 03 45 37.3 h = 93 km MB = 4.9
				D = 84.42 Az = 323.0 (USCGS)
				PV(A):1.6s 22.0nm MPV(A)=4.9
30.	eP	A	06 34 33.5	Off Coast of Northern Peru
				9.22 S 79.02 W
				H = 06 21 03.9 h = 56 km MB = 5.4
				D = 97.45 Az = 39.9 (USCGS)

Day	Phase		h m s	Remarks
30.	epP	A	16 26 00.5	Hokkaido, Japan Region 41.50 N 142.67 E
				H = 16 13 48.1 h = 50 km MB = 4.7
				D = 78.76 Az = 330.7 (USCGS)
30.	eP	A	17 55 43	Greece 38.80 N 20.7 E
	LmH	B	18 00.6	H = 17 52 30.4 h = 0 km
	LmV	B	02.0	D = 13.46 Az = 335 (ISC)
				LmH(B):15s 1.2/um MLH(B)=4.1
30.	eP	A	18 24 31	Greece 38.45 N 20.24 E
	e	B	27 35	H = 18 21 14.2 h = 6 km MB = 4.8
	e	BC	28 13	D = 13.64 Az = 336.1 (USCGS)
	e	BC	28 48	PV(A):0.9s 15.6nm
	LmH	B	29.4	LmH(B):14s 5.1/um MLH(B)=4.7
	LmV	B	30.9	LmV(B):11s 1.8/um
30.	ePKIKP	A	21 55 13	Tonga Islands 21.34 S 174.68 W
	ePKHKP	A	55 18	H = 21 35 32.9 h = 72 km MB = 4.6
	ePKP2	A	55 24.5	D = 150.33 Az = 351.9 (USCGS)
				PKHKPV(A):1.3s 28.4nm
30.	eP	A	22 18 25	Greece 38.82 N 20.58 E
	LmH	B	23.3	H = 22 15 13.8 h = 0 km MB = 4.2
	LmV	B	24.7	D = 13.41 Az = 335 (ISC)
				LmH(B):15s 1.2/um MLH(B)=3.9
				LmV(B):11s 0.5/um

Day	Phase		h m s	Remarks
1.	eP	A	01 43 20	Ionian Sea 38.3 N 20.1 E
	LmH	B	48.2	H = 01 40 06 (BCIS)
				D = 13.73
				LmH(B):15s 0.5/um MLH(B)=3.7
1.	eP	A	01 51 18	Greece 38.82 N 20.4 E
	LmH	B	56.2	H = 01 48 05 h = 0 km
				D = 13.35 Az = 335 (ISC)
				LmH(B):15s 1.3/um MLH(B)=4.1
1.	eP	A	03 29 25.5	Bonin Islands 28.08 N 139.28 E
				H = 03 17 25.5 h = 514 km MB = 4.8
				D = 89.01 Az = 329.7 (USNOAA)
				PV(A):1.6s 33.0nm MPV(A)=5.3
1.	e	A	04 47 43.5	Probably Timor (USNOAA)
1.	ePKP	A	13 03 17	Tonga Islands 17.12 S 175.02 W
				H = 12 44 03.6 h = 245 km MB = 4.7
				D = 146.13 Az = 352.4 (USNOAA)
				PKPV(A):1.9s 45.4nm
1.	eiP	A	15 55 22	Cyprus 35.29 N 31.23 E
	ei	A	55 34	H = 15 50 39.6 h = 36 km MB = 4.8
				D = 20.92 Az = 323.2 (USNOAA)
				PV(A):1.2s 56.9nm MPV(A)=4.8
1.	eiP	A	16 27 46	North Atlantic Ridge 23.77 N 45.57 W
	LmH	B	45.5	H = 16 18 42.8 h = normal MB = 5.1(USNOAA)
	LmV	B	45.5	D = 51.3
				PV(A):1.3s 32.8nm MPV(A)=5.1
				MLH(B):20s 1.3/um MLH(B)=4.8
				MLV(B):20s 1.6/um MLV(B)=5.1
1.	epP	A	21 40 56	Hokkaido, Japan Region 41.57 N 142.70 E
				H = 21 28 43.8 h = 51 km MB = 4.5
				D = 78.71 Az = 330.7 (USNOAA)

Day	Phase		h m s	Remarks
2.	eP	ABC	00 58 33	Near Coast of Peru 10.11 S 78.59 W
	epP	AB	58 46	H = 00 45 02.0 h = 62 km MB = 5.8
	eSKS	BC	01 09 09	D = 97.86 Az = 40.0 (USNOAA)
	eSKKS	BC	09 33	PV(A):1.5s 35.2nm MPV(A)=5.7
	LmH	B	40.7	LmH(B):21s 2.1/um MLH(B)=5.6
	LmV	B	40.7	LmV(B):20s 2.8/um MLV(B)=5.8
2.	iPKP	ABC	01 15 58	South of Australia 51.04 S 139.49 E
	eX1	A	16 00	H = 00 56 15.3 h = normal MB = 5.6
	eX2	AB	16 15	D = 147.65 Az = 290.1 (USNOAA)
	LmH	B	02 18.7	PKPV(A):1.6s 38.5nm
	LmV	B	26.0	X1V(A):1.4s 60.5nm
				X2V(A):2.2s 272.0nm
				X2V(B):7.0s 1.3/um
				LmH(B):20s 1.3/um MLH(B)=5.6
				LmV(B):21s 2.4/um MLV(B)=5.9
2.	eP1	A	02 29 22.5	Turkey 38.81 N 36.67 E
	eiP2	ABC	29 25.7	H = 02 24 35.8 h = 27 km MB = 4.8
	eS	BC	33 24	D = 21.26 Az = 312.0 (USNOAA)
	LmH	B	39.2	P1V(A):1.9s 26.5nm MPV1(A)=4.3
	LmV	B	39.3	P2V(A):1.9s 60.6nm MPV2(A)=4.6
				LmH(B):13s 2.4/um MLH(B) =4.8
				LmV(B):17s 3.0/um MLV(B) =4.9
2.	ePKHKP	A	07 37 07	Fiji Islands Region 21.78 S 179.37 W
	ePKP2	A	37 15.5	H = 07 18 22.5 h = 595 km MB = 4.9
	epPKP	A	39 30	D = 149.93 Az = 346.0 (USNOAA)
				PKHKPV(A):0.8s 25.0nm
2.	eP	A	07 53 23	Greece 38.73 N 20.51 E
	eX	A	53 34	H = 07 50 14.6 h = 34 km MB = 5.0
	e	B	57 04	D = 13.47 Az = 335.0 (USNOAA)
	LmH	B	58.3	XV(A):1.4s 51.1nm
	LmV	B	59.7	LmH(B):14s 4.0/um MLH(B)=4.6
				LmV(B):11s 2.7/um

Day	Phase		h m s	Remarks
2.	e	A	13 01 03.5	<u>Near Coast of Peru</u> 10.13 S 78.68 W
				H = 12 47 07.8 h = 60 km MB = 4.9
				D = 97.93 Az = 40.0 (USNOAA)
2.	eP	A	14 16 10	<u>Greece</u> 38.6 N 20.3 E
	LmH	B	21.0	H = 14 12 59 (BCIS)
				D = 13.55
				LmH(B):15s 1.0 μm MLH(B)=5.1
2.	eP	AB	19 26 30	<u>Northern Sumatra</u> 4.67 N 97.71 E
	e(PS)	B	37 40	H = 19 14 01.5 h = normal MB = 5.2
	LmH	B	20 09.5	D = 83.94 Az = 320.3 (USNOAA)
	LmV	B	09.5	PV(A):1.4s 25.6nm MPV(A)=5.2
				LmH(B):20s 0.5 μm MLH(B)=4.9
				LmV(B):19s 0.6 μm MLV(B)=5.0
3.	eP1	A	00 44 09.5	<u>Greece</u> 38.68 N 20.36 E
	eP2	A	44 12	H = 00 41 01.0 h = normal MB = 5.1
	e	B	47 52	D = 13.47 Az = 335.5 (USNOAA)
	LmH	B	49.2	P2V(A):0.7s 17.2nm MP2V(A)=5.1
	LmV	B	50.5	LmH(B):16s 2.9 μm MLH(B) =4.4
				LmV(B):10s 1.3 μm
3.	eP	A	10 01 31.5	<u>Northwest of Kurile Islands</u>
				49.85 N 150.62 E
				H = 09 50 33.5 h = 360 km MB = 5.1
				D = 73.93 Az = 334.2 (USNOAA)
				PV(A):0.8s 26.8nm MPV(A)=5.1
4.	eP	A	01 25 27	<u>North of Ascension Island</u>
	LmV	C	50.6	3.26 S 12.13 W
	LmH	C	51.0	H = 01 15 36.0 h = normal MB = 4.7
				D = 57.43 Az = 17.7 (USNOAA)
4.	iP	A	15 02 29	<u>Kurile Islands</u> 43.22 N 146.25 E
				H = 14 50 31.6 h = 57 km MB = 4.4
				D = 78.51 Az = 332.5 (USNOAA)

Day	Phase		h m s	Remarks
4.	LmH	B	20 17.3	**Banda Sea** 4.3 S 127.8 E
	LmV	B	20.6	H = 19 01 46.1 h = 59 km MB = 5.3(USNOAA)
				D = 109.8
				LmH(B):21s 0.6/um MLH=5.1
				LmV(B):20s 0.5/um MLV=5.1
4.	LmH	B	23 57.4	**Caucasus**
	LmV	B	58.5	H = 23 40 38 (UPP)
				LmH(B):14s 0.3/um
				LmV(B):14s 0.3/um
5.	eP	A	07 08 50	**Peru-Brazil Border Region**
	LmH	C	16.0	8.77 S 74.38 W
				H = 06 55 46.3 h = 140 km MB = 4.9
				D = 94.18 Az = 39.6 (USNOAA)
				PV(A):2.0s 47.0nm MPV(A)=5.5
5.	eP	A	14 25 55.5	**Mindanao, Philippine Islands**
	epP	A	26 09	7.43 N 126.85 E
	esP	A	26 16	H = 14 12 16.6 h = 59 km MB = 5.6
	e(pPP)	A	30 26	D = 99.79 Az = 324.2 (USNOAA)
	eSKS	BC	36 44	PV(A):1.8s 33.8nm MPV(A)=5.6
	ePS	C	39(12)	LmH(B):20s 1.6/um MLH(B)=5.5
	eSS	C	44(40)	LmV(B):20s 1.8/um MLV(B)=5.6
	LmH	B	15 14.5	
	LmV	B	14.5	
5.	ePKP	AB	24 08 54	**Tonga Islands** 17.43 S 173.43 W
	LmH	B	25 20.0	H = 23 49 13.5 h = 34 km MB = 4.8
	LmV	B	23.0	D = 146.61 Az = 354.2 (USNOAA)
				PKPV(A):1.8s 67.6nm
				LmH(B):18s 0.3/um MLH(B)=5.1
				LmV(B):18s 0.4/um MLV(B)=5.2
6.	eP1	A	03 44 47.5	**Kurile Islands** 43.28 N 147.64 E
	eP2	A	44 49.5	H = 03 32 48.8 h = 58 km MB = 4.7
	epP	A	45 00.5	D = 78.92 Az = 333.2 (USNOAA)
	LmH	C	04 17.0	

Day	Phase		h m s	Remarks
6.	ePKP	A	04 28 50	New Hebrides Islands 19.15 S 169.26 E
				H = 04 09 33.2 h = 148 km MB = 4.6
				D = 143.91 Az = 335.7 (USNOAA)
6.	eP	AB	05 57 07	South of Panama 4.03 N 78.23 W
	e	B	57.20	H = 05 44 23.3 h = normal MB = 4.5
	e	B	06 01 06	D = 86.82 Az = 39.6 (USNOAA)
	e	B	07 16	LmH(B):17s 0.6/um MLH(B)=5.1
	e	B	07 36	LmV(B):17s 0.6/um MLV(B)=5.1
	e	B	07 46	
	LmH	B	39.8	
	LmV	B	43.7	
6.	LmH	B	16 51.0	USSR-Mongolia Boder Region
	LmV	B	51.2	50.5 N 90.6 E H = 16 20 12 (ANUSSR)
				D = 47.8
				LmH(B):14s 0.3/um MLH(B)=4.4
				LmV(B):14s 0.3/um MLV(B)=4.5
7.	ePn1	A	06 06 18	Central Italy 42.83 N 13.00 E
	ePn2	A	06 20.5	H = 06 04 21.0 h = 17 km MB = 4.0
	e	A	06 30	D = 7.88 Az = 353.6 (USNOAA)
	ePg	A	06 56	Pn2V(A):1.1s 16.1nm
	iSn	A	07 44	LmH(B):8s 0.6/um MLH(B)=3.6
	eSg	ABC	08 44	
	LmH	BC	09.3	
7.	eiP	A	06 32 11	Kurile Islands 45.64 N 149.44 E
				H = 06 20 23.7 h = 80 km MB = 5.0
				D = 77.38 Az = 334.0 (USNOAA)
				PV(A):1.0s 33.5nm MPV(A)=5.2
7.	eP	A	08 06 01	North Atlantic Ridge 43.62 N 28.87 W
	eS	C	10 50	H = 08 00 11.7 h = normal MB = 4.8(USNOAA)
	LmH	B	14.4	D = 27.45
	LmV	B	15.4	PV(A):2.0s 21.4nm MPV(A)=4.4
				LmH(B):14s 0.4/um MLH(B)=4.2
				LmV(B):14s 0.4/um MLV(B)=4.3

Day	Phase		h m s	Remarks
7.	e(P)	A	22 23 56.5	Kurile Islands 44.38 N 148.0 E
				H = 22 11 58.9 h = 0 km MB = 5.0
				D = 78.06 Az = 333 (ISC)
8.	-iP1	AB	04 59 53	Virgin Islands 17.96 N 64.63 W
	+iP2	AC	59 54	H = 04 49 10.6 h = 150 km MB = 5.8(USNOAA)
	Pm	A	59 55.5	D = 67.6
	esP	BC	05 00 38	P1V(A):1.0s 43.3nm MP1V(A)=5.2
	ePP	BC	02 20	P2V(A):1.4s 65.0nm MP2V(A)=5.3
	eS	BC	08 38	PmV(A):2.2s 415.0nm MPmV(A)=5.9
	eScS	BC	09 36	P1V(B):6s 2.2 μm MP1V(B)=6.2
	e	A	09 40	PPV(B):5s 1.0 μm MPPV(B)=6.3
	esScS	BC	10 42	LmV(B):16s 1.1 μm
	eSSS	C	16 20	LmH(B):14.5s 1.1 μm
	LmV	B	23.9	
	LmH	B	24.7	
8.	ePg	A	14 00 44	Switzerland 46.3 N 7.7 E
	eSg	A	01 53	H = 13 59 06 (BCIS)
				D = 5.05
8.	ePg	A	14 09 09	Switzerland 46.3 N 7.8 E
	e	A	09 12	H = 14 07 33 (BCIS)
	eSg	A	10 15	D = 5.00
8.	eP	A	15 10 50	Hokkaido, Japan Region
	epP	A	11 13	42.04 N 142.51 E
	LmH	B	47.0	H = 14 58 56.0 h = 71 km MB = 4.8
	LmV	B	47.0	D = 78.23 Az = 330.6 (USNOAA)
				h = 91 km
				PV(A):1.5s 25.2nm MPV(A)=4.9
8.	eP	A	23 21(08)	Afghanistan-USSR Border Region
	LmH	B	43.0	37.09 N 66.05 E
	LmV	B	43.0	H = 23 13 29.7 h = normal MB = 5.1
				D = 40.61 Az = 307.3 (USNOAA)
				LmH(B):14s 0.1 μm MLH(B)=4.0
				LmV(B):12s 0.3 μm MLV(B)=4.4

Day	Phase		h m s	Remarks
9.	+iP	ABC	08 23 08.5	Kurile Islands Region 43.85 N 148.40 E
	eipP	A	23 20.5	H = 08 11 09.7 h = 51 km MB = 5.4
	LmH	B	55.8	D = 78.66 Az = 333.6 (USNOAA)
	LmV	B	09 02.1	h = 40 km
				PV(A):1.0s 51.2nm MPV(A)=5.5
				LmH(B):20s 1.2/um MLH(B)=5.2
				LmV(B):16s 0.9/um MLV(B)=5.3
9.	+iP1	ABC	11 36 38.3	Kurile Islands Region 43.89 N 148.53 E
	-eP2	A	36 39	H = 11 24 39.5 h = 41 km MB = 5.4
	eS	BC	46 32	D = 78.67 Az = 333.7 (USNOAA)
	esS	BC	46 48	P1V(A):0.7s 13.4nm MP1V(A)=5.0
	LmH	B	12 09.3	P2V(A):2.0s 145.0nm MP2V(A)=5.6
	LmV	B	15.6	LmH(B):19.5s 2.3/um MLH(B) =5.5
				LmV(B):16s 1.8/um MLV(B) =5.5
9.	eP	ABC	12 23 57	Kurile Islands Region 43.85 N 148.44 E
	Pm	A	23 59	H = 12 11 58.9 h = 48 km MB = 5.5
	eS	B	33 51	D = 78.67 Az = 333.6 (USNOAA)
	esS	B	34 09	PV(A):0.8s 61.5nm MPV(A)=5.6
	LmH	B	56.6	PmV(A):2.0s 214.0nm MPmV(A)=5.9
	LmV	B	13 03.0	LmH(B):20s 6.2/um MLH(B)=5.9
				LmV(B):15.5s 4.3/um MLV(B)=5.9
9.	iPg	A	17 05 23.5	Böhmischbruck, Federal Rep. Germany
	iSg	A	05 38	49°34.1'N 12°21.4'E
				explosion
				H c. 18 05 00 yield c. 5.0 to (Hannover)
				D c. 1.2
9.	eP	A	20 32 41	Kurile Islands Region 43.94 N 148.30 E
	LmH	B	21 05.0	H = 20 20 40.2 h = normal MB = 4.6
	LmV	B	11.7	D = 78.54 Az = 333.5 (USNOAA)
				PV(A):0.8s 13.5nm MPV(A)=5.1
				LmH(B):19s 0.3/um MLH(B)=5.7
				LmV(B):16s 0.3/um MLV(B)=5.7

Day	Phase		h m s	Remarks
9.	eP	A	20 34 47	Kurile Islands Region
				H = 20 22 46 (UPP)
				PV(A):0.6s 9.6nm
9.	eP	A	21 10 55	Rumania 45.71 N 26.47 E
	e	A	11 12	H = 21 08 18.5 h = 143 km MB = 4.6
				D = 11.08 Az = 301.8 (USNOAA)
10.	eP	A	01 23 21	Kurile Islands 44.13 N 148.36 E
	epP	A	23 33.5	H = 01 11 22.5 h = normal MB = 4.1
				D = 78.40 Az = 333.6 (USNOAA)
				h = 46 km
10.	ePKIKP	AB	09 44 32	Santa Cruz Islands 11.98 S 166.56 E
	ePP	AB	47 13	H = 09 25 24.4 h = 140 km MB = 5.3
				D = 136.34 Az = 337.0 (USNOAA)
				PKIKPV(A):2.3s 73.0nm
10.	eP	AB	13 27 48	Near Coast of Guerrero, Mexico
	eX	A	28 14	17.49 N 101.02 W
	ePP	B	31 25	H = 13 14 50.9 h = 46 km MB = 5.1
	eSKS	BC	38 12	D = 90.20 Az = 36.0 (USNOAA)
	e	BC	38 52	PV(A):1.8s 27.0nm MPV(A)=5.3
	ePS	BC	39 56	XV(A):1.6s 27.5nm
	eSS	C	44 46	LmH(B):17s 0.9/um MLH(B)=5.3
	LmH	B	14 11.5	LmV(B):18s 1.0/um MLV(B)=5.3
	LmV	B	11.5	
10.	eP	A	13 34 26	Turkey 40.88 N 35.58 E
				H = 13 29 57.0 h = 17 km MB = 4.6
				D = 19.28 Az = 308.5 (USNOAA)
				PV(A):2.4s 69.1nm MPV(A)=4.5
10.	ePg	A	14 05 56	Eschenlohe, Federal Rep. Germany
	e	A	06 02	47°37.9 N 11°08.7 E explosion
	eSg	A	06 39.5	H c. 14 05 00 yield c. 15 to (Hannover)
	eSg	A	06 40.5	D c. 3.0
	e	A	06 42.5	

Day	Phase		h m s	Remarks
10.	eP	A	14 21 17.5	Rumania 47.73 N 25.64 E
	e	A	21 25	H = 14 18 58.8 h = normal MB = 4.7
	e	AB	24 30	D = 9.64 Az = 292.8 (USNOAA)
	LmV	B	25.9	PV(A):1.1s 18.1nm
	LmH	B	26.0	LmV(B):7s 0.9 μm
				LmH(B):9s 0.8 μm MLH(B)=3.9
10.	eP	A	16 29 21	Ryukyu Islands 28.73 N 129.29 E
	LmH	B	17 11.5	H = 16 16 57.3 h = 63 km MB = 5.2
	LmV	B	11.5	D = 83.73 Az = 325.4 (USNOAA)
				PV(A):1.0s 17.7nm MPV(A)=5.1
				LmH(B):16.5s 0.6 μm MLH(B)=5.0
				LmV(B):17s 0.6 μm MLV(B)=5.1
10.	eP	AB	21 19 34.5	Mindoro, Philippine Islands
	e	A	19 41	13.93 N 120.45 E
	e	A	19 51	H = 21 06 38.3 h = 80 km MB = 5.6
	e	A	19 57	D = 90.87 Az = 322.9 (USNOAA)
	eS	BC	30 22	PV(A):1.5s 101.0nm MPV(A)=5.9
	e	B	30 50	LmH(B):20s 0.5 μm
	eSP	B	31 27	LmV(B):18s 0.3 μm
	e	BC	31 29	
	LmH	B	22 05.5	
	LmV	B	06.0	
11.	e	A	06 56 26	Poland 50.32 N 18.87 E
				H = 06 55 26.5 MB = 2.7 (ISC)
				D = 4.6
11.	e	A	13 38 22	Mediterranean Sea 34.41 N 22.24 E
	LmH	B	46.4	H = 13 33 51.9 h = 18 km MB = 4.6
	LmV	B	46.7	D = 17.98 Az = 337.3 (USNOAA)
				LmH(B):16s 0.4 μm MLH(B)=3.8
				LmV(B):16s 0.5 μm MLV(B)=4.0
11.	eP	A	14 40 31.5	Near East Coast of Honshu, Japan
	e	A	40 36	36.48 N 140.51 E
				H = 14 28 15.5 h = 67 km MB = 5.2
				D = 82.29 Az = 330.0 (USNOAA)
				PV(A):1.2s 24.4nm MPV(A)=5.1

Day	Phase		h m s	Remarks
11.	eP	A	21 29 03	Ryukyu Islands 28.33 N 129.44 E
	epP	A	29 15	H = 21 16 33.8 h = 35 km MB = 5.2
	eS	BC	39 24	D = 84.14 Az = 325.5 (USNOAA)
	LmH	B	22 04.2	h = 43 km
	LmV	B	11.0	PV(A):2.0s 30.0nm MPV(A)=4.9
				pPV(A):2.0s 34.2nm
				LmH(B):18s 1.4/um MLH(B)=5.4
				LmV(B):17s 0.6/um MLV(B)=5.1
11.	eP	ABC	22 47 15	Caspian Sea 37.55 N 48.98 E
	epP	A	47 28	H = 22 41 15.6 h = 65 km MB = 5.1
	eS	BC	52 12	D = 29.52 Az = 308.4 (USNOAA)
	e(SS)	B	54 00	h = 65 km
	LmH	B	23 05.0	PV(A):1.6s 22.0nm MPV(A)=4.6
	LmV	B	05.0	pPV(A):1.5s 25.1nm
				LmH(B):17s 1.5/um
				LmV(B):14.5s 1.2/um
11.	eP	A	23 32 28.5	Greece-Albania Border Region
	e	B	36 10	38.97 N 20.64 E
	LmH	B	37.3	H = 23 29 20.2 h = 41 km MB = 4.8
	LmV	B	38.7	D = 13.30 Az = 334.3 (USNOAA)
				PV(A):0.8s 26.9nm
				LmH(B):14.5s 2.6/um MLH(B)=4.4
				LmV(B):12s 1.1/um
12.	eP	ABC	09 31 25	Leyte, Philippine Islands
	e	A	31 48	10.83 N 125.39 E
	ePP	ABC	35 22	H = 09 17 59.0 h = 35 km MB = 5.5
	e(SKKS)	BC	42 20	D = 96.21 Az = 324.1 (USNOAA)
	e	BC	43 10	PV(A):1.2s 32.6nm MPV(A)=5.7
	LmH	B	10 11.6	LmH(B):24s 2.0/um MLH(B)=5.5
	LmV	B	19.3	LmV(B):18s 1.4/um MLV(B)=5.5
12.	ePKP	A	15 57 37	New Hebrides Islands 19.13 S 168.77 E
				H = 15 38 05.5 h = 28 km MB = -
				D = 143.69 Az = 335.3 (USNOAA)
				PKPV(A):0.7s 9.6nm

Day	Phase		h m s	Remarks
13.	eP	A	00 49 55	Greece 38.91 N 20.62 E
	eX	A	50 11	H = 00 46 47.4 h = normal MB = 4.7
	LmH	B	54.8	D = 13.35 Az = 334.4 (USNOAA)
	LmV	B	56.2	XV(A):1.6s 27.5nm
				LmH(B):14.5s 2.7/um MLH(B)=4.4
				LmV(B):10.5s 1.2/um
13.	ePKP	A	06 31 49.5	Tonga Islands 16.68 S 173.93 W
				H = 06 12 14.4 h = 47 km MB = 4.9
				D = 145.83 Az = 353.7 (USNOAA)
				PKPV(A):1.2s 16.3nm
13.	eP	A	16 11 34.5	Southern Alaska 60.43 N 152.04 W
	epP	A	12 00	H = 16 00 41.4 h = 104 km MB = 4.8
	e	A	12 05.5	D = 68.50 Az = 11.1 (USNOAA)
				h = 100 km
				PV(A):1.4s 16.3nm MPV(A)=4.7
14.	e	A	04 07 07	Probably Kermadec Islands (ISC)
14.	eP	ABC	18 11 33	Norwegian Sea 72.46 N 2.04 E
	Pm	A	11 38	H = 18 06 37.8 h = normal MB = 4.9
	eiX	A	11 43	D = 22.30 Az = 163.8 (USNCAA)
	eS	BC	15 40	PV(A):1.4s 34.9nm MPV(A)=4.6
	LmH	B	20.0	PmV(A):1.5s 65.4nm MPmV(A)=4.9
	LmV	B	21.8	XV(A):1.5s 90.5nm
				LmH(B):19s 1.1/um MLH(B)=4.3
				LmV(B):12s 0.7/um MLV(B)=4.4
15.	epPKP	A	15 13 42	New Hebrides Islands 20.79 S 169.85 E
				H = 14 53 46.8 h = 120 km MB = 4.9
				D = 145.63 Az = 335.3 (USNOAA)
16.	eP	A	01 53 13.5	Kurile Islands Region 43.88 N 148.33 E
	LmV	B	02 33.2	H = 01 41 13.9 h = 40 km MB = 4.7
	LmH	B	33.4	D = 78.61 Az = 333.6 (USNOAA)
				PV(A):0.9s 13.6nm MPV(A)=5.0
				LmV(B):12s 0.3/um MLV(B)=4.9
				LmH(B):16s 4.2/um MLH(B)=4.9

Day	Phase		h m s	Remarks
16.	ePKIKP	A	07 57 25	South of Fiji Islands 25.30 S 178.10 W
	ePKHKP	A	57 33	H = 07 38 00.8 h = 225 km MB = 5.2
	ePKP2	A	57 46.5	D = 153.61 Az = 346.0 (USNOAA)
	ei	A	57 51	h = 238 km
	epPKIKP	A	58(25)	PKIKPV(A):1.4s 14.0nm
	epPKHKP	A	58 34	
	ep PKP2	A	58 48	
16.	ePg	A	15 05 20	Hilders/Rhön, Federal Rep. Germany
	iPg	A	05 20.8	50°32.49'N 10°02.33'E explosion
	e	A	05 34	H = 15 05 00.74 yield 11.7 t (Hannover)
	iSg	A	05 36.3	D = 1.0
16.	ePKIKP	A	16 03 02.5	South of Fiji Islands 25.88 S 177.25 W
	ePKHKP	A	03 11	H = 15 43 20.2 h = 90 km MB = 5.3
	iPKP2	A	03 25	D = 154.35 Az = 346.9 (USNOAA)
16.	ePKIKP	A	21 37 26	Tonga Islands 19.19 S 173.45 W
	ePKHKP	A	37 30.5	H = 21 17 44.2 h = normal MB = 5.8
	e(PKP2)	AB	37 35	D = 148.35 Az = 353.8 (USNOAA)
	e	B	37 49	PKIKPV(A):1.7s 30.3nm
	e	B	38 34	PKHKPV(A):1.8s 203.0nm
	eSS	B	22 00 06	LmH(B):19s 2.7/um MLH(B)=6.0
	e	B	01 10	LmV(B):18s 2.6/um MLV(B)=6.0
	LmH	B	48.0	
	LmV	B	51.7	
16.	ePKP	A	23 43 41	Samoa Islands Region 15.62 S 172.89 W
	e	A	43 53	H = 23 24 02.8 h = normal MB = 4.7
				D = 144.88 Az = 355.0 (USNOAA)
17.	eP	A	07 44 11	Fox Islands, Aleutian Is.
	epP	A	44 23.5	51.05 N 171.31 W
				H = 07 32 10.1 h = normal MB = 4.9
				D = /8.65 Az = 358.1 (USNOAA)
				h = 46 km
				PV(A):1.1s 16.2nm MPV(A)=5.0
				pPV(A):0.9s 25.3nm

Day	Phase		h m s	Remarks
17.	eP	A	18 02 43.5	Ecuador 1.79 S 77.27 W
				H = 17 49 59.5 h = 182 km MB = 4.9
				D = 90.66 Az = 39.5 (USNOAA)
				PV(A):2.0s 42.8nm MPV(A)=5.1
17.	ePKIKP	AB	20 24 32.5	Tonga Islands Region 22.10 S 174.71 W
	+iPKHKP	A	24 38.5	H = 20 04 46.5 h = normal MB = 5.6
	ePKP2	A	24 46	D = 151.08 Az = 351.7 (USNOAA)
	ePP	B	28 12	PKIKPV(A):1.0s 23.6nm
	e	B	28 46	PKHKPV(A):1.6s 151.0nm
	ePPP	B	31 28	PKP2V(A):1.8s 263.0nm
	e	B	32 40	LmV(B):18s 6.0/um MLV(B)=6.4
	eSKKS	B	35 12	LmH(B):17s 5.2/um MLH(B)=6.3
	eSS	B	47 36	
	LmV	B	21 42.7	
	LmH	B	42.8	
18.	eiP	AB	02 00 34	Andreanof Islands, Aleutian Is.
	-i	A	00 39.5	51.39 N 178.52 W
	epP	A	00 56.5	H = 01 48 38.9 h = 46 km MB = 5.7
	esP	B	01 06	D = 77.97 Az = 353.4 (USNOAA)
	ePP	B	03 32	h = 90 km
	ePPP	B	05 22	PV(A):1.4s 100.0nm MPV(A)=5.6
	eS	B	10 25	PV(B):9s 1.5/um MPV(B)=6.0
	eSKS	B	10 37	PPV(B):9s 0.9/um MPPV(B)=5.9
	ePPS	B	11 29	LmH(B):19s 6.0/um MLH(B)=5.9
	eSS	B	15 40	LmV(B):20s 5.0/um MLV(B)=5.9
	LmH	B	39.2	
	LmV	B	39.3	
18.	ePKHKP	A	05 28 10	Tonga Islands 20.18 S 173.12 W
				H = 05 08 20.3 h = 31 km MB = 4.7
				D = 149.37 Az = 354.1 (USNOAA)
19.	eP	A	05 06 15	Iceland Region 62.91 N 24.68 W
	e	A	06 21	H = 05 01 11.6 h = normal MB = 4.0
	e	A	06 26.5	D = 23.00 Az = 105.3 (USNOAA)

Day	Phase		h m s	Remarks
19.	e	A	09 41 50.5	New Ireland Region 3.79 S 152.40 E
	ePP	AB	43 20	H = 09 22 40.1 h = 20 km MB = 5.5
	LmH	B	10 26.6	D = 122.88 Az = 331.4 (USNOAA)
	LmV	B	35.7	PPV(A):2.5s 77.0nm MPPV(A)=5.8
				LmH(B):23s 4.8/um MLH(B)=6.1
				LmV(B):19s 2.5/um MLV(B)=5.9
19.	eP	A	22 17 23	Central Mid-Atlantic Ridge
				0.87 S 22.82 W
				H = 22 07 22.1 h = normal MB = 4.6
				D = 59.11 Az = 24.8 (USNOAA)
19.	eP	A	23 43 44	Kurile Islands 44.62 N 148.15 E
				H = 23 31 47.5 h = normal MB = 4.7
				D = 77.89 Az = 333.4 (USNOAA)
21.	+iP	AB	01 25 51.5	Hindu Kush Region 36.47 N 70.47 E
	epP	A	26 36.5	H = 01 18 05.2 h = 210 km MB = 5.2
	ePP	A	27 30	D = 43.80 Az = 308.1 (USNOAA)
	LmH	B	53.7	h = 213 km
	LmV	B	57.6	PV(A):1.4s 60.5nm MPV(A)=4.9
				pPV(A):1.4s 23.3nm
				PPV(A):1.8s 33.8nm MPPV(A)=4.8
				LmH(B):18s 0.4/um
				LmV(B):16s 0.4/um
21.	+iP	A	03 10 43.5	Eastern Kazakh SSR 49.96 N 77.76 E
	ePn	A	12 14.5	H = 03 02 57.1 h = 0 km MB = 5.4
				D = 40.97 Az = 297.4 (USNOAA)
				Underground explosion (UPP)
				PV(A):0.7s 46.0nm MPV(A)=5.3
21.	ePKP	A	06 41 10	Tonga Islands 15.87 S 173.07 W
	epPKP	A	41 27	H = 06 21 33.5 h = normal MB = 5.0
	e(sPKP)	A	41 37	D = 145.10 Az = 354.8 (USNOAA)
				h = 57 km
				PKPV(A):1.5s 30.3nm
				pPKPV(A):1.7s 48.5nm
				sPKPV(A):1.3s 19.7nm

Day	Phase		h m s	Remarks
21.	epP	A	09 08 18	Tadzhik-Sinkiang Border Region 38.57 N 73.42 E H = 08 59 51.9 h = 120 km MB = 4.6 D = 44.39 Az = 306.7 (USNOAA) pPV(A):0.9s 9.7nm
21.	ePn	A	11 25 48	Switzerland 46.5 N 7.7 E
	ePg	A	26 10	H = 11 24 35 (BCIS)
	iSg	A	27 12.5	D = 4.9
	LmH	B	27.5	LmH(B):7.5s 1.4 um MLH(B)≠3.8
	LmV	B	27.6	LmV(B):9s 0.6 um
21.	ePKHKP	AB	19 04 08.5	Tonga Islands 19.20 S 173.42 W
	e	A	04 19	H = 18 44 30.2 h = 88 km MB = 5.1
	LmH	B	20 11.0	D = 148.37 Az = 353.9 (USNOAA)
	LmV	B	17.0	PKHKPV(A):1.4s 11.6nm
21.	eP	A	23 16 50	Unimak Islands Region 53.73 N 164.84 W H = 23 05 02 h = 18 km MB = 4.2 (ISC) D = 76.0
22.	ePKIKP	A	13 28 14.5	East New Guinea Region 5.53 S 147.27 E H = 13 09 39.8 h = 165 km MB = 5.3 D = 121.81 Az = 328.4 (USNOAA) PKIKPV(A):1.2s 10.2nm
23.	eP	A	05 31 24.5	Central Mid-Atlantic Ridge 6.11 N 33.60 W H = 05 21 31.4 h = normal MB = 4.8 D = 58.16 Az = 32.1 (USNOAA)
23.	eiPKP	A	16 04 41.5	Tonga Islands 15.71 S 173.85 W H = 15 45 14.6 h = 95 km MB = 5.4 D = 144.88 Az = 353.9 (USNOAA) PKPV(A):1.2s 32.5nm

Day	Phase		h m s	Remarks
24.	+iP	A	04 04 46.5	Eastern Kazakh SSR 49.80 N 78.17 E
	ePn	A	06 19.5	H = 03 56 57.4 h = 0 km MB = 5.3
				D = 41.28 Az = 297.7 (USNOAA)
				Underground explosion (UPP)
				PV(A):0.6s 76.6nm MPV(A)=5.6
24.	eP	A	08 14 10	Near Islands, Aleutian Is.
				52.14 N 171.45 E
				H = 08 02 23.6 h = normal MB = 5.0
				D = 76.18 Az = 346.9 (USNOAA)
25.	eP	A	11 16 47	Bonin Islands Region 28.85 N 139.04 E
				H = 11 04 43.7 h = 462 km MB = 4.9(USNOAA
				D = 88.2
25.	+iP1	AB	22 53 28.5	Kyushu, Japan 32.18 N 131.70 E
	eiP2	A	53 34	H = 22 41 10.7 h = 34 km MB = 6.1
	ePP	B	56 31	D = 82.06 Az = 326.2 (USNOAA)
	ePPP	B	58 15	P1V(A):1.7s 206.0nm MP1V(A)=6.0
	ePPPP	B	59 55	P1V(B):14s 20.0/um MP1V(B)=7.0
	iS	B	23 03 48	P2V(A):2.0s 760.0nm MP2V(A)=6.5
	iPS	B	04 38	PPV(B):12s 9.7/um MPPV(B)=7.0
	eSS	B	07 55	SH(B):13s 13.7/um MSH(B) =6.9
	ei	B	08 38	LmH(B):17s 440.0/um MLH(B) =7.9
	LmH	B	34.1	LmV(B):16s 566.0/um MLV(B) =8.1
	LmV	B	34.2	
26.	+iP	AB	07 22 54	Kyushu, Japan 32.19 N 131.81 E
	ePP	AB	26 04	H = 07 10 36.0 h = 35 km MB = 6.1
	ePPP	B	27 56	D = 82.11 Az = 326.3 (USNOAA)
	ePPPP	B	29 28	PV(A):2.0s 440.0nm MPV(A)=6.1
	eS	B	33 08	PV(B):12s 2.0/um MPV(B)=6.1
	eSKS	B	33 18	PPV(B):13s 1.1/um MPPV(B)=6.0
	ePS	B	33 58	SH(B):20s 2.4/um MSH(B)=6.0
	ePPS	B	34 12	LmV(B):16s 65.5/um MLV(B)=7.1
	eSS	B	38 22	LmH(B):14s 50.5/um MLH(B)=7.0
	eSSS	B	42 35	
	LmV	B	08 03.6	
	LmH	B	03.7	

Day	Phase		h m s	Remarks
27.	ePg	A	12 10 29	Switzerland 46.6 N 7.5 E
	e	A	10 38	H = 12 08 56 (BCIS)
	eSn	A	11 08	D = 4.9
	eSg	A	11 33	
	i	A	11 40	
27.	+ePKIKP	A	12 52 04.5	Off E. Coast of N. Island, N.Z.
	epPKIKP	A	52 26	37.83 S 177.61 E
				H = 12 31 18.7 h = 87 km MB = 5.7(USNOAA)
				D = 163.7 h = 75 km
				PKIKPV(A):1.2s 20.4nm
28.	ePKIKP	A	05 07 27	Tonga Islands 21.75 S 174.63 W
	ePKP2	A	07 39	H = 04 47 47.7 h = normal MB = 5.2
				D = 150.75 Az = 351.9 (USNOAA)
				PKP2V(A):1.6s 35.7nm
28.	eP	A	06 41 28	Hindu Kush Region 36.03 N 68.33 E
	ePP	A	43 08.5	H = 06 33 35.0 h = 60 km MB = 5.0
	LmH	B	07 02.5	D = 42.71 Az = 308.3 (USNOAA)
	LmV	B	02.9	PV(A):1.4s 20.9nm MPV(A)=4.7
				LmH(B):14s 0.2 µm
				LmV(B):10s 0.3 µm
28.	eP	A	19 39 50	Fox Islands, Aleutian Is.
	epP	A	40 10	54.11 N 165.95 W
				H = 19 28 13.8 h = 92 km MB = 4.8
				D = 75.60 Az = 1.6 (USNOAA)
				h = 80 km
28.	e	A	23 25 25	**Off East of Southern Chile**
	ePP	B	27 00	44.50 S 79.57 W
	e(PPP)	B	30 15	H = 23 06 22.9 h = normal MB = 4.8
	eSP	B	37 00	D = 123.24 Az = 49.6 (USNOAA)
	eSPP	B	38 32	LmH(B):19s 2.3 µm MLH(B)=5.8
	eSS	B	43 45	LmV(B):18s 2.2 µm MLV(B)=5.9
	eSSS	B	48 30	
	LmH	B	24 18.0	
	LmV	B	19.9	

Day	Phase		h m s	Remarks
29.	eP	AB	05 59 23.5	Southern Sinkiang Prov., China
	Pm	A	59 32	39.88 N 77.80 E
	ePP	B	06 01 16	H = 05 50 56.4 h = 13 km MB = 5.2
	e	B	03 00	D = 46.34 Az = 306.4 (USNOAA)
	eS	B	06 12	PV(A):1.2s 30.3nm MPV(A)=5.0
	e(ScS)	B	09 14	PmV(A):1.8s 81.3nm MPmV(A)=5.5
	eSS	B	09 20	PPV(B):9s 0.5μm MPPV(B)=5.4
	e	B	09 46	LmV(B):16.5s 16.2μm MLV(B)=6.1
	e	B	10 00	LmH(B):17s 13.4μm MLH(B)=6.0
	LmV	B	20.3	
	LmH	B	20.4	
29.	-iP	AB	10 27 04.5	Burma-India Border Region
	Pm	A	27 06	26.02 N 95.40 E
	e(sP)	AB	27 28	H = 10 16 19.3 h = 59 km MB = 6.5
	ePP	B	29 32	D = 66.52 Az = 316.4 (USNOAA)
	ePPP	B	31 12	PV(A):1.4s 1120.0nm MPV(A)=6.6
	e	B	31 24	PV(B):10.5s 10.3μm MPV(B)=6.7
	iS	B	35 50	PmV(A):1.4s 1400.0nm MPmV(A)=6.7
	e(PS)	B	36 24	PPV(B):9s 6.1μm MPPV(B)=6.9
	eSS	B	40 06	SH(B):10.5s 23.8μm MSH(B)=7.3
	eSSS	B	43 08	LmV(B):32s 127.4μm MLV(B)=7.0
	LmV	B	55.0	LmH(B):20.5s 50.2μm MLH(B)=6.7
	e	A	55.34	PKPPKPV(A):1.8s 88.0nm
	LmH	B	55.6	PKPPKPPKSV(A):2.1s 67.0nm
	ePKPPKP	A	55 38	
	eSKPPKP	A	58 54	
	ePKPPKS	A	59 06	
	eSKPP'P'	A	11 18 10	
	eP'P'PKS	A	18 19	
29.	+iP	A	10 41 55	Burma-India Border Region
				26.25 N 95.13 E
				H = 10 31 10.6 h = 48 km MB = 5.5
				D = 66.19 Az = 316.3 (USNOAA)
				PV(A):1.2s 54.9nm MPV(A)=5.4

Day	Phase		h m s	Remarks
29.	ePn	A	15 51 36.5	Austria 47.8 N 11.1 E
	ePg	A	51 42	H = 15 50 51 (BCIS)
	e	A	52 05	D = 2.85
	eSn	A	52 10	
	eSg	A	52 22.5	
29.	e	A	18 35 05	South Atlantic Ridge 14.01 S 14.31 W
				H = 18 23 58.3 h = normal MB = 5.2
				D = 68.27 Az = 17.4 (USNOAA)
30.	eP1	AB	00 59 01.5	Iran-USSR Border Region
	+iP2	A	59 06	37.82 N 55.88 E
	eS	B	01 04 28	H = 00 52 19.5 h = 19 km MB = 5.7
	eSS	B	06 14	D = 33.70 Az = 306.8 (USNOAA)
	e(SSS)	B	06 44	P1V(A):1.3s 80.8nm MP1V(A)=5.5
	e	B	08 25	P1V(B):5s 6.6/um MP1V(B)=6.8
	e	B	09 02.5	P2V(A):1.2s 240.0nm MP2V(A)=6.0
	LmH	B	18.4	SH(B):8s 10.4/um MSH(B) =6.6
	LmV	B	20.4	LmH(B):14s 58.9/um MLH(B) =6.4
				LmV(B):13s 39.6/um MLV(B) =6.4
30.	eP	A	02 08 45	Iran-USSR Border Region
	e	A	09 09	37.63 N 55.82 E
				H = 02 02 02.0 h = normal MB = 4.8
				D = 33.78 Az = 307.0 (USNOAA)
				PV(A):0.6s 13.4nm MPV(A)=5.1
30.	eP	A	02 27 07.5	Kenai Peninsula, Alaska
				60.58 N 148.57 W
				H = 02 16 08.8 h = 24 km MB = 4.7
				D = 67.99 Az = 13.5 (USNOAA)
				PV(A):0.6s 13.4nm MPV(A)=5.3
30.	eP	A	02 41 55	Iran-USSR Border Region
				37.48 N 55.76 E
				H = 02 35 11.5 h = normal MB = 4.9
				D = 33.83 Az = 307.2 (USNOAA)

Day	Phase		h m s	Remarks
30.	+eP1	A	05 07 26.5	Eastern Gulf of Aden 14.26 N 51.84 E
	+iP2	A	07 31	H = 04 58 43.8 h = normal MB = 5.5
	iP3	A	07 37	D = 48.70 Az = 326.8 (USNOAA)
	-eiP4	A	07 42.5	P1V(A):1.7s 57.6nm MP1V(A)=5.3
	+eiPP	A	09 33	P2V(A):2.0s 120.0nm MP2V(A)=5.6
				P3V(A):2.5s 600.0nm MP3V(A)=6.2
				P4V(A):2.1s 790.0nm MP4V(A)=6.4
				PPV(A):3.2s 2220.0nm MPPV(A)=6.5
30.	eP1	A	05 11 35	Socotra Region 13.9 N 52.1 E
	eP2	A	11 37	H = 05 02 56 h = 38 km MB = 5.1
	eP3	A	11 48	D = 49.17 Az = 327 (ISC)
	+eP4	A	11 54.5	P1V(A):1.6s 38.4nm MP1V(A)=5.1
	ePP	A	13 37	P2V(A):2.1s 135.0nm MP2V(A)=5.5
				P3V(A):1.7s 127.0nm MP3V(A)=5.6
				P4V(A):2.0s 522.0nm MP4V(A)=6.1
30.	e(PKP2)	A	06 02 41	Kermadec Islands Region 28.15 S 176.55 W
				H = 05 42 10.5 h = 48 km MB = 4.6(USNOAA)
				D = 157.1
				(PKP2V)(A):0.6s 11.5nm
30.	eP	A	07 25 01.5	North of Ascension Island
				0.87 S 13.12 W
				H = 07 15 26.8 h = normal MB = 4.9
				D = 55.49 Az = 18.9 (USNOAA)
				PV(A):0.6s 13.4nm MPV(A)=5.2
30.	ePKIKP	A	09 11 19	Near North Coast of New Guinea
	epPKIKP	A	11 48	4.45 S 143.96 E
				H = 08 52 40.7 h = 106 km MB = 5.4
				D = 119.14 Az = 327.4 (USNOAA)
				h = 109 km
30.	+iPKP	AB	19 03 40.8	Tonga Islands 15.55 S 174.00 W
	epPKP	A	04 12.5	H = 18 44 17.9 h = 122 km MB = 5.0
	esPKP	A	04 29	D = 144.70 Az = 353.8 (USNOAA)
				h = 124 km
				PKPV(A):1.4s 67.4nm

Day	Phase		h m s	Remarks
31.	eP	A	01 48 34	Kurile Islands 44.25 N 147.07 E
				H = 01 36 43.1 h = 79 km MB = 4.3
				D = 77.87 Az = 332.8 (USNOAA)
31.	eP	A	02 05 16	Kurile Islands 43.42 N 147.53 E
	e	A	05 24	H = 01 53 16.5 h = 49 km MB = 4.9
	epP	A	05 27	D = 78.76 Az = 333.2 (USNOAA)
	LmH	B	40.7	h = 41 km
	LmV	B	44.8	PV(A):1.1s 20.2nm MPV(A)=5.0
				pPV(A):1.4s 27.9nm
				LmH(B):16s 0.4/um MLH(B)=4.8
				LmV(B):16s 0.2/um MLV(B)=4.6
31.	ePKP	A	04 00 38	Tonga Islands 17.83 S 173.26 W
	eX1	AB	00 48	H = 03 40 57.1 h = normal MB = 4.8
	eX2	A	01 01	D = 147.03 Az = 354.3 (USNOAA)
				PKPV(A):2.0s 42.7nm
				X1V(A):2.0s 55.6nm
				X2V(A):1.8s 67.5nm
31.	eP1	A	13 21 56.5	Szechwan Province, China
	eP2	A	21 59	28.62 N 103.55 E
	LmH	B	49.0	H = 13 10 47.4 h = 25 km MB = 5.5
	LmV	B	55.5	D = 69.60 Az = 317.2 (USNOAA)
				P1V(A):1.3s 21.8nm MP1V(A)=5.1
				P2V(A):1.5s 35.2nm MP2V(A)=5.2
				LmH(B):22s 1.6/um MLH(B)=5.2
				LmV(B):12s 0.4/um MLV(B)=5.0
31.	ePKIKP	A	15 35 32	Easter Island Region 27.04 S 113.31 W
	eX	A	35 49	H = 15 16 18.7 h = normal MB = 5.3
	LmV	B	16 32.3	D = 132.36 Az = 44.9 (USNOAA)
	LmH	B	39.8	XV(A):2.5s 61.4nm
				LmV(B):20s 0.7/um MLV(B)=5.4
				LmH(B):17s 0.9/um MLH(B)=5.5

Day	Phase		h m s	Remarks
31.	-iP1	AB	17 19 47.5	Colombia 1.46 S 72.56 W
	+iP2	B	19 52	H = 17 08 05.4 h = 651 km MB = 7.1
	-iP3	B	20 10	D = 87.42 Az = 39.3 (USNOAA)
	eP4	B	20 31	P1V(B):6.5s 18.9/um MP1V(B)=7.0
	i	B	21 44	P2V(B):9.5s 60.1/um MP2V(B)=7.3
	e	B	22 33	P3V(B):11s 77.9/um MP3V(B)=7.4
	e	B	23 03	SH(B):10s 99.0/um MSH(B) =7.7
	e(PP)	B	23 45	XmV(A):2.8s 129.0nm
	iSKS	B	29 17	
	eiS	B	29 32	
	ei(sPPPP)	B	30 00	
	eiSP	B	30 38	
	eiSPP	B	31 12	
	eisS	B	33 43	
	eiSS	B	35 50	
	eisSS	B	39 22	
	e(SSSS)	B	42 42	
	eiPKPPKP	A	45 33	
	ei	A	47 47	
	e	A	18 05 35	
	eSKPP'P'	A	06 15	
	eiX	A	06 32	
	Xm	A	07 07	
31.	ePKP	A	21 05 40.5	Fiji Islands Region 14.96 S 176.98 W
				H = 20 46 05.5 h = normal MB = 5.2
				D = 143.73 Az = 350.7 (USNOAA)

Day	Phase		h m s	Remarks
1.	eP	A	01 30 56	Southern Sumatra 2.96 S 102.42 E
	epP	A	31 26	H = 01 17 56.3 h = 125 km MB = 5.4
	eX	A	34 19	D = 92.78 Az = 320.4 (USNOAA)
	ePP	A	34 39	PV(A):1.3s 17.5nm MPV(A)=5.1
	iSKS	B	40 15	XV(A):1.9s 22.7nm
	iS	B	40 49	PPV(A):2.0s 51.2nm MPPV(A)=5.6
	esS	A	42 37	SH(B):6.0s 0.7/um MSH(B)=6.0
	e	B	42 40	
1.	e	A	06 45 21	New Hebrides Islands 15.26 S 167.67 E
				H = 06 24 48.3 h = 180 km MB = 4.8
				D = 139.76 Az = 336.4 (USNOAA)
1.	eP	A	14 49 05	Greenland Sea 73.93 N 9.92 E
	LmH	B	58.5	H = 14 43 59.0 h = normal MB = 4.4
	LmV	B	58.5	D = 23.38 Az = 177.3 (USNOAA)
				PV(A):0.9s 19.5nm MPV(A)=4.6
				LmH(B):15s 0.2/um MLH(B)=3.8
				LmV(B):16s 0.2/um MLV(B)=3.8
2.	ePn	A	01 28 32	Upper Silesia
	ei(Sg)	A	29 37.5	rock burst (CLL)
				D c. 3.8
2.	+eiP	A	01 48 00.5	Kurile Islands 46.70 N 152.52 E
	epP	A	48 12.5	H = 01 36 10.6 h = 60 km MB = 5.0
	esP	A	48 19	D = 77.33 Az = 335.7 (USNOAA)
				h = 48 km
				PV(A):1.0s 37.4nm MPV(A)=5.3
2.	LmH	B	02 09.0	Off Coast of Mexico 8.6 N 102.6 W
	LmV	B	09.0	H = 01 13 59.7 h = normal MB = 4.6(USNOAA)
				D = 98.3
				LmH(B):20s 0.4/um MLH(B)=4.9
				LmV(B):20s 0.1/um MLV(B)=4.5
2.	ePKP	AB	13 35 57	Samoa Islands Region 16.39 S 172.90 W
	e	A	36 03	H = 13 16 19.9 h = normal MB = 4.7
	e	A	36 15	D = 145.63 Az = 354.9 (USNOAA)
				PKPV(A):1.8s 23.7nm

Day	Phase		h m s	Remarks
2.	-ePKP	AB	19 43 33	Samoa Islands Region 16.57 S 172.77 W
	eX1	A	43 37	H = 19 23 55.3 h = normal MB = 4.7
	eX2	A	43 49	D = 145.82 Az = 355.0 (USNOAA)
	e	B	44 00	PKPV(A):3.0s 184.0nm
	LmH	B	20 56.5	PKPV(B):8s 1.1 µm
	LmV	B	21 02.9	PX1V(A):1.8s 60.9nm
				PX2V(A):1.8s 88.0nm
				LmH(B):17.5s 0.7 µm MLH(B)=5.4
				LmV(B):17s 0.7 µm MLV(B)=5.5
3.	+eiPKP	AB	00 53 19	Tonga Islands 15.93 S 173.94 W
	e	A	53 42	H = 00 33 54.3 h = 120 km MB = 5.4
	ei	AB	53 50	D = 145.08 Az = 353.8 (USNOAA)
	LmH	B	02 02.9	PKPV(A):1.8s 270.0nm
	LmV	B	03.0	
3.	ePKP	A	03 53 11	Tonga Islands 16.21 S 174.62 W
	eX	A	53 16	H = 03 33 34.7 h = normal MB = 5.2
				D = 145.29 Az = 353.0 (USNOAA)
				PKPV(A):1.2s 12.4nm
				XV(A):1.6s 49.5nm
3.	ePKIKP1	AB	07 20 15	Solomon Islands 7.93 S 158.72 E
	ePKIKP2	A	20 16	H = 07 01 11.9 h = 67 km MB = 5.9
	epPKIKP	A	20 30	D = 129.44 Az = 333.4 (USNOAA)
	ePKS	B	23 38	h = 54 km
	epPKS	B	24 00	PKIKP1V(A):1.2s 10.2nm
	LmH	B	08 21.4	PKIKP2V(A):1.6s 38.5nm
	LmV	B	21.6	pPKIKPV(A):1.8s 74.3nm
				LmH(B):20s 1.9 µm
				LmV(B):20s 1.9 µm
3.	e(PKP)	A	13 35 03	Loyalty Islands Region
				22.61 S 170.60 E
				H = 13 15 18.4 h = 21 km MB = -
				D = 147.56 Az = 334.8 (USNOAA)

Day	Phase		h m s	Remarks
3.	-iP	AB	22 42 39	Northern Sumatra 2.60 N 97.98 E
	ipP	A	42 56	H = 22 30 02.5 h = 38 km MB = 5.9
	eSKS	B	52 59	D = 85.69 Az = 320.4 (USNOAA)
	e	B	53 00	h = 63 km
	e	B	54 50	PV(A):1.3s 80.8nm MPV(A)=5.8
	LmH	B	23 27.3	pPV(A):1.4s 144.0nm
	LmV	B	28.1	LmH(B):20s 0.4 µm
				LmV(B):18s 0.2 µm
4.	LmH	B	01 35.6	West New Guinea Region 4.6 S 134.0 E
	LmV	B	45.8	H = 00 32 06.4 h = normal MB = 5.2(USNOAA)
				D = 113.7
				LmH(B):18s 0.9 µm MLH(B)=5.4
				LmV(B):20s 0.3 µm MLV(B)=4.9
4.	eP	A	11 41 45.5	Fox Islands, Aleutian Is.
				53.33 N 166.86 W
				H = 11 29 59.6 h = 46 km MB = 4.8
				D = 76.39 Az = 1.0 (USNOAA)
4.	LmH	B	13 33.1	South of Honshu, Japan 31.8 N 139.3 E
	LmV	B	39.0	H = 12 41 41.9 h = 29 km MB = 4.8(USNOAA)
				D = 85.8
				LmH(B):13s 0.8 µm MLH(B)=5.3
				LmV(B):13s 0.5 µm MLV(B)=5.2
4.	eP	A	17 36 07	Greece 38.90 N 21.96 E
				H = 17 32 47.8 h = 65 km MB = 4.3(USNOAA)
				D = 13.9
4.	eiPKP	A	19 53 58.5	Tonga Islands 16.04 S 173.62 W
	eipPKP	A	54 12	H = 19 34 21.9 h = normal MB = 4.7
				D = 145.22 Az = 354.2 (USNOAA)
				h = 48 km
				PKPV(A):0.8s 23.1nm
				pPKPV(A):1.2s 24.4nm

Day	Phase		h m s	Remarks
5.	-ePn	A	04 29 09.3	Yugoslavia 44.11 N 15.95 E
	+iPn	A	29 10.3	H = 04 27 24.7 h = normal MB = 4.7
	e	A	29 37.5	D = 7.17 Az = 337.3 (USNOAA)
	eiSn	A	30 30.5	PnV(A):1.0s 94.5nm
	eSg	AB	31 16	LmH(B):6.0s 1.3/um MLH(B)=4.1
	LmH	B	31.4	
	LmV	B	33.0	
5.	eP	A	05 39 25	Near Coast of Northern Peru
				9.24 S 78.91 W
				H = 05 25 57.6 h = 69 km MB = 5.6(USNOAA)
				D = 97.4
				PV(A):1.3s 21.8nm MPV(A)=5.5
5.	-eP	AB	09 18 59.5	North Atlantic Ridge 11.86 N 43.74 W
	e	A	19 06	H = 09 08 59.4 h = normal MB = 5.2(USNOAA)
	LmV	B	39.7	D = 59.2
	LmH	B	40.5	PV(A):1.3s 43.7nm MPV(A)=5.4
				LmV(B):16s 0.4/um MLV(B)=4.7
				LmH(B):20s 0.4/um MLH(B)=4.5
5.	eP	A	13 34 43	Costa Rica 9.84 N 84.32 W
				H = 13 22 05.5 h = 56 km MB = 4.6(USNOAA)
				D = 86.2
5.	ePKIKP	A	13 48 04	Fiji Islands Region 21.3 S 176.5 W
				H = 13 28 38.5 h = 219 km MB = 4.3(USNOAA)
				D = 150.0
				PKIKPV(A):1.4s 11.6nm
6.	e	A	02 50 33	South of Honshu, Japan 31.7 N 139.4 E
	LmH	B	03 29.3	H = 02 37 49.0 h = 29 km MB = 4.7(USNOAA)
	LmV	B	29.5	D = 86.0
				LmH(B):14s 0.4/um MLH(B)=5.0
				LmV(B):14s 0.2/um MLV(B)=4.7
6.	iPn	A	13 55 50	Northern Italy 46.93 N 10.53 E
	iPg	A	56 07	H = 13 54 54.9 h = normal MB = - (USNOAA)

Day	Phase		h m s	Remarks
cont.				
6.	i	A	13 56 08	D = 3.8
	iSn	A	56 38.5	
	iSg	A	57 02	
	i	A	57 03.5	
6.	ePKHKP	AB	21 41 36	Tonga Islands Region 22.97 S 175.37 W
	ePKP2	A	41 48	H = 21 21 44.6 h = 50 km MB = 5.0(USNOAA)
	LmH	B	22 57.0	D = 151.8
	LmV	B	59.0	LmH(B):18s 0.4/um MLH(B)=5.2
				LmV(B):18s 0.3/um MLV(B)=5.1
6.	eP	A	22 17 19	Norwegian Sea 71.7 N 1.7 E
				H = 22 12 32 D = 21.6 (ANUSSR)
				Northeast of Jan Mayen, near 71 1/2 N 3 W
				H = 22 12 22 (UPP)
				D = 22.0
				PV(A):1.6s 11.0nm MPV(A)=4.0
7.	eP	A	01 55 20	Kurile Islands Region 43.77 N 148.35 E
	eX1	A	55 29	H = 01 43 19.0 h = normal MB = 5.0(USNOAA)
	eX2	A	55 31.5	D = 78.6
	LmH	B	02 28	PV(A):1.2s 21.0nm MPV(A)=5.1
	LmV	B	42	X2V(A):1.2s 32.5nm
				LmH(B):20s 0.5/um MLH(B)=4.8
7.	eP	A	04 57 26	Turkey 39.0 N 29.3 E
				H = 04 53 23 (BCIS)
				D = 17.1
7.	-eiPKIKP	AB	08 09 49.5	Fiji Islands Region 17.71 S 178.30 W
	+eiPKHKP	A	09 51.5	H = 07 51 12.0 h = 548 km MB = 5.5(USNOAA)
	eiPKP2	A	09 54.4	D = 146.2
	epPKP	AB	12 02.5	PKIKPV(A):1.6s 38.5nm
	eSKP	B	12 44	PKHKPV(A):1.6s 192.0nm
	ei(sSS)	B	35 00	pPKPV(A):1.6s 60.4nm

Day	Phase		h m s	Remarks
7.	ePKP	A	08 53 44	Fiji Islands Region 15.55 S 177.06 W
				H = 08 34 56.3 h = 416 km MB = 4.6(USNOAA)
				D = 144.2
				PKPV(A):1.1s 10.1nm
7.	eP	A	13 52 25	Kurile Islands Region 43.63 N 148.40 E
				H = 13 40 23.4 h = normal MB = 4.5
				D = 78.86 Az = 333.6 (USNOAA)
7.	eP	A	14 45 57.5	Kurile Islands 48.12 N 153.36 E
				H = 14 34 18.4 h = 110 km MB = 4.8(USNOAA)
				D = 76.2
				PV(A):1.1s 20.1nm MPV(A)=4.8
7.	eP	AB	16 46 30	Bonin Islands Region 27.29 N 141.74 E
	e	A	46 47	H = 16 33 29.2 h = normal MB = 5.4(USNOAA)
	ePP	A	50 06	D = 90.7
	e	A	50 14	PV(A):2.0s 42.7nm MPV(A)=5.4
	eSKS	B	57 00	LmH(B):15s 1.9/um MLH(B)=5.6
	eS	B	57(24)	LmV(B):13s 1.5/um MLV(B)=5.6
	eSP	B	58 25	
	ePS	B	58 30	
	LmH	B	17 27.1	
	LmV	B	33.7	
7.	ePKHKP	A	19 25 35	South of Fiji Islands 22.56 S 179.09 E
				H = 19 06 49.3 h = 583 km MB = 4.7(USNOAA)
				D = 150.3
8.	ePn	A	04 28 56	Northern Italy 44.67 N 12.82 E
	ePg	A	29 28	H = 04 27 27.4 h = 27 km MB = 4.2(USNOAA)
	iSn	A	30 05	D = 6.0
	eSg	A	30 49	
	LmH	B	31.9	
	LmV	B	31.9	
8.	eP	A	09 13 19	Kyushu, Japan 30.63 N 129.99 E
	e	A	14 04	H = 09 01 08.3 h = 120 km MB = 5.1(USNOAA)
	e(PS)	B	24 20	D = 82.5

Day	Phase		h m s	Remarks
cont.				
8.	LmH	B	09 49.0	PV(A):1.8s 67.5nm MPV(A)=5.1
	LmV	B	55.0	
8.	eX	A	11 55 02	Northern Sinkiang Prov.,China
	LmH	B	12 11.1	44.31 N 81.21 E
	LmV	B	14.5	H = 11 46 31.3 h = normal MB = 4.7
				D = 45.88 Az = 303.8 (USNOAA)
				XV(A):1.2s 20.4nm
				LmH(B):7.5s 0.7/um MLH(B)=5.0
8.	e	A	12 16 45	Probably Southern Greece (USNOAA)
	e	A	16 50.5	
	e	A	16 56	
8.	eP	A	14 39 47	Kurile Islands 44.56 N 148.46 E
				H = 14 27 51.2 h = normal MB = 4.4(USNOAA)
8.	ePKP	A	17 17 10.5	Fiji Islands Region 17.53 S 179.09 W
				H = 16 58 32.7 h = 558 km MB = 4.1(USNOAA)
				D = 145.8
8.	eP	A	21 18 12	Molucca Passage 1.15 N 126.11 E
	e	A	18 18	H = 21 04 05.9 h = 24 km MB = 5.8(USNOAA)
	e	B	18 24	D = 104.5
	ePP	AB	22 34	PV(A):1.7s 30.3nm MPV(A)=5.9
	e	A	22 40	SH(B):9s 0.65/um MSH(B)=6.1
	e	A	22 44.5	LmH(B):21s 2.9/um MLH(B)=5.8
	eSKS	B	28 46	LmV(B):16s 1.8/um MLV(B)=5.7
	eS	B	30 00	
	LmH	B	22 04.8	
	LmV	B	11.3	
9.	LmH	B	11 57.4	South Pacific Cordillera
	LmV	B	57.5	62.80 S 160.51 W
				H = 10 16 04.6 h = normal MB = 5.2(USNOAA)
				D = 167.0
				LmH(B):20s 0.7/um MLH(B)=5.5
				LmV(B):20s 0.4/um MLV(B)=5.3

Day	Phase		h m s	Remarks
9.	e	A	20 11 27	United Kingdom 54.47 N 2.39 W
	eSn	A	12 58.5	H = 20 09 00.5 h = normal MB = 4.0
				D = 9.35 Az = 108.5 (USNOAA)
10.	-iPKP	AB	11 53 28.5	Tonga Islands 16.47 S 174.05 W
				H = 11 33 55.7 h = 60 km MB = 5.0
				D = 145.61 Az = 353.6 (USNOAA)
				PKPV(A):1.2s 87.3nm
10.	ePKP	AB	15 34 33	New Hebrides Islands 13.87 S 166.79 E
	e	A	34 41.5	H = 15 15 19.5 h = 46 km MB = 6.0
	ePP	AB	37 32	D = 138.15 Az = 336.4 (USNOAA)
	eiSKP	AB	38 20	SKPV(A):1.0s 51.1nm
	ePS	B	47 45	SS2H(B):24s 14.6/um
	e	B	50 00	PSPS2H(B):25s 27.4/um
	e(SS2)	B	16 08 52	LmH(B):20s 6.0/um MLH(B)=6.3
	e(PSPS2)	B	13 09	LmV(B):18s 6.6/um MLV(B)=6.4
	LmH	B	37.7	
	LmV	B	44.1	
10.	-iPn	A	16 40 29	Yugoslavia 44.15 N 16.15 E
	ePg	A	40 57	H = 16 38 43.2 h = 27 km MB = 4.2
	iSn	A	41 51	D = 7.19 Az = 336.3 (USNOAA)
	e(Sg)	A	42 24	PnV(A):0.8s 100.0nm
	e	A	42 32	
	LmH	B	42.7	
	LmV	B	42.7	
11.	eP	A	01 15 21	Kashmir-Tibet Border Region
				34.10 N 79.33 E
				H = 01 06 19.3 h = normal MB = 4.8
				D = 50.90 Az = 310.6 (USNOAA)
11.	+eP1	AB	03 58 28.5	North of Ascension Island
	iP2	A	58 36.5	1.11 S 13.87 W
	ePP	AB	04 00 29	H = 03 48 52.4 h = normal MB = 5.4
	e	A	00 37	D = 55.96 Az = 19.3 (USNOAA)
	ePPP	B	01 38	P1V(A):1.6s 121.0nm MP1V(A)=5.7
	eS	B	06 08	P1V(B):12s 1.1/um MP1V(B)=5.8

Day	Phase		h m s	Remarks
cont.				
11.	ei	B	04 06 19	P2V(A):1.5s 226.0nm MP2V(A)=6.0
	LmH	B	21.7	PPV(B):10s 0.9 µm MPPV(B)=5.8
	LmV	B	23.4	SH(B):13s 5.9 µm MSH(B) =6.3
				LmH(B):19s 16.8 µm MLH(B) =6.2
				LmV(B):18s 15.4 µm MLV(B) =6.2
11.	ePKP	A	08 53 35.5	<u>Fiji Islands Region</u> 17.74 S 178.74 W
				H = 08 35 01.0 h = 609 km MB = 4.1
				D = 146.14 Az = 348.2 (USNOAA)
				PKPV(A):1.0s 11.6nm
11.	ePKIKP	A	10 28 54	<u>New Hebrides Islands</u> 14.16 S 166.61 E
				H = 10 09 27.9 h = normal MB = 5.1
				D = 138.34 Az = 336.1 (USNOAA)
				PKIKP(A):1.4s 18.6nm
11.	ePKIKP	AB	10 41 35	<u>New Hebrides Islands</u> 14.12 S 166.65 E
	eX1	A	41 39	H = 10 22 20.0 h = normal MB = 6.2
	iX2	A	41 48	D = 138.33 Az = 336.2 (USNOAA)
	ei	B	42 10	X1V(A):1.6s 30.2nm
	eiPP	AB	44 32	X2V(A):3.0s 526.0nm
	iPKS	A	45 21	LmH(B):20s 27.0 µm MLH(B)=6.9
	i	B	45 24	LmV(B):21s 30.3 µm MLV(B)=7.0
	i	A	45 29	
	ePS	B	54 48	
	LmH	B	11 44.8	
	LmV	B	45.0	
11.	LmH	B	18 27.0	Probably <u>West New Guinea Region</u> (USNOAA)
				LmH(B):20s 0.2 µm
11.	eP	A	18 36 43.5	<u>Southwestern Ryukyu Islands</u>
	e	A	36 49	25.26 N 125.75 E
	e	A	37 02	H = 18 24 13.8 h = 60 km MB = 5.0
	e	A	37 11	D = 84.75 Az = 324.3 (USNOAA)
				PV(A):1.8s 33.8nm MPV(A)=5.1

Day	Phase		h m s	Remarks
11.	eP	A	20 11 15.5	Off Coast of Costa Rica 8.92 N 84.64 W
	epP	A	11 21.5	H = 19 58 32.4 h = normal MB = 4.8
				D = 87.12 Az = 39.3 (USNOAA)
				PV(A):2.0s 25.6nm MP1V(A)=5.2
				pPV(A):1.9s 37.9nm
11.	ePP	AB	20 30 30	South Sandwich Islands Region
	eSP	B	40 16	60.61 S 25.40 W
	ePS	B	40 28	H = 20 10 52.5 h = normal MB = 6.0(USNOAA)
	eSPP	B	41 06	D = 114.8
	ePPS	B	41 16	PPV(A):1.9s 68.2nm MPPV(A)=6.1
	LmV	B	21 13.9	LmV(B):20s 0.8/um MLV(B) =5.3
	LmH	B	14.0	LmH(B):20s 0.7/um MLH(B) =5.3
12.	ePKP	B	01 00 04	New Hebrides Islands 13.93 S 166.51 E
	e	A	00 12	H = 00 40 42.9 h = 39 km MB = 5.4
	ePP	B	02 50	D = 138.09 Az = 336.1 (USNOAA)
	eSKP	A	03 38	SKPV(A):1.5s 25.1nm
	LmH	B	59.2	LmH(B):20s 1.2/um MLH(B)=5.6
	LmV	B	02 09.0	LmV(B):20s 1.9/um MLV(B)=5.8
12.	ePKIKP	A	01 58 50	New Hebrides Islands 13.86 S 166.49 E
	e	AB	58 58	H = 01 39 36.7 h = 43 km MB = 5.8
	e	A	59 00	D = 138.02 Az = 336.2 (USNOAA)
	e	A	59 11	LmH(B):19.5s 4.5/um MLH(B)=6.2
	ePP	AB	02 01 44	LmV(B):18s 4.7/um MLV(B)=6.3
	e	A	02 31	
	LmH	B	03 07.6	
	LmV	B	08.0	
12.	ePKIKP	A	08 40 47	New Hebrides Islands 13.91 S 166.67 E
	eSKP	A	44 20	H = 08 21 24.4 h = 42 km MB = 5.1
				D = 138.14 Az = 336.3 (USNOAA)
				PKIKPV(A):1.5s 15.1nm
				SKPV(A):1.4s 18.6nm
12.	ePKHKP	A	09 25 10	New Hebrides Islands 14.15 S 166.69 E
	ePKIKP	A	25 16	H = 09 05 48.9 h = 37 km MB = 5.2
	eSKP	A	28 47	D = 138.36 Az = 336.2 (USNOAA)

Day	Phase		h m s	Remarks
12.	eP	AB	09 36 49	Nicaragua 12.02 N 86.54 W
	e	A	36 14	H = 09 24 11.5 h = normal MB = 5.9
	e	A	36 25	D = 85.92 Az = 39.2 (USNOAA)
	ePP	A	40 17	PV(A):4.0s 435.0nm MPV(A)=6.1
	e	AB	40 30	LmH(B):18.5s 25.8/um MLH(B)=6.6
	e	A	40 41	LmV(B):18.5s 32.5/um MLV(B)=6.8
	eS	B	47 20	
	eSSS	B	56 30	
	LmH	B	10 14.5	
	LmV	B	14.7	
12.	eP	A	10 37 03.5	Nicaragua 12.05 N 86.54 W
	e	A	37 16.5	H = 10 24 23.9 h = normal MB = 5.6
	e	A	37 34.5	D = 85.89 Az = 39.2 (USNOAA)
12.	ePKIKP	A	12 54 16	New Hebrides Islands 14.10 S 166.37 E
				H = 12 34 51.4 h = 28 km MB = 4.9
				D = 138.19 Az = 336.0 (USNOAA)
12.	eP	A	18 55 40	Northeast of Taiwan 24.96 N 124.67 E
	LmH	B	19 31.5	H = 18 43 14.9 h = 87 km MB = 5.2
	LmV	B	38.6	D = 84.42 Az = 323.9 (USNOAA)
				PV(A):1.0s 17.7nm MPV(A)=5.0
12.	+eP	A	22 58 13	Kurile Islands 44.07 N 147.84 E
				H = 22 46 19.0 h = 69 km MB = 5.2
				D = 78.28 Az = 333.3 (USNOAA)
				PV(A):0.9s 37.0nm MPV(A)=5.3
13.	ePKHKP	A	00 42 33.5	South Of Fiji Islands 25.82 S 176.98 W
	ePKP2	A	42 47	H = 00 22 43.0 h = 90 km MB = 5.1
				D = 154.35 Az = 347.3 (USNOAA)
13.	eP diff	A	04 36 43.5	Sumbawa Island Region 8.88 S 117.98 E
	e	A	40 02	H = 04 22 38.5 h = 117 km MB = 6.0
	ePKIKP	A	40 48	D = 107.19 Az = 320.3 (USNOAA)
	e(PP)	A	41 12	PV(A):1.6s 19.2nm MPV(A)=6.0
	e	B	46 06	PcPPKPV(A):1.8s 33.8nm
	ePcPPKP	A	56 15	
	e	A	56 25	

Day	Phase		h m s	Remarks
13.	e	A	07 11 29	Burma-India Border Region 24.70 N 93.92 E H = 07 00 41.6 h = 40 km MB = 4.7 D = 66.55 Az = 316.5 (USNOAA)
13.	eP	A	15 53 19	North Atlantic Ridge 48.97 N 28.37 W
	eX	A	53 28	H = 15 47 50.5 h = normal MB = 4.5 D = 25.65 Az = 71.0 (USNOAA) XV(A):1.6s 16.5nm
13.	eP	A	19 36 24	Lake Baikal Region 51.82 N 105.49 E
	epP	A	36 31.5	H = 19 26 55.5 h = normal MB = 4.7
	LmH	B	20 02.7	D = 54.70 Az = 308.9 (USNOAA)
	LmV	B	02.7	PV(A):1.0s 15.7nm MPV(A)=5.0 pPV(A):1.8s 40.5nm LmH(B):14s 2.2/um MLH(B)=5.3 LmV(B):14.5s 2.7/um MLV(B)=5.5
13.	eP	A	22 56 19.5	Near East Coast of Kamchatka 53.66 N 160.36 E H = 22 44 52.3 h = normal MB = 4.6 D = 72.79 Az = 339.8 (USNOAA) PV(A):0.8s 11.5nm MPV(A)=4.9
13.	LmH	B	24 25.6	Mariana Islands 14.1 N 146.5 E
	LmV	B	37.5	H = 23 27 05.7 h = 46 km MB = 5.4(USNOAA) D = 104.5 LmH(B):16.5s 0.8/um MLH(B)=5.3 LmV(B):14.5s 0.8/um MLV(B)=5.6
14.	eP	A	03 50 05.5	Central Alaska 64.89 N 147.80 W
	epP	A	50 12.5	H = 03 39 33.5 h = 19 km MB = 5.0 D = 63.70 Az = 14.5 (USNOAA) PV(A):1.4s 23.3nm MPV(A)=5.2
14.	e	A	06 41 30	Probably Northern California (USNOAA)

Day	Phase		h m s	Remarks
14.	ePKP	A	13 27 09	<u>Tonga Islands</u> 16.25 S 172.97 W H = 13 07 31.5 h = normal MB = 4.6 D = 145.49 Az = 354.8 (USNOAA) PKPV(A):1.7s 27.3nm
14.	ePn	A	13 30 15	<u>Czechoslovakia, explosion</u>
	i(Sg)	A	30 38.5	50.62 N 14.35 E
	i	A	30 43	H = 13 h 29 min yield 7.6 to D = 1.74 Az = 272 (PRU)
15.	ePKP	A	02 01(25)	<u>Fiji Islands Region</u> 16.71 S 177.11 W
	e(pPKP)	A	01 30	H = 01 41 48.3 h = 50 km MB = 4.7 D = 145.43 Az = 350.2 (USNOAA) (pPKP)V(A):1.2s 16.2nm
15.	ePKP	A	05 02 32.5	<u>Fiji Islands Region</u> 16.98 S 177.18 W
	e	A	02 39.5	H = 04 42 55.1 h = normal MB = 5.3 D = 145.68 Az = 350.1 (USNOAA) PKPV(A):1.4s 23.2nm
15.	ePKHKP	A	07 20 24	<u>South of Fiji Islands</u> 25.88 S 177.08 W
	ePKP2	A	20 35	H = 07 00 32.4 h = 99 km MB = 4.6 D = 154.38 Az = 347.1 (USNOAA)
16.	ePn	A	10 48 27.5	<u>Ionian Sea</u> 37.94 N 16.49 E
	ei	A	48 38	H = 10 45 21.4 h = 52 km MB = 4.3
	i	A	48 49	D = 13.17 Az = 346.3 (USNOAA)
	e	A	51 29	
	LmH	B	54.0	
	LmV	B	56.0	
17.	e	A	02 45 26	<u>Albania</u> 41.6 N 19.9 E
	eSn	A	47 12	H = 02 42 36 (BCIS)
	e	A	48 02	D = 10.75
18.	ePn	A	04 26 51	<u>Switzerland</u> 46.4 N 7.5 E
	e	A	27 00	H = 04 25 33 (BCIS)
	eiPb	A	27 06	D = 5.09

Day	Phase		h m s	Remarks
cont.				
18.	eiPg	A	04 27 14	LmH(B):7.0s 0.8/um MLH(B)=3.5
	eSg	AB	28 10	LmV(B):7.0s 1.0/um
	i	AB	28 14	
	LmH	B	28.85	
	LmV	B	28.85	
17.	e	A	04 27 41	Albania 41.6 N 19.9 E
				H = 04 22 19 (BCIS)
				D = 10.75
18.	-eiP	A	17 05 11	Celebes Sea 4.54 N 123.02 E
	ePP	AB	09 29	H = 16 52 25.3 h = 561 km MB = 5.7
				D = 99.85 Az = 323.0 (USNOAA)
				PV(A):1.3s 39.3nm MPV(A)=5.7
				PPV(A):1.3s 39.3nm MPPV(A)=5.5
18.	eP	A	17 43 29	Greece 39.21 N 21.85 E
	e	A	43 38	H = 17 40 16.4 h = 31 km MB = 4.6
	e	A	44 14	D = 13.53 Az = 331.1 (USNOAA)
	LmH	B	49.3	LmH(B):13s 1.0/um MLH(B)=4.1
	LmV	B	49.3	LmV(B):12s 1.1/um
18.	+eiP1	A	18 03 02.5	Southern Alaska 60.70 N 145.38 W
	iP2	AB	03 03	H = 17 52 06.3 h = 16 km MB = 5.6
	e	AB	03 10	D = 67.47 Az = 15.6 (USNOAA)
	e	AB	05 31	P2V(A):1.0s 181.0nm MP2V(A)=6.3
	ePP	B	05 40	P2V(B):4s 1.04/um MP2V(B)=6.4
	eS	B	11 57	PPV(B):10s 1.02/um MPPV(B)=6.2
	e	B	12 08	LmH(B):17s 3.8/um MLH(B) =5.7
	eSS	B	16 24	LmV(B):15s 3.0/um MLV(B) =5.7
	ePKPPKP	A	31 29	
	LmH	B	38.4	
	LmV	B	39.0	
18.	eSn	A	18 45 50	Italy 43.3 N 11.6 E
	e	A	47 36	H = 18 42 45 (BCIS)
				D = 7.40

Day	Phase		h m s	Remarks
19.	eP1	A	02 04 30	Albania 41.10 N 19.77 E
	iP2	A	04 33.5	H = 02 01 53.1 h = normal MB = 5.2
	iP3	A	04 40	D = 11.10 Az = 332.0 (USNOAA)
	iP4	AB	04 43.5	P1V(A):1.0s 27.6nm MP1V(A)=5.4
	ei	A	05 32	P2V(A):1.0s 59.0nm MP2V(A)=5.8
	eS	B	06 31	P3V(A):0.9s 80.0nm MP3V(A)=6.0
	LmH	B	08.4	P4V(A):1.2s 173.0nm MP4V(A)=6.2
	LmV	B	09.6	LmH(B):17s 22.8/um MLH(B) =5.2
				LmV(B):16s 18.2/um
19.	e	A	03 16 46	Near East Coast of Honshu, Japan
				39.47 N 142.91 E
				H = 03 04 29.2 h = normal MB = 4.7
				D = 80.62 Az = 331.0 (USNOAA)
19.	e	A	05 17 31.5	Probably Poland (ISC)
19.	eP	A	10 09 08	Off East Coast of Kamchatka
				52.86 N 159.43 E
				H = 09 57 40.2 h = 55 km MB = 4.7
				D = 73.35 Az = 339.3 (USNOAA)
				PV(A):1.2s 10.2nm MPV(A)=4.7
19.	-eP	A	12 21(42)	Central Italy 43.24 N 11.06 E
				H = 12 19 54.5 h = normal MB = 5.1
				D = 7.42 Az = 2.7 (USNOAA)
20.	ePKHKP	A	13 12(46)	Loyalty Islands Region
	ePKP2	A	12 51	22.31 S 171.49 E
				H = 12 53 13.0 h = 100 km MB = 5.2
				D = 147.63 Az = 335.8 (USNOAA)
20.	e(P)	A	15 36 41.5	Southern Iran 29.28 N 51.62 E
	LmH	B	58.0	H = 15 29 52.2 h = 33 km MB = 4.4(USNOAA)
	LmV	B	58.0	D = 36.7
21.	+iP	A	00 55 54	Kurile Islands 45.82 N 150.06 E
	e	A	56 00	H = 00 44 06.4 h = 80 km MB = 5.2

Day	Phase		h m s	Remarks
cont.				
21.	e	A	00 56 06	D = 77.40 Az = 334.4 (USNOAA) PV(A):1.1s 50.4nm MPV(A)=5.3
21.	ePKP	A	08 57 57	Samoa Islands Region 14.62 S 175.67 W H = 08 38 22.3 h = normal MB = 5.3 D = 143.58 Az = 352.2 (USNOAA)
21.	eP	A	12 09 42	Southern Alaska 60.81 N 142.50 W
	e	A	09 48	H = 11 58 50.0 h = 14 km MB = 4.5 D = 66.96 Az = 17.6 (USNOAA)
21.	eP	A	15 33 47.5	Hindu Kush Region 36.12 N 68.46 E H = 15 25 56.8 h = 98 km MB = 5.2 D = 42.74 Az = 308.2 (USNOAA)
22.	eP1	A	11 38 44	Off East Coast of Kamchatka
	eP2	A	38 46	53.55 N 161.35 E H = 11 27 15.3 h = normal MB = 5.1 D = 73.10 Az = 340.4 (USNOAA) P1V(A):0.8s 23.1nm MP1V(A)=5.2 P2V(A):0.9s 38.9nm MP2V(A)=5.4
22.	eP	A	12 08 05	Greece 38.07 N 20.00 E
	e	A	08 15	H = 12 04 48.2 h = normal MB = 4.6
	LmH	B	13.3	D = 13.92 Az = 337.3 (USNOAA)
	LmV	B	14.4	LmH(B):13s 0.6 μm MLH(B)=3.9 LmV(B):12s 0.7 μm
23.	eP	A	03 45 14.5	Ascension Island Region 7.11 S 11.85 W H = 03 35 01.3 h = normal MB = 4.9 D = 61.00 Az = 16.9 (USNOAA) PV(A):1.2s 12.2nm MPV(A)=4.9
23.	ePKIKP	A	05 17 08.5	Fiji Islands Region 17.79 S 178.78 W
	ePKHKP	A	17 10	H = 04 58 31.5 h = 560 km MB = 5.0 D = 146.18 Az = 348.1 (USNOAA) PKHKPV(A):1.2s 34.6nm

Day	Phase		h m s	Remarks
23.	eP1	AB	11 13 12.5	North Atlantic Ocean 53.14 N 35.10 W
	eP2	A	13 14.5	H = 11 07 18.4 h = normal MB = 5.0
	ePP	B	14 50	D = 28.54 Az = 75.9 (USNOAA)
	eS	B	17 50	P1V(A):1.4s 14.0nm MP1V(A)=4.5
	ei	B	18 04	P2V(A):1.6s 27.5nm MP2V(A)=4.7
	LmH	B	24.3	LmH(B):16s 2.7/um MLH(B) =5.0
	LmV	B	24.3	LmV(B):16s 2.6/um MLV(B) =5.0
23.	ePn	A	11 40(40)	Albania 40.9 N 20.2 E
	e(Sg)	A	44 03	H = 11 37 45 (BCIS)
				D = 11.50
23.	eP	A	19 05 42.5	Southern Sumatra 1.83 S 100.18 E
	e	A	05 52	H = 18 52 46.1 h = 69 km (ISC)
				D = 90.5
24.	eX	A	07 14 02	West of Macquarie Island 58.99 S 149.3 E
				H = 06 53 31 h = 15 km
				D = 154.60 Az = 270 (ISC)
				XV(A):1.0s 7.9nm
24.	ePKHKP	A	09 52 46.5	South of Fiji Islands 25.81 S 177.12 W
	ePKP2	A	53 00	H = 09 32 52.6 h = 63 km MB = 5.2
	ePP	A	56 46	D = 154.31 Az = 347.1 (USNOAA)
				PKHKPV(A):1.6s 11.0nm
				PKP2V(A):1.6s 30.1nm
24.	ePKIKP	AB	12 50 20	South Pacific Cordillera
	ePKHKP	A	50(42)	56.59 S 142.48 W
	-iPKP2	AB	51 13.5	H = 12 30 19.5 h = normal MB = 5.9
	PKP2max	A	51 17	D = 163.58 Az = 100.3 (USNOAA)
	ePP	AB	54 57	PKIKPV(A):1.5s 27.6nm
	ePPP	AB	58 49	PKHKPV(A):2.8s 193.0nm
	e	A	13 00 41	PKP2maxV(A):2.4s 456.0nm
	eX	A	01 30.5	PPV(A):2.5s 322.0nm
	iSS	B	15 32	XV(A):1.2s 20.3nm
	LmH	B	14 10.8	

Day	Phase		h m s	Remarks
24.	eP	A	16 41(10)	Turkey 38.32 N 39.31 E
				H = 16 36 03.2 h = normal MB = 4.4
				D = 23.13 Az = 311.1 (USNOAA)
25.	ePn1	A	01 42 17.5	Yugoslavia 43.29 N 18.41 E
	ePn2	A	42 21.5	H = 01 40 09.6 h = 10 km MB = 5.2
	e	A	43 41.5	D = 8.70 Az = 330.1 (USNOAA)
	iSn	A	43 47.5	Pn1V(A):0.8s 50.0nm
	i	A	44 22.5	Pn2V(A):0.9s 151.5nm
	i	A	44 26.5	LmH(B):12s 8.0/um MLH(B)=4.5
	LmH	B	45.2	LmV(B):10s 3.8/um
	LmV	B	45.8	
25.	eP	A	03 54 54	Kurile Islands 45.52 N 149.99 E
				H = 03 43 01.6 h = 50 km MB = 4.8
				D = 77.65 Az = 334.3 (USNOAA)
				PV(A):1.0s 13.8nm MPV(A)=4.9
25.	ePKP	A	05 08 32	Samoa Islands Region 16.56 S 172.84 W
				H = 04 48 53.0 h = 22 km MB = 4.7
				D = 145.81 Az = 355.0 (USNOAA)
				PKPV(A):1.2s 26.4nm
25.	ePKP2	A	12 05 50.5	South of Fiji Islands 24.35 S 179.93 E
				H = 11 46 38.7 h = 488 km MB = 5.0
				D = 152.24 Az = 343.9 (USNOAA)
				PKP2V(A):0.6s 15.3nm
26.	eP	AB	15 24 38.5	Luzon, Philippine Islands
	eipP	A	24 50	18.02 N 120.48 E
	ei	AB	24 55	H = 15 11 54.9 h = 58 km MB = 5.5
	ei	A	25 07.5	D = 87.64 Az = 323 (ISC)
	e	B	28 20	PV(A):2.4s 96.7nm MPV(A)=5.6
	eSKS	B	35 00	LmV(B):13s 1.9/um MLV(B)=5.6
	eiS	B	35 15	LmH(B):17s 2.5/um MLH(B)=5.7
	eSS	B	41 10	
	LmV	B	16 06.7	
	LmH	B	07.1	

Day	Phase		h m s	Remarks
26.	ePKIKP	A	18 33(08)	Fiji Islands Region 20.09 S 178.08 W
	eiPKHKP	A	33 12.5	H = 18 14 26.5 h = 550 km MB = 4.7
	ePKP2	A	33 17.5	D = 148.55 Az = 348.1 (USNOAA)
				PKHKPV(A):0.8s 17.3nm
26.	eP	A	21 07 11	Off East Coast of Honshu, Japan
	e	A	07 20	34.78 N 141.60 E
	LmH	B	51.0	H = 20 54 43.0 h = 45 km MB = 5.0
	LmV	B	51.0	D = 84.20 Az = 330.6 (USNOAA)
				PV(A):1.4s 9.3nm MPV(A)=4.7
				LmH(B):15s 1.1/um MLH(B)=5.4
				LmV(B):14s 0.8/um MLV(B)=5.3
27.	ePKHKP	A	13 20 20	Fiji Islands Region 21.80 S 179.63 W
	ePKP2	A	20 27	H = 13 01 38.7 h = 640 km MB = 4.7
				D = 149.88 Az = 345.7 (USNOAA)
27.	eP	A	16 41 32	Philippine Islands Region
				19.92 N 121.76 E
				H = 16 28 47.1 h = 33 km MB = 5.1(USNOAA)
				D = 86.8
27.	ePKP	AB	16 41 59.5	Tonga Islands 15.22 S 173.28 W
	eX	A	41 09	H = 16 22 24.7 h = 23 km MB = 5.4
	LmH	B	17 49.9	D = 144.44 Az = 354.6 (USNOAA)
	LmV	B	50.0	PKPV(A):2.0s 68.4nm
				PKPV(B):10s 1.3/um
				XV(A):1.8s 81.0nm
				LmH(B):18s 2.2/um
				LmV(B):20s 1.0/um
27.	eP1	AB	19 57 36	Near Coast of Oaxaca, Mexico
	eP2	A	57 42	15.41 N 95.60 W
	ePP	AB	20 01 02.5	H = 19 44 42.0 h = 31 km MB = 5.5
	e	B	01 11	D = 88.75 Az = 37.5 (USNOAA)
	eSKS	B	08 04	P1V(B):8s 0.7/um MP1V(B)=6.0
	eiS	B	08 25	P2V(A):2.0s 128.0nm MP2V(A)=5.9

Day	Phase		h m s	Remarks
cont.				
27.	LmH	B	20 39.0	PPV(B):7s 0.33 μm MPPV(B)=5.9
	LmV	B	40.0	SH(B):11.5s 2.1 μm MSH(B)=6.1
				LmH(B):17.5s 2.2 μm MLH(B)=5.6
				LmV(B):19s 2.4 μm MLV(B)=5.7
27.	ePKP	A	20 31 25.5	<u>Tonga Islands</u> 15.21 S 173.22 W
	e	A	31 33	H = 20 11 52.0 h = normal MB = 5.1
				D = 144.45 Az = 354.7 (USNOAA)
28.	eP	A	01 18 12	<u>New Ireland Region</u> 4.57 S 153.06 E
	+iPKIKP	AB	21 39	H = 01 02 49.0 h = 88 km MB = 5.9(USNOAA)
	+i	A	21 41.8	D = 123.8
	ePP	AB	23 17	PKIKPV(A):2.0s 128.0nm
	eSKS	B	28 30	PPV(B):9s 2.0 μm MPPV(B)=6.2
	e	B	30 44	LmH(B):22s 8.8 μm MLH(B) =6.4
	ePS	B	33 03	LmV(B):21s 7.2 μm MLV(B) =6.3
	eSS	B	40 12	
	LmH	B	02 13.5	
	LmV	B	17.8	
28.	eP	A	01 34 46	<u>India-East Pakistan Border Region</u>
	eX	A	34 57	24.66 N 91.67 E
				H = 01 24 04.4 h = 17 km MB = 4.9
				D = 65.17 Az = 316.3 (USNOAA)
				PV(A):0.9s 9.7nm MPV(A)=5.0
				XV(A):1.0s 11.8nm
28.	ePKIKP	A	10 25 58.5	<u>South of Kermadec Islands</u>
	ePKHKP	A	26 11.5	33.86 S 179.82 W
	ePKP2	A	26 42	H = 10 06 08.8 h = 90 km MB = 5.6
	e	A	26 55	D = 161.25 Az = 336.9 (USNOAA)
	e	A	29 40	PKIKPV(A):2.0s 42.8nm
	ePP	A	30 25	
	e	A	30 36	
28.	ePKP	A	14 21 03.5	<u>Tonga Islands</u> 15.12 S 173.40 W
	e	A	21 11.5	H = 14 01 29.9 h = normal MB = 5.3
				D = 144.33 Az = 354.5 (USNOAA)
				PKPV(A):1.3s 21.8nm

Day	Phase		h m s	Remarks
28.	ePKP	A	14 36 02	Tonga Islands 15.06 S 173.17 W H = 14 16 23.1 h = 33 km MB = 5.0 D = 144.30 Az = 355 (ISC)
28.	eP	A	18 34 50	Luzon, Philippine Islands 18.63 N 120.99 E H = 18 22 02.9 h = 23 km MB = 5.1 D = 87.45 Az = 323.0 (USNOAA) PV(A):1.9s 30.3nm MPV(A)=5.2
28.	ePb	A	20 52 52.5	Italy 44.6 N 10.9 E
	ePg	A	53 05.5	H = 20 50 59 (BCIS)
	eSn	A	53 34	D = 6.12
	e	A	53 40	
	eSg	A	54 17	
	e	A	54 20	
29.	-iP	AB	01 54 52.8	Near West Coast of Honshu, Japan
	epP	A	56 04	36.99 N 136.75 E
	ePP	A	57 58	H = 01 43 12.2 h = 284 km MB = 5.2 D = 80.29 Az = 328.1 (USNOAA) PV(A):1.2s 85.4nm MPV(A)=5.4
29.	eP	A	03 11 56	Venezuela 7.59 N 71.93 W H = 02 59 43.0 h = 26 km MB = 4.6 D = 80.08 Az = 40.0 (USNOAA)
29.	ePn	A	10 44(48)	Albania 41.45 N 19.42 E
	e	A	44 55	H = 10 42 17.5 h = 38 km MB = 4.4
	eSn	A	46 47	D = 10.67 Az = 332.1 (USNOAA)
29.	+eP1	A	15 10 20	Eastern Russia 51.10 N 135.26 E
	iP2	A	10 22	H = 14 59 22.6 h = normal MB = 5.4
	LmH	B	39.2	D = 67.90 Az = 325.1 (USNOAA)
	LmV	B	45.3	P1V(A):1.2s 48.8nm MP1V(A)=5.5 P2V(A):1.3s 78.5nm MP2V(A)=5.6 LmH(B):16s 0.9/um MLH(B)=5.1 LmV(B):12s 0.7/um MLV(B)=5.1

Day	Phase		h m s	Remarks
29.	iPg	A	16 00 21	Explosion Bischhofsheim/Rhön,Fed.Rep.Germany
	iSg	A	00 36	50°25.29'N 10°01.71'E
	i	A	00 37	H = 16 00 01.51 yield 5.1 to (Hannover)
				D = 1.1
29.	ePKP	A	20 41 55.5	Tonga Islands 16.77 S 174.05 W
				H = 20 22 22.5 h = 64 km MB = 4.7
				D = 145.90 Az = 353.6 (USNOAA)
				PKPV(A):1.2s 32.5nm
30.	ePKIKP	A	00 47 53	New Ireland Region 4.77 S 153.37 E
				H = 00 29 00.1 h = 60 km MB = 5.0
				D = 124.19 Az = 331.5 (USNOAA)
				PKIKPV(A):1.3s 21.8nm
30.	+eP	AB	00 50 15	Off East Coast of Kamchatka
	ePcP	B	50 28	52.09 N 159.65 E
	eS	B	59 46	H = 00 38 40.1 h = normal MB = 5.2
	e	B	59 54	D = 74.12 Az = 339.5 (USNOAA)
	LmH	B	01 28.2	PV(A):2.5s 161.0nm MPV(A)=5.6
	LmV	B	28.2	PV(B):5s 0.4/um MPV(B)=5.7
				LmH(B):16s 2.3/um MLH(B)=5.6
				LmV(B):15s 1.7/um MLV(B)=5.5
30.	+ePKP	A	01 03 33	Samoa Islands Region 16.14 S 172.52 W
	LmV	B	31.0	H = 00 43 57.4 h = normal MB = 5.3
	LmH	B	31.2	D = 145.42 Az = 355.4 (USNOAA)
				PKPV(A):2.7s 230.0nm
				LmV(B):12s 0.9/um MLV(B)=5.8
				LmH(B):15.5s 1.0/um MLH(B)=5.7
30.	eP	A	03 25 32.5	Off East Coast of Kamchatka
				52.17 N 159.73 E
				H = 03 13 58.4 h = normal MB = 4.8
				D = 74.06 Az = 339.5 (USNOAA)
30.	ePn	A	03 51 39.5	Yugoslavia 43.4 N 18.6 E
	e	A	53 42.5	H = 03 49 36 (BCIS)
	eSg	A	54 13	D = 8.72

Day	Phase		h m s	Remarks
30.	iP	A	16 24 15.5	**Iran** 37.38 N 56.04 E H = 16 17 31.0 h = normal MB = 5.1 D = 34.07 Az = 307.3 (USNOAA) PV(A):0.9s 17.5nm MPV(A)=5.0
30.	iP1	A	17 56 30.5	**Sea of Okhotsk** 52.38 N 151.60 E
	iP2	B	56 32	H = 17 46 09.0 h = 645 km MB = 6.6
	ipP	B	58 44	D = 71.91 Az = 334.5 (USNOAA)
	e	B	59 48	SH(B):15s 118.0/um MSH(B)=7.2
	esP	B	59 56	St = Strainseismograph
	esPP	B	18 02 31	
	esPPP	B	03 56	
	eiS	B	04 56	
	iSKS	B	05 32	
	esS	B	08 55	
	esPS	St	09 16	
	eSS	B	09 52	
	esSS	B	13 12	
	e	B	16 16	
	e	A	23 29	
	e	A	43 42	
30.	eP	A	20 47 53	**Off East Coast of Kamchatka** 52.14 N 159.71 E H = 20 36 19.2 h = normal MB = 4.6 D = 74.09 Az = 339.5 (USNOAA) PV(A):1.1s 10.1nm MPV(A)=4.7
31.	eP	A	04 25 46	**Albania** 40.71 N 19.90 E H = 04 22 58.1 h = 67 km MB = 4.3 D = 11.49 Az = 332.6 (USNOAA) PV(A):0.8s 7.7nm MPV(A)=4.5
31.	eX	A	13 00 51.5	**Off East Coast of Kamchatka** 52.23 N 159.62 E H = 12 49 03.0 h = normal MB = 4.6 D = 73.99 Az = 339.5 (USNOAA) XV(A):1.0s 11.8nm

Day	Phase		h m s	Remarks
1.	eP	A	01 10 02	Greece 38.03 N 20.16 E
	LmH	B	15.3	H = 01 06 41.8 h = 16 km MB = 4.8
	LmV	B	16.4	D = 14.00 Az = 337.0 (USNOAA)
				PV(A):1.0s 19.7nm
				LmH:14s 7.2/um MLH=4.9
				LmV:13s 4.8/um
1.	eP	AB	05 25 06	Mariana Islands 17.72 N 147.62 E
	e	A	25 24	H = 05 11 16.1 h = 40 km MB = 6.3
	iPP	AB	29 16.3	D = 101.73 Az = 333.1 (USNOAA)
	eSKS	B	35 42	PV(A):2.0s 94.0nm MPV(A)=6.1
	eSS	B	43 40	PV(B):9s 1.85/um MPV(B)=6.7
	LmH	B	06 13.1	PPV(A):2.3s 1084.0nm MPPV(A)=6.9
	LmV	B	15.6	PPV(B):6.5s 7.2/um MPPV(B)=7.2
				LmH:16s 14.4/um MLH =6.6
				LmV:18s 14.8/um MLV =6.5
1.	eP	A	07 33 09	North Atlantic Ridge 15.42 N 46.12 W
				H = 07 23 17.7 h = normal MB = 4.4
				D = 57.89 Az = 39.5 (USNOAA)
				PV(A):1.8s 27.0nm MPV(A)=5.0
1.	eP	A	15 36 14.5	North of Ascension Island
	LmH	B	58.7	1.65 S 12.75 W
	LmV	B	16 01.7	H = 15 26 36.9 h = normal MB = 5.0
				D = 56.10 Az = 18.8 (USNOAA)
				PV(A):2.0s 81.1nm MPV(A)=5.4
				LmH:15s 0.92/um MLH =5.0
				LmV:16s 0.94/um MLV =5.0
1.	eP	A	16 15 37	North of Ascension Island
				1.69 S 12.32 W
				H = 16 06 02.8 h = normal MB = 4.7
				D = 56.01 Az = 18.1 (USNOAA)
1.	eP	A	16 27 37	North of Ascension Island
	LmH	B	38.1	1.60 S 12.70 W

Day	Phase		h m s	Remarks
cont. 1.	LmV	B	16 41.1	H = 16 17 59.4 h = normal MB = 4.8
				D = 56.04 Az = 18.4 (USNOAA)
				PV(A):1.2s 10.2nm MPV(A)=4.7
				LmH:14s 0.62/um MLH =4.8
				LmV:18s 0.83/um MLV =4.9
2.	iP	A	01 22 56	Near East Coast of Kamchatka
	esP	A	23 09	53.13 N 159.79 E
	ePcP	A	23 14	H = 01 11 29.3 h = 60 km MB = 4.9(USNOAA)
				D = 73.17 Az = 339 (ISC)
				PV(A):1.0s 39.4nm MPV(A)=5.3
2.	eP	A	04 03 10.5	Off East Coast of Kamchatka
	epP	A	03 25	52.23 N 159.60 E
				H = 03 51 39.1 h = 60 km MB = 4.7(USNOAA)
				D = 73.99 Az = 339 (ISC)
				PV(A):1.0s 15.7nm MPV(A)=4.9
2.	eP	A	07 55 20	Off W. Coast of Northern Sumatra
				1.80 N 94.59 E
				H = 07 42 49.3 h = 33 km MB = 5.2(USNOAA)
				D = 84.2
				PV(A):1.8s 23.6nm MPV(A)=5.0
3.	LmH		00 44.5	LmH:23s 0.6/um
3.	eP	AB	05 37 04	Turkey 39.62 N 38.70 E
	Pm	A	37 10	H = 05 32 09.7 h = 23 km MB = 5.1
	èS	B	41 11	D = 21.93 Az = 309.1 (USNOAA)
				PV(A):1.6s 99.0nm MPV(A)=5.0
				PmV(A):1.7s 225.0nm MPmV(A)=5.2
3.	ePKP	A	09 51 45	New Hebrides Islands 16.92 S 167.83 E
	e	A	53 31	H = 09 32 23.0 h = 44 km MB = 5.5
	LmH	B	10 57.1	D = 141.33 Az = 335.7 (USNOAA)
	LmV	B	11 10.8	

Day	Phase		h m s	Remarks
3.	ePKHKP	A	19 30 08	Tonga Islands Region 22.90 S 175.00 W
	ePKP2	A	30 18	H = 19 10 15.7 h = normal MB = 4.7
	epPKP2	A	30 27.5	D = 151.83 Az = 351.1 (USNOAA)
4.	eP	A	13 19 43	Hindukush Region 36.58 N 70.11 E
	eipP	A	20 37.5	H = 13 12 02.2 h = 290 km MB = 4.9
				D = 43.50 Az = 307.9 (USNOAA)
				pPV(A):1.6s 44.0nm MPV(A)=4.5
4.	ePg	A	15 30 32	Federal Rep. Germany explosion 4.8 t
	eSg	A	30 52.5	51.19 N 9.24 E
				H = 15 30 00.78 (HAN)
				D = 1.55
				phi = $51^{o}11.31'$N, lambda = $09^{o}14.73'$E
4.	eP	A	23 31 08	Yugoslavia 43.78 N 18.10 E
				H = 23 29 08.0 h = 38 km MB = 4.4
				D = 8.16 Az = 329.6 (USNOAA)
				PV(A):0.7s 7.7nm
5.	eP	A	03 55 40	Taiwan 23.99 N 121.25 E
	LmH	B	04 37	H = 03 43 16.5 h = 43 km MB = 4.8(USNOAA)
	LmV	B	37	D = 83.3
				PV(A):(1.8)s 27.0nm MPV(A)=5.0
5.	-iP	A	08 02 55.2	Sea of Okhotsk 52.23 N 151.43 E
	ipP	AB	04 55	H = 07 52 27.9 h = 580 km MB = 5.7
	iPP	B	05 48	D = 72.00 Az = 334.4 (USNOAA)
	eiS	B	11 32	PV(A):1.2s 214.0nm MPV(A)=5.6
	isS	B	14 58	pPV(A):1.6s 385.0nm
	LmH	B	21.7	LmH:16.5s 3.3/um
	LmV	B	28.7	LmV:13.5s 2.3/um
	ePKPPKP	A	30 42	ei(B) 15 00 e(A) 32 55
	epPKPPKP	A	32 48	
5.	LmH	B	10 07.8	LmH:17s 0.6/um
	LmV	B	11.5	LmV:18s 0.6/um

Day	Phase		h m s	Remarks
5.	eP	A	11 47 33	Arabian Sea 14.83 N 53.76 E
	epP	AB	47 40	H = 11 38 46.1 h = normal MB = 4.9
	i(sP)	A	47 45	D = 49.25 Az = 325.7 (USNOAA)
	eS	B	54 41	PV(A):1.1s 24.2nm MPV(A)=5.1
	eSP	B	54 49	pPV(A):1.8s 67.6nm MPV(A)=5.4
	LmH	B	12 14.0	PV(B):5s 0.44/um MPV(B)=5.7
	LmV	B	15.0	SH:17s 2.32/um MSH(B)=5.8
				LmH:20s 3.3/um MLH =5.3
				LmV:16s 1.9/um MLV =5.3
5.	eP	A	13 48 20	Szechwan Province, China
	epP	A	48 24.5	32.04 N 101.16 E
	esP	A	48 35.5	H = 13 37 34.9 h = normal MB = 5.0
	LmH	B	14 13.7	D = 65.70 Az = 305.7 (USNOAA)
	LmV	B	18.4	PV(A):1.0s 11.8nm MPV(A)=4.9
				LmH:19.5s 1.9/um MLH =5.3
				LmV:15s 1.2/um MLV =5.3
5.	ePKHKP	A	17 27 31	Tonga Islands Region 17.45 S 171.90 W
	ePKP2	A	27 34	H = 17 07 50.7 h = normal MB = 5.2
				D = 146.76 Az = 355.9 (USNOAA)
				PKHKPV(A):1.5s 42.7nm
				PKP2V(A):1.6s 88.0nm
5.	ePKP	A	17 34 56.5	Tonga Islands 17.45 S 171.90 W
				H = 17 15 15.9 h = normal MB = 5.0
				D = 146.77 Az = 355.9 (USNOAA)
				PKPV(A):1.5s 30.2nm
5.	eP	A	19 17 22	Greenland Sea 73.65 N 8.70 E
				H = 19 12 18.6 h = normal MB = 4.6
				D = 23.13 Az = 175.3 (USNOAA)
				PV(A):1.5s 25.1nm MPV(A)=4.5
5.	eP	A	19 34 24	Afghanistan-USSR Border Region
	eX	A	34 31.5	37.04 N 71.38 E
	epP	A	34 48	H = 19 26 25.6 h = 112 km MB = 5.1
				D = 44.02 Az = 307.7 (USNOAA)
				eX(A):1.9s 30.3nm MPV(A)=4.7

Day	Phase		h m s	Remarks
6.	+iP	A	04 10 46.5	**Eastern Kazakh SSR** 49.77 N 78.09 E
	ePP	A	12 19	H = 04 02 57.4 h = 0 km MB = 5.6
				D = 41.25 Az = 297.7 (USNOAA)
				PV(A):0.9s 87.6nm MPV(A)=5.5
				Underground explosion, MB = 6.4 (UPP)
6.	eP	A	15 54 16	**Southeastern Alaska** 60.11 N 141.16 W
				H = 15 43 18.1 h = 5 km MB = 4.7
				D = 67.43 Az = 18.4 (USNOAA)
				PV(A):1.0s 11.8nm MPV(A)=5.1
7.	ePKHKP	A	03 51 02	**Fiji Islands Region** 21.01 S 178.42 W
				H = 03 32 09.1 h = 497 km MB = 4.4
				D = 149.39 Az = 347.4 (USNOAA)
				PKHKP(A):1.8s 27.0nm
7.	ePKP2	A	06 47 40	**Tonga Islands Region** 18.41 S 172.48 W
				H = 06 27 56.1 h = normal MB = 4.5(USNOAA)
				D = 147.4
7.	eP	A	12 55 58.5	**Yugoslavia** 43.96 N 15.96 E
	LmH	B	58.3	H = 12 54 13.0 h = 36 km MB = 4.6(USNOAA)
	LmV	B	58.9	D = 7.35
				PV(A):0.9s 77.7nm
				LmH:6s 1.3 μm
				LmV:10s 0.6 μm
7.	eP	A	14 04 14	**Central Italy** 43.01 N 12.89 E
	iSn	A	05 34	H = 14 02 23.3 h = 36 km MB = 4.3(USNOAA)
	LmH	B	07.2	D = 7.75
	LmV	B	07.2	PV(A):1.0s 27.6nm
				LmH:7.0s 1.9 μm
				LmV:14.0s 1.4 μm
7.	eP	A	18 36 37	**Morocco** 34.16 N 4.90 W
				H = 18 31 55 h = 0
				D = 20.34 Az = 31 (ISC)

Day	Phase		h m s	Remarks
7.	+eiPn	AB	21 00 39	Yugoslavia 43.95 N 16.06 E
	eiSn	A	01 57	H = 20 58 49.8 h = 5 km MB = 5.5
	LmH	B	03.0	D = 7.35 Az = 337,3 (USNOAA)
	LmV	B	03.6	PV(A):1.0s 17630.0nm
				LmH:6.0s 27.5 /um
				LmV:9.0s 18.4 /um
7.	ePn	A	23 27 06	Yugoslavia 44.15 N 15.67 E
	iPx	A	27 07	H = 23 25 23.2 h = normal (USNOAA)
	eSn	A	28 30.5	D = 7.08
	eSg	A	29 14	PV(A):0.6s 15.3nm
				ei(A) 28 30.5 e(A) 28 50 e(A) 29 17
8.	ePKP	A	04 21 22	Tonga Islands 15.93 S 173.07 W
	esPKP	A	21 32	H = 04 01 44.8 h = 25 km MB = 4.4
				D = 145.16 Az = 354.8 (USNOAA)
				PKP(A):1.0s 11.8nm
8.	e(Pn)	A	07 32 11	Albania 41.08 N 19.57 E
	e	A	34 41	H = 07 29 23.2 h = 25 km MB = 4.1
	e	A	35 04.5	D = 11.05 Az = 332.6 (USNOAA)
	eSg	A	35 21	
8.	ePKIKP	A	14 46 10	South of Fiji Islands 23.78 S 179.24 E
	ePKHKP	A	46 16.5	H = 14 27 21.6 h = 524 km MB = 4.9(USNOAA)
				D = 151.5
				PKHKPV(A):1.1s 14.1nm
8.	eP	A	17 02 27	Hokkaido, Japan Region 42.20 N 143.05 E
				H = 16 50 33.5 h = 70 km MB = 4.8
				D = 78.27 Az = 330.8 (USNOAA)
				PV(A):0.8s 13.5nm MPV(A)=4.9
9.	eP	A	00 17 35	Cyprus 34.76 N 32.67 E
				H = 00 12 41.6 h = normal MB = 4.6
				D = 22.05 Az = 322.5 (USNOAA)
				PV(A):1.3s 19.7nm MPV(A)=4.4

Day	Phase		h m s	Remarks
9.	e(P)	A	05 30 03	**Turkey** 38.97 N 29.52 E
				H = 05 25 58.8 h = 0 km MB = 4.8 (ISC)
				D = 16.6
9.	eP	A	19 31 16	**Ryukyu Islands Region** 29.85 N 131.77 E
	ePcP	A	31 21	H = 19 18 44.1 h = 13 km MB = 5.1
	esP	A	31 28	D = 84.03 Az = 326.4 (USNOAA)
	LmH	B	20 13.9	sPV(A):1.8s 73.3nm MPV(A)=5.6
	LmV	B	13.8	LmH:15.5s 1.9 µm MLH =5.6
				LmV:15.5s 1.9 µm MLV =5.6
10.	eP	A	02 51 00	**West Pakistan** 29.90 N 70.39 E
				H = 02 42 20.9 h = 20 km MB = 4.9
				D = 48.01 Az = 312.9 (USNOAA)
10.	erKP	A	03 36 15	**Fiji Islands Region** 17.80 S 178.59 W
				H = 03 17 40.5 h = 586 km MB = 4.7
				D = 146.23 Az = 348.3 (USNOAA)
10.	-iPg	A	13 47 32.8	**Federal Rep. Germany**, possibly artificial
	iSg	A	47 42.8	49.5 N 11.6 E
				H = 13 47 24 (BCIS)
				D = 1.17
11.	ePKP	AB	01 23 49.5	**Eastern Island Cordillera**
	epPKP	A	23 57	50.05 S 114.47 W
	e	A	24 24	H = 01 04 12.4 h = normal MB = 5.2
	LmH	B	02 25.2	D = 146.23 Az = 67.8 (USNOAA)
	LmV	B	24.9	pPKPV(A):1.2s 32.5nm
				LmH:20s 2.0 µm MLH=5.8
				LmV:20s 2.2 µm
11.	ePKP	A	11 03 53.5	**Loyalty Islands Region** 21.86 S 170.6 E
				H = 10 44 22 h = 86 km
				D = 146.88 Az = 335 (ISC)
12.	ePn	A	08 41 28	**Yugoslavia** 44.23 N 15.66 E
	iX	A	41 30	H = 08 39 45.9 h = 43 km MB = 4.5
	eSn	A	42 51	D = 6.98 Az = 338.3 (USNOAA)

Day	Phase		h m s	Remarks
cont.				
12.	LmH	B	08 43.7	PnV(A):1.0s 25.6nm
	LmV	B	44.0	XV(A):0.7s 99.6nm
				LmH:7.0s 0.7/um MLH=3.7
12.	eP	A	14 43 26	Southern California 34.26 N 117.55 W
	ePcP	A	43 30	H = 14 30 51.9 h = 9 km MB = 5.4
	e	A	43 38	D = 84.31 Az = 29.7 (USNOAA)
	LmH	B	15 22.7	PV(A):1.4s 27.9nm MPV(A)=5.3
	LmV	B	23.4	PcP(A):1.5s 45.2nm MP2V(A)=5.5
				LmH:15s 1.5/um MLH =5.5
				LmV:15s 2.2/um MLV =5.7
12.	eP	A	15 44 50.5	Kurile Islands 45.40 N 149.81 E
				H = 15 32 59.1 h = 60 km MB = 4.9
				D = 77.71 Az = 334.3 (USNOAA)
13.	eP	A	08 41 29	Northern Sumatra 0.94 N 100.19 E
				H = 08 28 58.1 h = 197 km MB = 4.8
				D = 88.38 Az = 320.5 (USNOAA)
				PV(A):1.6s 16.5nm MPV(A)=4.7
13.	ePP	A	14 04 56	Volcano Islands Region 25.64 N 142.77 E
	LmH	B	46.0	H = 13 47 56.6 h = 26 km MB = 4.9(USNOAA)
	LmV	B	48.0	D = 92.7
				PPV(A):1.6s 16.5nm MPPV(A)=5.2
13.	eP	A	16 24 45.5	North Atlantic Ridge 49.76 N 28.96 W
				H = 16 19 14.6 h = normal MB = 4.5
				D = 25.76 Az = 72.3 (USNOAA)
				PV(A):1.2s 12.2nm MPV(A)=4.4
13.	eP	A	21 22 41	Off Coast of Northern California
	ePP	A	25 44	40.22 N 125.14 W
	LmH	B	59.9	H = 21 10 23.0 h = normal MB = 5.4
	LmV	B	22 00.9	D = 81.90 Az = 26.1 (USNOAA)
				PV(A):1.2s 12.2nm MPV(A)=4.8
				PPV(A):1.7s 18.2nm MPPV(A)=5.2
				LmH:17s 1.2/um MLH =5.3
				LmV:15s 1.0/um MLV =5.4

Day	Phase		h m s	Remarks
14.	eP	A	01 07 02.5	Northern Celebes 0.57 S 121.83 E
	ePP	A	11 07.5	H = 00 53 05.1 h = normal MB = 5.6
	e	A	11 19	D = 103.15 Az = 322.2 (USNOAA)
	LmH	B	02 01	PPV(A):1.9s 30.3nm MPPV(A)=5.5
	LmV	B	02	
14.	eP	A	09 51 54.5	Southern Sinkiang 39.94 N 77.02 E
	e	A	51 59	H = 09 43 33.5 h = normal MB = 5.2
	epP	A	52 02	D = 45.83 Az = 306.2 (USNOAA)
				PV(A):0.8s 13.5nm MPV(A)=4.9
14.	+iP	AB	09 57 05	Near East Coast of Honshu, Japan
	ePP	B	10 00 09	38.74 N 142.17 E
	ePPP	B	01 59.5	H = 09 44 53.7 h = 44 km MB = 5.6
	eS	B	07 06	D = 80.98 Az = 330.7 (USNOAA)
	eSKS	B	07 20	PV(A):2.5s 445.0nm MPV(A)=5.9
	eSS	B	12 06.5	PV(B):8s 2.6/um MPV(B)=6.1
	LmH	B	34.4	PPV(B):10.5s 1.1/um MPPV(B)=6.0
	LmV	B	36.2	LmH:17s 15.9/um MLH =6.4
				LmV:17s 16.6/um MLV =6.5
				e(B) 00 16 e(B) 02 14.5 e(B) 07 08
				e(B) 12 10 i(B) 12 21
14.	eP	A	12 55 57.5	Kuril Islands Region 43.50 N 147.98 E
	e	A	56 08	H = 12 43 55.6 h = 30 km MB = 4.6
	LmH	B	13 35	D = 78.84 Az = 333.4 (USNOAA)
	LmV	B	35	PV(A):1.0s 11.8nm MPV(A)=4.9
14.	ePKIKP	A	15 55 23	Near Coast of Central Chile
	ePP	A	56 18	33.97 S 72.17 W
				H = 15 36 51.3 h = 31 km MB = 5.6
				D = 111.82 Az = 43.0 (USNOAA)
				PKPV(A):1.2s 10.2nm
14.	eP	A	16 19 00	Hindu Kush Region 36.42 N 70.07 E
				H = 16 11 15.6 h = 219 km MB = 5.1
				D = 43.57 Az = 308.1 (USNOAA)
				PV(A):1.4s 11.6nm MPV(A)=4.1

Day	Phase		h m s	Remarks
14.	eP	A	18 29 04	Kurile Islands 43.64 N 147.60 E H = 18 17 03.1 h = normal MB = 4.3 D = 78.59 Az = 333.2 (USNOAA)
14.	e	A	18 52 01	Kurile Islands 43.37 N 147.63 E H = 18 39 46.3 h = normal MB = 4.2 D = 78.84 Az = 333.2 (USNOAA)
14.	eP	A	19 56 33	Kurile Islands 43.46 N 147.90 E H = 19 44 31.5 h = 30 km MB = 5.1 D = 78.85 Az = 333.4 (USNOAA) PV(A):1.2s 20.3nm MPV(A)=5.0
15.	eP e	A A	04 07 29.5 07 35.5	Greenland Sea 73.29 N 5.80 E H = 04 02 29.7 h = normal MB = 4.6 D = 22.87 Az = 170.4 (USNOAA) PV(A):0.8s 17.3nm MPV(A)=4.5
15.	eSg	A	07 56 43	Poland 50.28 N 18.95 E H = 07 54 10.2 MAG = 2.9 (ISC, WAR) D = 4.65
15.	ePKIKP eiPKHKP ePKP2	A A A	09 54 46 54 51 54 56.5	Fiji Islands Region 20.53 S 178.84 W H = 09 36 10.5 h = 615 km MB = 5.1 D = 148.83 Az = 347.1 (USNOAA) PKHKPV(A):1.5s 72.9nm
15.	LmH LmV	B B	12 08.6 09.8	LmH:18.5s 1.1/um LmV:16s 0.9/um
15.	ePKIKP ePKHKP ePKP2 LmH LmV	A A A B B	21 02 53 03 04 03 28.5 22 29 25	Kermadec Islands 30.21 S 177.61 W H = 20 42 59.1 h = 34 km MB = 5.2(USNOAA) D = 158.51 Az = 344 (ISC) PKP2V(A):1.5s 37.7nm

Day	Phase		h m s	Remarks
15.	iPn	A	22 48 46.5	Federal Rep. Germany 48.4 N 11.1 E
	e	A	48 54	H = 22 48 09 h = 0 km
	e(Pg)	A	48 57.5	D = 2.31 Az = 8 (ISC)
	eSg	A	49 37	
15.	eP	A	23 27 24	Mozambique Channel 23.49 S 37.29 E
				H = 23 15 30.9 h = normal MB = 4.9
				D = 77.22 Az = 343.6 (USNOAA)
				PV(A):1.6s 13.7nm MPV(A)=4.7
16.	ePKP2	A	00 55 59	Kermadec Islands 30.17 S 177.66 W
	epPKP2	A	56 05	H = 00 35 29.6 h = normal MB = 5.0(USNOAA
				D = 158.3
				PKPV(A):1.3s 15.3nm
16.	e	A	02 06 37	Mariana Islands 13.01 N 144.43 E
	e	B	06(45)	H = 01 49 20.5 h = 47 km MB = 6.0(USNOAA)
	ePP	B	07 50	D = 104.5
	eSKS	B	13 57	PPV(A):2.5s 158.0nm MPPV(A)=6.0
	e(PS)	B	16 40	LmH:18.5s 4.2/um MLH =6.0
	LmH	B	55.6	LmV:19s 0.7/um MLV =5.2
	LmV	B	55.3	
16.	eP	A	03 17 37.5	Taiwan Region 22.25 N 121.34 E
	e	A	17 51.5	H = 03 05 05.4 h = normal MB = 5.0
	epP	A	17 57	D = 84.76 Az = 323.0 (USNOAA)
				PV(A):1.2s 20.3nm MPV(A)=5.2
16.	e	A	04 16 49	Poland 50.31 N 18.93 E
				H = 04 14 31.5 MAG=3.1 (ISC, WAR)
				D = 4.7
17.	ePKP	A	05 02 53.5	New Hebrides Islands 17.96 S 168.93 E
	epPKP	A	03 03	H = 04 43 21.7 h = 27 km MB = 5.0
				D = 142.69 Az = 336.1 (USNOAA)
17.	eP	A	11 36 20	North Atlantic Ridge 26.56 N 22.76 W
	esP	A	36 36	H = 11 29 23.7 h = normal MB = 4.6
				D = 35.55 Az = 38.2 (USNOAA)

Day	Phase		h m s	Remarks
17.	ePKP	A	23 33 19	Chile-Argentina Border Region
				31.81 S 69.95 W
				H = 23 15 02.1 h = 118 km MB = 5.3
				D = 108.97 Az = 41.8 (USNOAA)
18.	iP1	AB	02 11 25.4	Jan Mayen Island Region 71.16 N 7.69 W
	iP2	A	11 34.6	H = 02 06 30.4 h = normal MB = 5.1
	eP1	B	11 35	D = 22.40 Az = 146.5 (USNOAA)
	eS	B	15 28	P1mV(A):1.6s 65.9nm MP1V(A)=4.9
	LmH	B	21.0	P2mV(A):1.4s 181.0nm MP2V(A)=5.3
	LmV	B	22.8	SH:18s 2.3 µm MSH =5.3
				LmH:15s 3.5 µm MLH =4.9
				LmV:13s 3.0 µm MLV =5.0
18.	eP1	AB	16 17 35.5	North Atlantic Ridge 51.12 N 29.55 W
	eP2	A	17 39.5	H = 16 12 07.1 h = normal MB = 5.2
	eS	B	22 14	D = 25.74 Az = 74.8 (USNOAA)
	e	B	22 36	P1V(A):0.5s 23.1nm MP1V(A)=5.0
	LmH	B	26.8	P2V(A):1.7s 72.7nm MP2V(A)=5.0
	LmV	B	27.4	LmH:15s 3.0 µm MLH =4.9
				LmV:12s 2.0 µm MLV =5.0
18.	eP	A	16 58 07.5	Crete 34.34 N 26.25 E
	e	A	58 12.5	H = 16 53 40.0 h = 25 km MB = 4.5
				D = 19.49 Az = 331.2 (USNOAA)
19.	eP	A	00 55 55.5	South of Honshu, Japan 32.41 N 137.71 E
	epP	A	57 28	H = 00 44 01.2 h = 365 km MB = 5.1
				D = 84.60 Az = 328.9 (USNOAA)
				PV(A):0.9s 11.7nm MPV(A)=4.7
19.	eP	A	01 16 06.5	Mongolia 48.36 N 89.30 E
	LmH	B	34.7	H = 01 07 22.3 h = normal MB = 4.7
	LmV	B	36.7	D = 48.31 Az = 303.6 (USNOAA)
				PV(A):1.3s 15.3nm MPV(A)=4.9
				LmH:13.5s 0.94 µm MLH =4.9
				LmV:10s 0.65 µm MLV =5.0

Day	Phase		h m s	Remarks
19.	LmH	B	07 43.6	LmH:18s 1.0/um
	LmV	B	44.0	LmV:18s 1.3/um
19.	eP	A	21 02 08	Jan Mayen Island Region 71.15 N 7.90 W H = 20 57 11.6 h = normal MB = 4.5 D = 22.43 Az = 146.1 (USNOAA) PV(A):1.5s 22.6nm MPV(A)=4.4
19.	eP	A	21 37 46.5	Jan Mayen Island Region 70.99 N 8.47 W H = 21 32 48.4 h = normal MB = 4.2 D = 22.40 Az = 145.0 (USNOAA)
19.	eP	A	22 03 03	Jan Mayen Island Region 70.93 N 8.11 W H = 21 58 02.0 h = normal MB = 3.9 D = 22.29 Az = 145.5 (USNOAA)
19.	ePKIKP	A	24 00 46	Near Coast of Southern Chile 51.93 S 74.14 W H = 23 41 48.0 h = normal MB = 4.8 D = 125.20 Az = 51.0 (USNOAA)
20.	eP	A	10 50 40	South of Honshu, Japan 29.51 N 141.33 E
	ePP	A	54 05	H = 10 37 48.5 h = normal MB = 5.0 D = 88.67 Az = 330.7 (USNOAA)
20.	ePKP	A	12 10 03.5	Samoa Islands Region 15.98 S 172.92 W
	ePKP2	A	10 06.5	H = 11 50 27.7 h = normal MB = 4.5 D = 145.23 Az = 354.9 (USNOAA) PKPV(A):1.1s 12.1nm
21.	ePKHKP	A	01 31 03.5	South of Fiji Islands 24.48 S 176.38 W
	ePKP2	A	31 14	H = 01 11 08.7 h = normal MB = 4.9 D = 153.15 Az = 348.7 (USNOAA) PKHKPV(A):0.5s 17.3nm PKP2V(A):1.4s 11.6nm
21.	e(P)	A	11 59 22.5	Turkey 39.8N 38.9 E
	e	A	59 27	H = 11 54 22 h = 15 km (ANUSSR) D = 21.9

Day	Phase		h m s	Remarks
21.	iPKP2	A	15 59 54	Kermadec Islands Region 30.32 S 179.80 W H = 15 40 04.3 h = 367 km MB = 4.4(USNOAA) D = 157.9
21.	eP	A	19 12 07.5	Turkey 38.85 N 29.91 E H = 19 08 02.3 h = 32 km MB = 4.2 D = 17.49 Az = 318.3 (USNOAA)
22.	eP	A	06 49 25	Greece 38.01 N 20.10 E
	eS	A	52 10	H = 06 46 08.9 h = normal MB = 4.4
	LmH	B	55.9	D = 14.01 Az = 337.2 (USNOAA)
	LmV	B	55.9	PV(A):0.7s 9.6nm MPV(A)=4.6
				LmH:12s 1.0/um MLH =4.1
				LmV:12s 1.2/um
23.	eiP	A	12 17 46	Alaska Peninsula 58.41 N 155.37 W
	epP	A	18 18	H = 12 06 39.6 h = 100 km MB = 4.6
	e	A	18 28	D = 70.78 Az = 8.7 (USNOAA)
	e	A	18 40	PV(A):0.6s 34.5nm MPV(A)=5.4
23.	ePKIKP	A	12 23 54	Solomon Islands 6.47 S 154.62 E
	epPKP	A	24 06	H = 12 04 54.2 h = 39 km MB = 5.7
	ePP	C	25 52	D = 126.27 Az = 331.6 (USNOAA)
	eSKKKS	A	32 48	LmH:21s 8.5/um MLH=6.4
	eSKSP	C	35 39	LmV:18s 2.3/um MLV=5.9
	ePKKS2	A	37 05	e(A) 32 03.5 e(C) 43 00
	ePPS	C	37 20	
	eSS	C	42 35	
	LmH	B	13 22.0	
	LmV	B	31	
23.	eP	A	21 14 49.5	Andreanof Islands, Aleutian Is. 51.35 N 179.41 W H = 21 02 54.6 h = 43 km MB = 5.2 D = 77.95 Az = 352.8 (USNOAA) PV(A):1.2s 26.4nm MPV(A)=5.1

Day	Phase		h m s	Remarks
23.	eP	A	23 01 10.5	<u>Near Coast of Guatemala</u> 13.51 N 90.09 W H = 22 48 33.8 h = 80 km MB = 4.7 D = 86.94 Az = 38.6 (USNOAA)
23.	eP	A	23 08 37	<u>Near Coast of Guatemala</u> 13.52 N 90.08 W H = 22 56 00.1 h = 83 km MB = 4.6 D = 86.93 Az = 38.6 (USNOAA)
23.	ePKIKP	AC	23 30 58	<u>Solomon Islands</u> 6.55 S 154.68 E
	ePP	B	32 55	H = 23 11 58.5 h = 47 km MB = 5.3
	eSKSP	B	42(40)	D = 126.37 Az = 331.6 (USNOAA)
	ePPS	B	44 20	PPV:10s 0.81 /um MPPV=5.9
	LmH	B	30.9	LmH:18s 2.1 /um MLH =5.8
	LmV	B	30.9	LmV:18s 2.4 /um MLV =5.9
23.	iP	ABC	16 56 03.5	<u>Near East Coast of Kamchatka</u>
	eS	BC	17 05 20	54.66 N 162.83 E
	e	BC	05 28	H = 16 44 40.0 h = 34 km MB = 5.3
	LmH	B	32.2	D = 72.34 Az = 341.2 (USNOAA)
	LmV	B	34	PV(A):1.8s 98.0nm MPV(A)=5.5 PV(B):8s 0.49 /um MPV(B)=5.6 SH:14s 0.42 /um MSH =5.5 LmH:15s 2.0 /um MLH =5.5 LmV:14s 2.1 /um MLV =5.6
24.	eP	A	18 45 27	<u>Svalbard Region</u> 79.91 N 5.06 E H = 18 39 24.3 h = normal MB = 4.6 D = 29.47 Az = 170.5 (USNOAA)
26.	e(P)	A	06 17 34	<u>Greece</u> 38.98 N 21.89 E H = 06 14 08.8 h = normal MB = 4.3 D = 13.74 Az = 331.4 (USNOAA)
26.	eP	AB	12 15 05	<u>Near West Coast of Colombia</u>
	eisP	A	15 09	6.20 N 77.59 W
	ePP	BC	18 08	H = 12 02 29.3 h = 8 km MB = 6.1
	eSKS	BC	25 22	D = 84.74 Az = 39.7 (USNOAA)

Day	Phase		h m s	Remarks
cont. 26.	eiS	BC	12 25 32	PmV(A):1.8s 122.0nm MPV(A)=5.8
	eiPS	C	26 25	sPmV(A):2.0s 385.0nm MPV(A)=6.3
	eSS	B	31 23	PV(B):10s 9.7/um MPV(B)=7.0
	eSSS	C	35 44	PPV:10s 3.5/um MPPV =6.7
	eSSSS	C	37 20	SH:14s 28.0/um MSH =7.1
	LmH	B	48.5	LmH:19.5s 21.5/um MLH =6.5
	LmV	B	48.5	LmV:20s 23.4/um MLV =6.6
				ei(B) 26 29
26.	iPn	A	16 44 03.4	<u>Northern Italy</u> 44.33 N 12.30 E
	ePg	A	44 34.5	H = 16 42 28 h = 12 km
	iSn	A	45 12.5	D = 6.33 Az = 356 (ISC)
	e	A	45 24	
	e	A	45 40	
	eSg	A	46 07	
27	eP	A	03 51 11	<u>Near West Coast of Colombia</u>
	+iPcP	AB	51 14	6.43 N 77.41 W
	ePP	BC	54 40	H = 03 38 36.2 h = 8 km MB = 5.8
	eS	BC	04 01 40	D = 84.46 Az = 39.8 (USNOAA)
	ePS	C	02 15	PmV(A):3.0s 921.0nm MPV(A)=6.5
	eiPPS	BC	03 03	PV(B):9s 3.7/um MPV(B)=6.6
	eSS	B	06 45	PPV:10s 2.36/um MPPV =6.7
	eSSS	C	10 55	SH:13s 8.0/um MSH =6.6
	eLQ	C	13 44	LmH:25s 43.1/um MLH =6.7
	ePKPPKP	A	17 34	LmV:24s 53.0/um MLV =6.9
	eLR	C	18 00	e(B,C) 00 53 ei(B) 02 38
	LmH	B	22.9	
	LmV	B	23.0	
27.	ePKP	A	09 14 05	<u>New Britain Region</u> 5.08 S 152.81 E
				H = 08 55 10.3 h = 58 km MAG=4.9 (ISC)
				D = 124.2
27.	eP	A	10 25 02	<u>Philippine Islands Region</u>
				20.65 N 122.23 E
				H = 10 12 19.4 h = normal MB = 4.7(USNOAA)
				D = 86.5

Day	Phase		h m s	Remarks
27.	eP	A	15 59 36	Greece-Albania Border Region 39.30N 20.14E
	LmH	B	16 04.2	H = 15 56 32.2 h = 23 km MB = 4.2
	LmV	B	05.8	D = 12.84 Az = 334.9 (USNOAA)
27.	eP	A	16 36 55.5	Greece-Albania Border Region
	LmH	B	41	39.29 N 20.25 E
	LmV	B	42	H = 16 33 53.8 h = 40 km MAG=4.2
				D = 12.88 Az = 335 (ISC)
27.	ePn	A	19 20 49	Southern Italy 41.71 N 13.83 E
	eSn	A	22 31	H = 19 18 36.7 h = 27 km MB = 4.4
				D = 9.06 Az = 351.0 (USNOAA)
				PV(A):0.8s 7.7nm MPV(A)=(5.0)
28.	eP	A	04 33 26.5	Unimak Island Region 54.48 N 164.49 W
				H = 04 21 49.8 h = 77 km MB = 5.0
				D = 75.19 Az = 2.6 (USNOAA)
				PV(A):1.1s 16.1nm MPV(A)=4.8
28.	eP	A	09 21 03	Near East Coast of Kamchatka
				55.11 N 162.64 E
				H = 09 09 41.5 h = normal MB = 4.4(USNOAA)
				D = 71.8
28.	eP	A	11 44 47	North Atlantic Ridge 57.17 N 33.35 W
	e	A	44 50.5	H = 11 39 09.0 h = normal MB = 4.6
	e	A	44 54	D = 26.87 Az = 84.6 (USNOAA)
	LmH	B	55	PV(A):1.6s 16.5nm MPV(A)=4.4
	LmV	B	55	
28.	eP	A	15 45 26	Kurile Islands 44.05 N 148.99 E
	ePcP	A	45 40	H = 15 33 25.7 h = normal MB = 4.6
				D = 78.67 Az = 333.9 (USNOAA)
				PV(A):1.4s 11.6nm MPV(A)=4.7
28.	-eiP	ABC	17 33 29.5	Near East Coast of Kamchatka
	eS	B	42 44	53.33 N 158.74 E
	eSPP	BC	43 30	H = 17 22 12.1 h = 118 km MB = 5.4

Day	Phase		h m s	Remarks
cont. 28.	esS	BC	17 43 40	D = 72.77 Az = 338.8 (USNOAA)
				PV(A):1.3s 264.0nm MPV(A)=5.9
28.	eP	A	19 58 24	Turkey 37.10 N 28.60 E
	LmH	B	20 06	H = 19 54 09.4 h = 36 km MB = 4.2
	LmV	B	06	D = 18.20 Az = 323.4 (USNOAA)
				PV(A):1.8s 27.0nm MPV(A)=4.1
28.	eP	A	23 55 10.5	North Atlantic Ocean 57.20 N 33.45 W
	e	A	55 42	H = 23 49 32.0 h = normal MB = 4.8(USNOAA)
				D = 26.89 Az = 85 (ISC)
				PmV(A):2.0s 51.2nm MPmV(A)=4.8
28.	eP	A	24 02 42.5	North Atlantic Ocean 57.28 N 33.30 W
	e	A	02 57	H = 23 57 04.1 h = normal MB = 4.9
	LmH	B	13.4	D = 26.83 Az = 84.9 (USNOAA)
	LmV	B	13.8	PV(A):1.0s 15.8nm MPV(A)=4.6
				PmV(A):1.7s 42.4nm MPmV(A)=4.8
				LmH:15.5s 1.5 um MLH =4.6
				LmV:15 s 1.4 um MLV =4.8
29.	iP	AC	04 55 04.5	Nicaragua 11.47 N 85.48 W
	iPcP	A	55 12	H = 04 42 46.6 h = 192 km MB = 5.4(USNOAA)
	ipP	AC	55 52.5	D = 85.66 Az = 39 (ISC)
	eS	C	05 05 22	LmH:16s 0.4 um
	ePS	BC	06 36	LmV:16s 0.5 um
	LmH	B	28.4	
	LmV	B	28.5	
29.	ePKP	ABC	06 22 45	New Hebrides Islands 13.51 S 166.55 E
	ePP	A	25 36	H = 06 03 26.0 h = 59 km MB = 5.8(USNOAA)
	eSS	BC	43 40	D = 137.7
	LmH	B	07 28.3	PKPV(A):1.7s 21.2nm
	LmV	B	28.2	LmH:20s 4.1 um MLH=6.1
				LmV:21.5s 5.2 um MLV=6.2
				e(BC) 25 40 e(AB) 25 49 e(A) 26 02
				e(BC) 26 42 e(C) 48 28

Day	Phase		h m s	Remarks
29.	ePn	A	10 36 35	Yugoslavia 44.1 N 15.7 E
	eSn	A	37 57.5	H = 10 34 56 h = 59 km
	eSg	A	38 40	D = 7.10 Az = 338 (ISC)
				PnV(A):0.9s 15.6nm MPnV(A)=4.6
29.	ePP	A	23 30 21	Solomon Islands 9.59 S 159.34 E
				H = 23 09 33.6 h = 548 km MB = 4.9(USNOAA)
				D = 131.1
				PPV(A):1.6s 22.0nm MPPV(A)=5.1
30.	eP	A	10 05 03.5	Philippine Islands Region
	esP	A	05 16	20.59 N 122.02 E
	e	B	05 32	H = 09 52 22.7 h = normal MB = 5.1
	ePP	A	08 26.5	D = 86.47 Az = 323.3 (USNOAA)
	e	A	08 38	PV(A):1.5s 20.1nm MPV(A)=5.2
	eS	BC	15 36	PPV(A):1.7s 30.3nm MPPV(A)=5.5
	LmH	B	44.9	LmH:12.5s 3.1/um MLH =5.9
	LmV	B	49.2	LmV:16s 2.9/um MLV =5.8
30.	+eiP	A	18 25 42.5	Alaska Peninsula 58.38 N 155.28 W
				H = 18 14 35.9 h = 101 km MB = 4.5
				D = 70.80 Az = 8.8 (USNOAA)
				PV(A):0.8s 19.2nm MPV(A)=5.0

Day	Phase		h m s	Remarks
1.	eP	A	09 53 28	Sakhalin Island 46.85 N 143.64 E
	e(PP)	A	56 08	H = 09 42 28.6 h = 385 km MB = 5.0
				D = 74.44 Az = 330.6 (USNOAA)
1.	eP	A	22 25 28.5	Greece 38.13 N 22.76 E
	LmH	B	31.8	H = 22 21 54.9 h = 24 km MB = 4.7
	LmV	B	31.8	D = 14.81 Az = 331.2 (USNOAA)
				PV(A):2.0s 30.0nm MPV(A)=4.1
				LmH(B):11s 2.1/um MLH(B)=4.5
				LmV(B):9.5s 2.2/um MLV(B)=4.8
1.	eP	A	22 42 03.5	Greece 38.04 N 22.79 E
	LmH	B	47.1	H = 22 38 35.3 h = 29 km MB = 5.0
	LmV	B	48.4	D = 14.90 Az = 331.3 (USNOAA)
				PV(A):1.8s 40.5nm MPV(A)=4.2
				LmH(B):12s 1.8/um MLH(B)=4.4
				LmV(B):10s 1.8/um MLV(B)=4.7
2.	eP	A	00 35 31.5	Algeria 35.57 N 6.18 E
	LmH	B	41.3	H = 00 31 48.7 h = 36 km MB = 4.5
	LmV	B	42.0	D = 15.57 Az = 13.0 (USNOAA)
				PV(A):1.2s 24.4nm MPV(A)=4.2
				LmH(B):13s 0.5/um MLH(B)=3.9
				LmV(B):12s 0.4/um MLV(B)=4.0
2.	ePKIKP	A	06 34 32.5	Solomon Islands 6.82 S 154.91 E
	LmH	B	22.2	H = 06 15 32.8 h = 54 km MB = 5.4
	LmV	B	28.7	D = 126.71 Az = 331.7 (USNOAA)
				PKIKPV(A):1.5s 30.2nm
				LmH(B):21s 2.5/um MLH(B)=5.9
				LmV(B):21s 1.6/um MLV(B)=5.7
2.	ePKP2	A	10 00 43	Kermadec Islands Region
	LmH	B	11 15.7	28.95 S 177.05 W
	LmV	B	21.4	H = 09 40 21.2 h = 59 km MB = 5,5 (USNOAA)
				D = 157.3
				PKP2V(A):1.4s 27.9nm
				LmH(B):20s 1.8/um MLH(B)=5.8
				LmV(B):18s 1.4/um MLV(B)=5.8

Day	Phase		h m s	Remarks
2.	iP	A	14 45 20	Explosion Hilders/Rhön,Fed.Rep.Germany
				$50^{o}32.52'N$ $10^{o}02.39'E$
				H = 14 45 00.87 yield 21.4 t (Hannover)
				D = 1.0
3.	iP	A	08 23 35.5	Gulf of Alaska 58.36 N 150.52 W
	ipP	A	23 41.5	H = 08 12 20.8 h = 25 km MB = 5.0
				D = 70.38 Az = 12.0 (USNOAA)
3.	LmH	B	11 38.5	Probably New Britain Region (USNOAA)
	LmV	B	41.7	LmH(B):22s 2.9 µm
3.	eP	A	14 29 28	Near West Coast of Colombia
				6.39 N 77.51 W
				H = 14 16 55.3 h = 25 km MB = 5.2(USNOAA)
				D = 84.5
				PV(A):1.8s 40.5nm MPV(A)=5.4
4.	ePKHKP	A	00 55 38	Fiji Islands Region 20.84 S 178.76 W
	ePKP2	A	55 45	H = 00 36 57.9 h = 631 km MB = 4.7
				D = 149.15 Az = 347.1 (USNOAA)
				PKHKPV(A):0.8s 17.3nm
4.	eP	A	02 24 15	North Atlantic Ocean 58.3 N 31.4 W
				H = 02 18 48.4 h = 33 km MB = 4.0
				D = 25.77 Az = 88 (ISC)
4.	ePKIKP	A	17 15 23	Banda Sea 6.83 S 130.41 E
				H = 16 56 54.3 h = 92 km MB = 5.6
				D = 113.31 Az = 322.6 (USNOAA)
4.	eP	A	17 52 22	Gulf of California 30.63 N 113.65 W
	LmH	B	18 32.0	H = 17 39 45.4 h = normal MB = 5.0
	LmV	B	32.0	D = 85.75 Az = 31.4 (USNOAA)
				PV(A):1.2s 10.2nm MPV(A)=4.9

Day	Phase		h m s	Remarks
5.	eP	A	10 51 08.5	Kirgiz-Sinkiang Border Region
				40.16 N 77.09 E
				H = 10 42 49.1 h = normal MB = 5.0
				D = 45.74 Az = 306.0 (USNOAA)
				PV(A):0.8s 11.5nm MPV(A)=4.9
5.	ePKIKP	A	14 44 51.5	Fiji Islands 22.55 S 178.22 W
	+iPKHKP	A	44 57.5	H = 14 26 09.8 h = 545 km MB = 4.7
	ePKP2	A	45 03	D = 148.97 Az = 347.8 (USNOAA)
				PKHKPV(A):1.3s 35.0nm
5.	eP	A	14 58 36	Jordan-Syria Region 35.12 N 38.90 E
				H = 14 53 11.7 h = 33 km MB = 4.8
				D = 25.12 Az = 316.6 (USNOAA)
				PV(A):1.4s 32.5nm MPV(A)=4.7
5.	+iPKP	A	20 00 42.8	Fiji Islands Region 15.81 S 177.66 W
				H = 19 41 58.2 h = 456 km MB = 4.8
				D = 144.46 Az = 349.8 (USNOAA)
				PKPV(A):1.2s 40.6nm
5.	-iPn	AB	23 26 08.4	Yugoslavia 44.03 N 15.78 E
	ePg	A	26 36.5	H = 23 24 23.1 h = 49 km MB = 5.0
	iSn	A	27 29	D = 7.20 Az = 338.3 (USNOAA)
	e	A	27 59	LmH(B):11s 1.6 µm MLH(B)=3.9
	iSg	B	28 16	LmV(B):12s 1.3 µm
	LmH	B	29.1	
	LmV	B	29.1	
6.	eP	A	04 27 52	Kurile Islands 49.60 N 156.07 E
	e	A	28 28	H = 04 16 10.8 h = 46 km MB = 4.9
	LmH	B	05 03.0	D = 75.61 Az = 337.5
	LmV	B	03.6	
6.	e	A	14 59 37.5	ČSSR, explosion, yield 11.5 t (PRU)
				50°47.4'N 14°31.5'E

Day	Phase		h m s	Remarks
6.	e(P)	A	15 50 45	Near West Coast of Colombia
				5.95 N 77.60 W
				H = 15 38 06.8 h = 38 km MB = 4.5
				D = 84.94 Az = 39.7 (USNOAA)
6.	eP	A	21 37 54	Near West Coast of Colombia
	e	A	38 00.5	6.18 N 77.62 W
	LmH	B	22 14.5	H = 21 25 21.0 h = 33 km MB = 5.2
	LmV	B	14.5	D = 84.79 Az = 39.7 (USNOAA)
				LmH(B):18s 0.3/um MLH(B)=4.8
				LmV(B):20s 0.6/um MLV(B)=5.1
6.	eP1	A	22 14 20.5	Tadzhik SSR 39.07 N 71.62 E
	iP2	A	14 22.5	H = 22 06 26.8 h = 68 km MB = 5.2
	ePn	A	15 57	D = 42.96 Az = 306.0 (USNOAA)
	ePP	A	16 02	P1V(A):1.0s 15.7nm MP1V(A)=4.8
				P2V(A):1.0s 39.3nm MP2V(A)=5.1
6.	eP	A	23 47 23.5	Hokkaido, Japan Region
				41.51 N 142.10 E
				H = 23 35 28.6 h = 71 km MB = 5.0
				D = 78.54 Az = 330.4 (USNOAA)
				PV(A):1.4s 20.9nm MPV(A)=4.8
7.	e(P)	A	01 24 03.5	Near West Coast of Colombia
				5.98 N 77.58 W
				H = 01 11 23.1 h = 34 km MB = 4.8
				D = 84.91 Az = 39.7 (USNOAA)
				(P)V(A):1.5s 17.6nm
7.	e(pP)	A	02 28 24	Southern Iran 27.78 N 56.50 E
				H = 02 20 36.7 h = 43 km MB = 5.0
				D = 40.78 Az = 316.5 (USNOAA)
7.	eP	A	10 51 10.5	Western Caucasus 43.36 N 44.04 E
	e	A	51 25	H = 10 46 03.8 h = normal MB = 4.5
	LmH	C	59.0	D = 23.14 Az = 299.7 (USNOAA)

Day	Phase		h m s	Remarks
cont. 7.	LmV	C	11 00.9	PV(A):0.8s 13.5nm MPV(A)=4.5 LmH(C):20s 0.4/um MLH(C)=3.9 LmV(C):14s 0.5/um MLV(C)=4.3
7.	ePn	A	14 47 02	D = 2.9
	e	A	47 07	
	e(Sn)	A	47 34.5	
	eSg	A	47 48	
7.	eP	A	16 13 09	<u>Rat Islands, Aleutian Is.</u> 50.50 N 177.66 E H = 16 01 10.3 h = 41 km MB = 4.9 D = 78.53 Az = 351.0 (USNOAA) PV(A):1.1s 16.1nm MPV(A)=49
7.	ePKP	A	19 02 41.5	<u>Samoa Islands Region</u> 16.44 S 172.43 W
	eipPKP	A	02 46.5	H = 18 43 01.2 h = 14 km MB = 5.1 D = 145.73 Az = 355.4 (USNOAA) h = 18 km PKPV(A):1.1s 30.2nm
8.	eP	A	01 54 19.5	<u>Northwest of Kurile Islands</u> 47.93 N 146.75 E H = 01 43 30.4 h = 490 km MB = 4.6 D = 74.50 Az = 332.2 (USNOAA) PV(A):1.0s 19.7nm MPV(A)=4.6
8.	P1	A	02 51 16.5	<u>Dead Sea Region</u> 31.7 N 35.3 E
	P2	A	51 19.5	H = 02 45 45 h = 0 km D = 25.82 Az = 324 (ISC) P1V(A):0.6s 13.2nm MP1V(A)=4.9 P2V(A):1.3s 28.4nm MP2V(A)=4.7
8.	ePKP	A	04 07 57.5	<u>Tonga Islands</u> 19.27 S 173.47 W
	e	A	08 20	H = 03 48 13.3 h = 40 km MB = 5.2
	e	A	08 40	D = 148.44 Az = 353.8 (USNOAA) PKPV(A):1.6s 30.2nm

Day	Phase		h m s	Remarks
8.	+iP	ABC	05 04 46	Near East Coast of Kamchatka
	eS	BC	14 04	53.83 N 160.45 E
	ePS	C	14 32	H = 04 53 21.8 h = 53 km MB = 5.6
	ePPS	BC	14 57	D = 72.66 Az = 339.8 (USNOAA)
	e	BC	15 05	PV(A):1.2s 287.0nm MPV(A)=6.1
	LmV	B	35.3	PV(B):4.5s 7.7/um MPV(B)=5.9
	LmH	B	35.8	LmV(B):26s 1.7/um MLV(B)=5.2
				LmH(B):25s 2.4/um MLH(B)=5.4
8.	ePKIKP	A	07 17 52	Fiji Islands Region 21.63 S 179.05 W
	-iPKHKP	A	17 57.5	H = 06 59 10.7 h = 577 km MB = 5.0
	ePKP2	A	18 05.5	D = 149.85 Az = 346.4 (USNOAA)
				PKHKPV(A):1.1s 76.6/um
8.	ePKP2	A	10 59 09.5	Kermadec Islands 29.30 S 176.96 W
				H = 10 38 44.5 h = 49 km MB = 4.7(USNOAA)
				D = 157.7
				PKP2V(A):1.2s 12.2nm
8.	ePKP2	A	11 28 07.5	Kermadec Islands Region 29.37 S 176.89 W
				H = 11 07 43.5 h = 47 km MB = 4.8(USNOAA)
				D = 157.7
8.	eP	A	13 14 06.5	Andreanof Islands, Aleutian Is.
	e	A	14 12.5	50.45 N 176.20 W
	e	A	14 18.5	H = 13 02 04.7 h = 38 km MB = 5.1
	e	A	14 26.5	D = 79.07 Az = 354.9 (USNOAA)
				PV(A):1.4s 46.5 MPV(A)=5.3
8.	ePKP2	A	15 30 17.5	Tonga Islands Region 22.50 S 175.36 W
				H = 15 10 13.1 h = normal MB = 4.5
				D = 151.38 Az = 350.7 (USNOAA)
				PKP2V(A):1.2s 10.2nm
8.	eP1	A	22 17 37	Greece 38.28 N 20.25 E
	eP2	A	17 39.5	H = 22 14 23.5 h = 57 km MB = 4.5
	e(S)	A	20 22	D = 13.80 Az = 336.4 (USNOAA)

Day	Phase		h m s	Remarks
cont. 8.	LmH	B	22 24.3	P2V(A):1.0s 19.7nm MP2V(A)=4.2
	LmV	B	24.3	LmH(B):11s 0.7/um
				LmV(B):12s 0.9/um
8.	eP	A	23 24 43.5	<u>Kurile Islands</u> 43.59 N 146.89 E
				H = 23 12 44.8 h = normal MB = 4.7
				D = 78.39 Az = 332.8 (USNOAA)
8.	+iP1	ABC	23 48 11	<u>Kurile Islands</u> 43.80 N 147.43 E
	iP2	A	48 13	H = 23 36 09.7 h = 15 km MB = 5.8
	e	A	48 21	D = 78.38 Az = 333.1 (USNOAA)
	eS	BC	58 00	P1V(A):1.3s 107.0nm MP1V(A)=5.8
	e	B	58 20	P2V(A):1.1s 187.5nm MP2V(A)=6.1
	eSS	C	24 03.0	LmV(B):17.5s 4.3/um MLV(B)=5.9
	LmV	B	27.5	LmH(B):16s 4.1/um MLH(B)=5.9
	LmH	B	27.6	
9.	eP	A	01 02 05	<u>Ionian Sea</u> 37.85 E 19.94 E
	LmH	B	08.7	H = 00 58 43.6 h = 14 km MB = 4.6
	LmV	B	08.7	D = 14.10 Az = 337.8 (USNOAA)
				PV(A):1.3s 13.1nm MPV(A)=4.5
				LmH(B):11s 1.6/um MLH(B)=4.4
				LmV(B):12s 1.3/um
9.	eP1	A	01 26 36	<u>Tadzhik SSR</u> 39.00 N 71.64 E
	iP2	A	26 37.9	H = 01 18 44.1 h = 81 km MB = 5.1
	ePP	A	28 17.5	D = 43.02 Az = 306.1 (USNOAA)
	e	A	28 29	P2V(A):1.2s 34.6nm MP2V(A)=5.1
	e	A	28 52	LmH(B):8s 0.5/um
	e	A	29 11.5	LmV(B):11s 0.4/um
	LmH	B	43.9	
	LmV	B	47.3	
9.	eP1	A	07 37 23	<u>Mediterranean Sea</u> 34.96 N 13.67 E
	eP2	A	37 26	H = 07 33 39.2 h = normal MB = 4.3
	P m	A	37 28	D = 15.75 Az = 355.2 (USNOAA)
	LmH	B	44.4	P2mV(A):2.0s 42.7nm MP2mV(A)=4.2

Day	Phase		h m s	Remarks
cont. 9.	LmV	B	07 45.6	LmH(B):14s 0.8 µm MLH(B)=5.0
				LmV(B):13s 0.4 µm MLV(B)=5.0
9.	LmH	C	11 10.0	Probably Mindanao, Philippine Islands
	LmV	C	20.0	(USNOAA)
9.	eP	A	11 19 16	Andreanof Islands, Aleutian Is.
				51.38 N 178.42 W
				H = 11 07 20.2 h = 41 km MB = 5.2
				D = 78.00 Az = 353.5 (USNOAA)
				PV(A):1.2s 16.3nm MPV(A)=4.9
9.	eP1	A	13 56 48	Tadzhik SSR 39.10 N 71.67 E
	eP2	A	56 50.2	H = 13 48 52.6 h = 46 km MB = 5.2
	ePP	A	58 28.5	D = 42.97 Az = 306.0 (USNOAA)
	LmH	B	14 15.8	P1V(A):0.9s 19.5nm MP1V(A)=4.9
	LmV	B	18.5	P2V(A):1.2s 36.6nm MP2V(A)=5.0
				LmH(B):14s 0.4 µm MLH(B) =4.4
				LmV(B):13.6s 0.3 µm MLV(B) =4.5
9.	e(Sn)	A	17 37 42	Yugoslavia 46.6 N 15.17 E
				H = 17 35 25 (ISC)
				D = 5.7
10.	eP	A	00 28 52	Andreanof Islands, Aleutian Is.
				50.18 N 178.59 W
				H = 00 16 49.1 h = normal MB = 4.6
				D = 79.18 Az = 353.4 (USNOAA)
				PV(A):0.8s 9.6nm MPV(A)=4.9
10.	eiP1	ABC	09 05 28.5	South Indian Ocean 3.59 S 86.16 E
	iP2	A	05 38.2	H = 08 53 04.8 h = normal MB = 5.9
	iP3	A	05 48.7	D = 83.03 Az = 321.8 (USNOAA)
	ePP	BC	08 44	P1V(A):1.3s 100.0nm MP1V(A)=5.8
	e	B	08 47	P1V(B):6s 1.2 µm MP1V(B)=6.3
	eS	B	15 40	P3V(A):1.5s 261.0nm MP3V(A)=6.1
	iS	C	15 46	PPV(B):12s 0.97 µm MPPV(B)=6.1
	eSS	C	20 56	SH(B):15s 3.45 µm MSH(B) =6.3

Day	Phase		h m s	Remarks
cont. 10.	eSSS	C	09 24 44	LmV(B):16s 5.4/um MLV(B)=6.0
	eSSSS	C	27 25	LmH(B):17.5s 5.7/um MLH(B)=6.0
	LmV	B	47.6	
	LmH	B	47.8	
10.	eP	A	10 49 58.5	Kurile Islands 46.49 N 153.32 E
	e	A	50 02.5	H = 10 38 02.8 h = normal MB = 4.8
				D = 77.75 Az = 336.2 (USNOAA)
				PV(A):1.0s 11.8nm MPV(A)=4.9
10.	eiP	A	13 51 41.5	Ionian Sea 38.03 N 19.89 E
	eipP	A	51 47.5	H = 13 48 23.5 h = 18 km MB = 4.6
	LmH	B	58.4	D = 13.92 Az = 337.6 (USNOAA)
	LmV	B	58.4	PV(A):0.8s 21.2nm
				LmH(B):10s 1.6/um MLH(B)=4.4
				LmV(B):11s 1.6/um
10.	ePKIKP	BC	22 19 40	Kermadec Islands Region
	e	A	19 42	31.90 S 177.93 W
	eiPKP2	AB	20 19	H = 21 59 42.9 h = normal MB = 5.9
	ei	B	20 50	D = 159.97 Az = 342.0 (USNOAA)
	ePP	C	24 00	PPV(B):8.5s 1.63/um MPPV(B)=6.2
	ePP	B	24 04	LmV(B):18s 6.5/um MLV(B) =6.5
	ePS	C	34 24	LmH(B):17s 3.5/um MLH(B) =6.2
	eiSPP	BC	37 26	
	e(PPS)	C	37 31	
	eSS	C	44 08	
	ei	B	44 28	
	eiSSS	B	50 00	
	LmV	B	23 48.1	
	LmH	B	49.7	
11.	ePKP2	A	01 23 33	South of Kermadec Islands
				32.03 S 177.90 W
				H = 01 02 58.0 h = normal MB = 4.6(USNOAA)
				D = 160.2
				PKP2V(A):1.5s 20.1nm

Day	Phase		h m s	Remarks
11.	eP	A	02 38 47.5	Greece 38.19 N 20.06 E
	LmV	B	45.4	H = 02 35 29.2 h = 21 km MB = 4.7
	LmH	B	45.6	D = 13.83 Az = 337.0 (USNOAA)
				PV(A):0.8s 34.6nm
				LmV(B):8s 1.0/um
				LmH(B):8s 0.9/um MLH(B)=4.2
11.	ePKIKP	ABC	03 36 47	Kermadec Islands Region 31.81 S 178.13 W
	iPKP2	ABC	37 25	H = 03 16 49.6 h = normal MB = 5.6
	ePP	BC	41 00	D = 159.84 Az = 341.8 (USNOAA)
	ePP	A	41 05	PPV(B):9s 2.0/um MPPV(B)=6.3
	ePS	C	51 32	LmV(B):15.5s 10.5/um MLV(B) =6.8
	eSPP	C	54 30	LmH(B):14.5s 7.6/um MLH(B) =6.6
	eSS	C	04 01 08	
	eSSS	C	06 15	
	e	C	07 00	
	LmV	B	26.2	
	LmH	B	26.4	
11.	eP	A	03 42 41	Ryukyu Islands 26.74 N 129.69 E
				H = 03 30 01.9 h = 27 km MB = 5.3
				D = 85.57 Az = 325.7 (USNOAA)
				PV(A):2.3s 110.0nm MPV(A)=5.6
11.	eP	A	05 11 43.5	Ryukyu Islands 29.29 N 130.43 E
	e	A	11 54.5	H = 04 59 18.3 h = 51 km MB = 5.1
				D = 83.84 Az = 325.9 (USNOAA)
11.	eiP	A	05 41 16	Kurile Islands 43.48 N 147.66 E
	eipP	A	41 28	H = 05 29 17.3 h = 50 km MB = 5.2
				D = 78.75 Az = 333.2 (USNOAA)
				h = 44 km
11.	ePKIKP	A	05 58 05	South of Kermadec Islands
	ePKP2	A	58 41.5	32.05 S 177.82 W
	eSKKS2	BC	06 13 32	H = 05 38 06.0 h = 32 km MB = 5.6
	ePPS	BC	15 52	D = 160.15 Az = 342.1 (USNOAA)
	LmH	B	07 12.2	LmH(B):20s 3.3/um MLH(B)=6.1
	LmV	B	16.0	LmV(B):20s 3.1/um MLV(B)=6.1

Day	Phase		h m s	Remarks
11.	ePKP2	A	07 05 45	Kermadec Islands Region 31.94 S 177.80 W H = 06 45 07.7 h = normal MB = 4.8(USNOAA) D = 160.1
11.	eP	A	08 57 55	Near East Coast of Kamchatka 51.47 N 157.17 E H = 08 46 21.3 h = normal MB = 4.8 D = 74.14 Az = 338.0 (USNOAA) PV(A):0.8s 7.7nm MPV(A)=4.8
11.	eP	A	10 38 01	Near East Coast of Kamchatka
	LmH	B	11 20.4	53.80 N 160.73 E
	LmV	B	27.7	H = 10 26 34.5 h = normal MB = 5.0 D = 72.74 Az = 340.0 (USNOAA) PV(A):0.9s 19.5nm MPV(A)=5.1 LmH(B):18s 1.3 µm MLH(B)=5.2 LmV(B):14s 1.4 µm MLV(B)=5.4
11.	e	A	10 44 35	Ryukyu Islands 26.56 N 129.73 E
	e	A	44 45	H = 10 31 33.1 h = normal MB = 4.9(USNOAA) D = 85.7
11.	ePKP2	A	11 47 17.5	South of Kermadec Islands
	e	A	47 43.5	32.31 S 177.77 W H = 11 26 40.3 h = normal MB = 5.0(USNOAA) D = 160.3 PKP2V(A):1.6s 22.0nm
11.	e	A	16 00 31	Probably Oaxaca, Mexico (USNOAA)
11.	ePKP2	A	20 46 48	Kermadec Islands Region 31.45 S 178.09 W H = 20 26 14.3 h = normal MB = 4.3(USNOAA) D = 159.7
11.	ePKP2	A	22 49 01	South of Kermadec Islands 32.28 S 178.02 W H = 22 28 25.0 h = normal MB = 4.5(USNOAA) D = 160.3

Day	Phase		h m s	Remarks
12.	e	A	03 59 10	Upper Silesia / Poland, rockburst (CLL)
12.	ePKHKP ePKP2	A A	07 20(11) 20 35	South of Kermadec Islands 32.03 S 178.04 W H = 06 59 58.3 h = normal MB = 5.1(USNOAA) D = 160.1 PKP2V(A):1.2s 26.4nm
12.	ePKP2 LmH LmV	A B B	07 31 13.5 08 49.5 49.5	South of Kermadec Islands 32.23 S 178.04 W H = 07 10 36.3 h = normal MB = 4.8(USNOAA) D = 160.3 PKP2V(A):1.2s 20.3nm LmH(B):16s 0.5/um MLH(B)=5.3 LmV(B):16s 0.6/um MLV(B)=5.5
12.	ePKP2 e	A A	07 32 58.5 33 03	South of Kermadec Islands H = 07 12 21 (UPP)
12.	eiP epP esP	A A A	09 44 11 46 08 47 05	E.Russia-N.E. China Border Region 42.75 N 131.04 E H = 09 33 36.6 h = 555 km MB = 5.2 D = 73.07 Az = 324.6 (USNOAA) PV(A):1.8s 132.0nm MPV(A)=5.1 pPV(A):1.7s 60.6nm
12.	ePKP2	A	11 51 04	South of Kermadec Islands 32.3 S 178.0 W H = 11 30 27.0 h = normal MB = 4.7(USNOAA) D = 160.3
12.	eP LmH LmV	A B B	20 18 14.5 20 54.5 56.5	Gulf of California 30.07 N 113.38 W H = 20 05 34.5 h = normal MB = 5.2 D = 86.11 Az = 31.5 (USNOAA) PV(A):1.2s 12.2nm MPV(A)=5.0 LmH(B):20s 0.5/um MLH(B)=4.9 LmV(B):18s 0.5/um MLV(B)=5.0

Day	Phase		h m s	Remarks
13.	eP1	A	00 58 29	Turkey 38.28 N 36.98 E
	eP2	A	58 32	H = 00 53 37.3 h = 34 km MB = 4.6
	LmV	B	01 09.8	D = 21.80 Az = 313 (ISC)
	LmH	B	11.0	P2V(A):1.6s 44.0nm MPV(A)=4.4
13.	ePKP	AB	04 19 12.5	Fiji Islands Region 18.85 S 175.99 E
				H = 03 59 35.9 h = normal MB = 5.3
				D = 145.92 Az = 342.2 (USNOAA)
				PKPV(A):2.3s 158.0nm
13.	ePKP	A	07 13 05.5	Fiji Islands Region 18.78 S 176.00 E
	e	A	13 13.5	H = 06 53 28.6 h = normal MB = 5.1
				D = 145.86 Az = 342.2 (USNOAA)
				PKPV(A):1.3s 50.2nm
13.	ePKP2	A	08 32 40	South of Kermadec Islands
				32.3 S 177.9 W
				H = 08 12 03.4 h = normal MB = 4.8(USNOAA)
				D = 160.4
				PKP2V(A):1.4s 16.3nm
13.	ePKP	A	14 22 02	Fiji Islands Region 18.54 S 176.12 E
	e	A	22 09	H = 14 02 25.8 h = normal MB = 5.1
				D = 145.66 Az = 342.4 (USNOAA)
				PKPV(A):2.0s 68.4nm
13.	ePKIKP	A	19 12 05.5	New Guinea 4.06 S 143.03 E
	e	BC	13 20	H = 18 53 30.0 h = 120 km MB = 5.7
	ePP	A	13 28.5	D = 118.32 Az = 327.2 (USNOAA)
	e	B	14 10	PKIKPV(A):1.2s 20.3nm
	ePKKP	A	22 21	PPV(A):1.4s 48.8nm MPPV(A)=5.3
	LmH	B	56.0	PKKPV(A):2.2s 32.7nm
	LmV	B	20 04.0	LmH(B):19s 1.2 μm
				LmV(B):15s 0.7 μm
13.	eP	A	23 34 16.5	Near Coast of Guatemala 13.76 N 90.05 W
	e	A	34 31	H = 23 21 42.5 h = 100 km MB = 4.8
				D = 86.72 Az = 38.6 (USNOAA)
				PV(A):1.3s 13.1nm MPV(A)=4.8

Day	Phase		h m s	Remarks
14.	e(P)	A	00 45 36.5	India-West Pakistan Border Region
				31.23 N 74.34 E
				H = 00 36 34.5 h = normal MB = 5.2
				D = 49.63 Az = 312.0 (USNOAA)
14.	+iP	ABC	06 06 03.0	Novaya Zemlya 73.32 N 55.15 E
	iPn	B	06 30	H = 05 59 57.1 h = 0 km MB = 6.7
	+iPcP	B	09 10	D = 29.35 Az = 243.4 (USNOAA)
	eiS	B	10 56	PV(A):1.4s 1083.0nm MPV(A)=6.5
	eiSn	B	11 35	PV(B):3.0s 3.1 µm MPV(B)=6.6
	LmH	B	17.8	LmH(B):13s 24.2 µm MLH(B)=6.0
	LmV	B	20.8	LmV(B):8s 15.3 µm MLV(B)=6.1
	e(P'P')	A	39 25	
14.	eP	A	07 26 23.5	Kurile Islands 43.79 N 146.88 E
				H = 07 14 26.2 h = 40 km MB = 4.6
				D = 78.22 Az = 332.8 (USNOAA)
				PV(A):1.3s 13.1nm MPV(A)=4.8
14.	eP	A	07 39 19	Southern Sinkiang Prov., China
	LmH	B	08 02.7	40.93 N 89.43 E
	LmV	B	02.7	H = 07 29 58.6 h = 0 km MB = 4.6
				D = 52.77 Az = 308.6 (USNOAA)
				Atmospheric nuclear explosion (UPP)
				LmH(B):14s 1.3 µm MLH(B)=5.1
				LmV(B):14s 1.6 µm MLV(B)=5.3
14.	eP	A	07 56 46	Mindanao, Philippine Islands
	epP	A	56 57.5	7.03 N 123.84 E
	ePP	A	08 00 56	H = 07 43 12.6 h = 59 km MB = 5.4
				D = 98.35 Az = 323.4 (USNOAA)
				PV(A):1.2s 8.1nm MPV(A)=5.1
14.	ePKP2	A	09 26 35	South of Kermadec Islands
	e	A	26 41	32.62 S 177.48 W
	e	A	26 44	H = 09 05 52.6 h = normal MB = 4.4(USNOAA)
	e	A	26 58	D = 160.8
				PKP2V(A):1.0s 9.8nm

Day	Phase		h m s	Remarks
14.	+ePKIKP	AB	10 59 31	Fiji Islands Region 18.06 S 178.46 W
	+iPKHKP	A	59 33.7	H = 10 40 58.0 h = 609 km MB = 5.3
	eiPKP2	A	59 35.8	D = 146.51 Az = 348.4 (USNOAA)
	e	A	11 05 48	PKIKPV(A):1.4s 34.9nm
				PKHKPV(A):1.2s 97.6nm
14.	+eP	A	14 42 18.5	Southern Nevada 37.07 N 116.01 W
	LmV	C	15 20.5	H = 14 30 00.0 h = 0 km MB = 5.5
				D = 81.26 Az = 30.7 (USNOAA)
				37°04'14.4"N 116°00'18.5"W
				Nevada test site "Tijeras" (USAEC)
				PV(A):1.3s 43.6nm MPV(A)=5.4
				LmV(C):16s 0.2/um MLV(C)=4.5
14.	+iP	A	16 12 35	Kurile Islands Region 43.40 N 148.04 E
	ipP	A	12 48	H = 16 00 34.1 h = 42 km MB = 5.2
	LmH	B	45.0	D = 78.94 Az = 333.4 (USNOAA)
	LmV	B	52.3	h = 48 km
				PV(A):1.2s 38.6nm MPV(A)=5.3
				pPV(A):1.5s 75.4nm
				LmH(B):20s 1.6/um MLH(B)=5.4
				LmV(B):16s 1.6/um MLV(B)=5.5
14.	epP	A	17 02 56	Burma 21.90 N 94.25 E
	e	A	03 05	H = 16 51 37.9 h = 71 km MB = 5.2
				D = 68.79 Az = 317.4 (USNOAA)
				h = 78 km
14.	eP	A	18 18 01	Kurile Islands 43.51 N 147.85 E
	epP	A	18 13.5	H = 18 05 59.9 h = 33 km MB = 5.1
				D = 78.78 Az = 333.3 (USNOAA)
				h = 46 km
				PV(A):1.6s 35.7nm MPV(A)=5.1
				pPV(A):1.4s 65.0nm
14.	+iP	AB	18 27 38.7	Kurile Islands Region 43.52 N 147.97 E
	+ipP	A	27 51.5	H = 18 15 37.3 h = 30 km MB = 5.5
	eS	BC	37 32	D = 78.82 Az = 333.4 (USNOAA)
	eSS	BC	42 48	h = 47 km

Day	Phase		h m s	Remarks
cont. 14.	LmH	B	19 07.4	PV(A):1.3s 69.8nm MPV(A)=5.5
	LmV	B	07.5	PV(B):10s 2.2/um MPV(B)=6.2
				pPV(A):2.0s 265.0nm
				SH(B):13s 3.2/um MSH(B)=6.2
				LmH(B):15s 17.8/um MLH(B)=6.5
				LmV(B):16s 21.7/um MLV(B)=6.6
14.	eP	A	18 38 36.5	<u>Kurile Islands</u> 43.56 N 147.68 E
	ePcP	A	38 46	H = 18 26 37.3 h = 45 km MB = 4.9
	epP	A	38 49	D = 78.68 Az = 333.2 (USNOAA)
				h = 46 km
14.	eP	A	18 47 16	<u>Kurile Islands Region</u> 43.44 N 148.05 E
				H = 18 35 16.2 h = 52 km MB = 4.8
				D = 78.91 Az = 333.4 (USNOAA)
				PV(A):1.5s 20.1nm MPV(A)=4.9
14.	+iP	AB	21 25 59.3	<u>Kurile Islands</u> 43.49 N 146.98 E
	eS	BC	35 50	H = 21 14 01.0 h = 41 km MB = 5.4
	eSS	C	40 45	D = 78.51 Az = 332.8 (USNOAA)
	LmV	B	22 05.5	PV(A):1.1s 52.4nm MPV(A)=5.4
	LmH	B	05.6	PV(B):8s 1.2/um MPV(B)=5.9
				SH(B):8s 1.2/um MSH(B)=6.0
				LmV(B):17s 6.3/um MLV(B)=6.1
				LmH(B):14s 5.7/um MLH(B)=6.1
15.	eP	A	00 38 27.5	<u>Kurile Islands</u> 43.88 N 147.80 E
	epP	A	38 40	H = 00 26 28.2 h = normal MB = 4.7
				D = 78.44 Az = 333.3 (USNOAA)
				h = 46 km
15.	eP	A	01 56 08.5	<u>Kurile Islands</u> 43.70 N 147.80 E
	epP	A	56 20	H = 01 44 07.8 h = normal MB = 4.6
				D = 78.60 Az = 333.3 (USNOAA)
				h = 43 km
15.	eP	A	04 03 38.5	<u>Southern Sinkiang Prov., China</u>
				39.76 N 77.23 E

Day	Phase		h m s	Remarks
cont.				
15.	LmH	B	04 24.3	H = 03 55 16.1 h = normal MB = 4.9
	LmV	B	24.3	D = 46.06 Az = 306.4 (USNOAA)
				PV(A):1.0s 5.9nm MPV(A)=4.6
				LmH(B):12s 0.6 µm MLH(B)=4.7
				LmV(B):12s 0.8 µm MLV(B)=5.0
15.	LmH	B	05 11.4	<u>Southern Sinkiang Prov., China</u>
	LmV	B	11.4	39.8 N 72.2 E
				H = 04 42 19.0 h = normal MB = 4.6(USNOAA)
				D = 46.1
				LmH(B):12s 0.3 µm MLH(B)=4.5
				LmV(B):13s 0.5 µm MLV(B)=4.7
				P is superimposed by calibration pulses
15.	ePKIKP	A	12 32 00	<u>Fiji Islands Region</u> 17.61 S 178.84 W
	+iPKHKP	A	32 01.8	H = 12 13 23.6 h = 564 km MB = 5.1
				D = 145.99 Az = 348.1 (USNOAA)
				PKIKPV(A):1.2s 14.2nm
				PKHKPV(A):1.4s 65.1nm
15.	eP	A	15 37 21.5	<u>Hokkaido, Japan Region</u> 42.00 N 144.29 E
	LmH	B	16 16.0	H = 15 25 22.3 h = 56 km MB = 4.7
	LmV	B	16.0	D = 78.90 Az = 331.5 (USNOAA)
				PV(A):1.5s 20.1nm MPV(A)=4.9
				LmH(B):16s 0.6 µm MLH(B)=5.0
				LmV(B):17.5s 0.8 µm MLV(B)=5.1
16.	eiP1	ABC	05 38 21	<u>Honshu, Japan</u> 39.26 N 140.70 E
	eP2	A	38 24	H = 05 26 13.3 h = 24 km MB = 5.9
	eP3	A	38 26.5	D = 79.95 Az = 329.9 (USNOAA)
	ePP	B	41 28	P1V(A):1.4s 65.0nm MP1V(A)=5.5
	ePP	C	41 30	P1V(B):9s 1.9 µm MP1V(B)=6.1
	eS	C	48 22	P2V(A):1.6s 148.5nm MP2V(A)=5.8
	LmH	B	06 11.4	P3V(A):2.2s 392.0nm MP3V(A)=6.1
	LmV	B	20.5	PPV(B):10s 0.9 µm MPPV(B)=5.9
				LmH(B):18s 13.4 µm MLH(B)=6.3
				LmV(B):14.5s 8.1 µm MLV(B)=6.2

Day	Phase		h m s	Remarks
16.	eP	A	10 55 56.5	Honshu, Japan 39.32 N 140.72 E H = 10 43 49.0 h = 25 km MB = 4.7 D = 79.91 Az = 329.9 (USNOAA)
16.	ePKP	A	12 38 50	New Hebrides Islands 19.18 S 168.83 E H = 12 19 19.9 h = 33 km MB = 4.5 D = 143.77 Az = 335.3 (USNOAA) PKPV(A):0.7s 9.6nm
16.	eP	A	16 23 38	Fox Islands, Aleutian Is. 53.01 N 167.66 W H = 16 11 50.5 h = 45 km MB = 4.2 D = 76.72 Az = 0.5 (USNOAA)
17.	eP	A	01 54 53	Turkey 40.47 N 35.83 E
	LmH	B	02 03.5	H = 01 50 23.9 h = normal MB = 4.2
	LmV	B	05.5	D = 19.68 Az = 309.2 (USNOAA)
17.	eP	A	03 23 21.5	Taiwan Region 23.99 N 122.05 E
	LmH	B	04 05.0	H = 03 10 54.8 h = 51 km MB = 4.8
	LmV	B	05.0	D = 83.78 Az = 323.1 (USNOAA) PV(A):1.2s 10.2nm MPV(A)=4.7 LmH(B):16s 0.8 µm MLH(B)=5.2 LmV(B):16s 0.8 µm MLV(B)=5.2
17.	eP	A	05 41 39.5	Kirgiz-Sinkiang Border Region 41.43 N 79.24 E H = 05 33 15.2 h = normal MB = 5.0 D = 46.33 Az = 305.5 (USNOAA) PV(A):1.0s 15.7nm MPV(A)=5.0
17.	eP	A	13 51 50.5	Kurile Islands 43.71 N 147.59 E H = 13 39 50.4 h = normal MB = 4.7 D = 78.52 Az = 333.2 (USNOAA)
18.	ePKIKP	A	01 26 07.5	South of Fiji Islands 25.69 S 178.59 E
	ePKHKP	A	26 15.5	H = 01 07 21.0 h = 572 km MB = 4.9
	ePKP2	A	26 30	D = 153.14 Az = 341.5 (USNOAA)

Day	Phase		h m s	Remarks
cont. 18.	epPKHKP	A	01 28 20	
	epPKP2	A	28 36	
18.	eP	A	06 18 14.5	Southern Iran 27.34 N 55.05 E
	LmH	B	37.0	H = 06 10 39.1 h = 40 km MB = 4.8
	LmV	B	37.0	D = 40.22 Az = 317.3 (USNOAA)
				LmH(B):16s 0.6 /um MLH(B)=4.5
				LmV(B):16s 0.7 /um MLV(B)=4.7
18.	-iPKP	A	09 39 11.5	Samoa Islands Region 16.74 S 172.09 W
	ei	A	39 17.5	H = 09 19 33.2 h = 35 km MB = 5.5
	LmH	B	10 58.5	D = 146.04 Az = 355.8 (USNOAA)
	LmV	B	11 00.0	PKPV(A):1.3s 129.0nm
18.	e(PKHKP)	A	16 33 56.5	South of Fiji Islands 22.88 S 176.01 W
	e	A	34 02	H = 16 14 07.0 h = 30 km MB = 5.0
	e(PKP2)	A	34 07.5	D = 151.65 Az = 349.8 (USNOAA)
18.	ePKP2	A	18 49 08.5	South of Kermadec Islands
				33.3 S 179.7 W
				H = 18 28 40.6 h = 115 km MB = 5.1(USNOAA)
				D = 160.7
				PKP2V(A):1.4s 18.6nm
18.	ePKIKP	A	20 51 09	New Britain Region 5.11 S 152.09 E
				H = 20 32 17.0 h = 68 km MB = 5.4
				D = 123.88 Az = 330.8 (USNOAA)
				PKIKPV(A):1.2s 20.3nm
18.	eP	A	21 32 16.5	Andaman Islands Region 10.79 N 93.51 E
	e	A	32 21	H = 21 20 35.7 h = 95 km MB = 5.3
				D = 76.61 Az = 319.6 (USNOAA)
19.	ePKHKP	A	05 22 10	Tonga Islands Region 22.86 S 174.84 W
	ePKP2	A	22 20	H = 05 02 15.5 h = normal MB = 5.1
				D = 151.81 Az = 351.3 (USNOAA)
				PKHKPV(A):1.7s 42.4nm

Day	Phase		h m s	Remarks
19.	eP	A	05 48 49.5	West Pakistan 26.87 N 66.96 E H = 05 40 12.2 h = normal MB = 5.4 D = 47.93 Az = 315.1 (USNOAA) PV(A):0.9s 19.5nm MPV(A)=5.2
19.	eP epP	A A	08 20 38.5 20 50.5	Kurile Islands 43.34 N 147.53 E H = 08 08 38.6 h = 47 km MB = 4.9 D = 78.83 Az = 333.2 (USNOAA) h = 45 km PV(A):1.3s 28.4nm MPV(A)=5.1
19.	eP	A	10 14 04.5	Southern Sinkiang Province, China 38.6 N 75.6 E H = 10 05 47 MB = 4.5 (ANUSSR) D = 45.7
19.	eP epP esP	A A A	18 23 07.5 23 23 23 31	South of Honshu, Japan 31.98 N 140.73 E H = 18 10 29.3 h = 63 km MB = 5.2 D = 86.27 Az = 330.2 (USNOAA) h = 67 km
20.	ePKHKP ePKP2	A A	00 40 27.5 40 32.5	Tonga Islands 18.36 S 173.29 W H = 00 20 43.6 h = normal MB = 4.8 D = 147.55 Az = 354.2 (USNOAA) PKHKPV(A):1.5s 32.7nm
20.	ePKIKP iX	A A	08 44 54 45 02	New Hebrides Islands 15.37 S 167.41 E H = 08 25 45.1 h = 112 km MB = 5.5 D = 139.76 Az = 336.2 (USNOAA) XV(A):1.4s 60.4nm
20.	ePn e eSg	A A A	13 47 37.5 49 13 49 39.5	Yugoslavia 44.82 N 17.23 E H = 13 45 55.0 h = 17 km MB = 4.4 D = 6.95 Az = 329.0 (USNOAA) PnV(A):0.7s 9.6nm
20.	eiPn iSn	A A	20 21 08.5 22 34.5	Yugoslavia 44.84 N 17.33 E H = 20 19 23.9 h = 8 km MB = 5.0

Day	Phase		h m s	Remarks
cont. 20.	eiSg	A	20 23 10	D = 6.97 Az = 328.5 (USNOAA) PnV(A):1.0s 47.2nm
20.	eP	A	23 34 53.5	Greenland Sea 74.74 N 9.13 E H = 23 29 38.8 h = normal MB = 4.5 D = 24.20 Az = 176.1 (USNOAA) PV(A):1.5s 40.2nm MPV(A)=4.8
21.	eP	A	01 26 59.5	Near East Coast of Honshu, Japan
	epP	A	27 12	40.01 N 142.86 E H = 01 14 52.4 h = 48 km MB = 4.7 D = 80.13 Az = 330.9 (USNOAA) h = 46 km pPV(A):1.2s 20.3nm
21.	-eiP	AB	08 19 27.5	Greenland Sea 74.62 N 8.41 E H = 08 14 14.1 h = normal MB = 5.5(USNOAA) D = 24.13 PV(A):1.8s 872.0nm MPV(A)=6.1 PV(B):2.0s 1.4/um MPV(B)=6.2
21.	eP	A	12 47 42.5	Andreanof Islands, Aleutian Is. 51.44 N 174.25 W H = 12 35 43.7 h = 21 km MB = 5.1 D = 78.17 Az = 356.2 (USNOAA) PV(A):1.0s 19.7nm MPV(A)=5.1
21.	eP	ABC	16 00 04	Central Mid-Atlantic Ridge 7.66 N 37.60 W
	eS	BC	08 10	H = 15 50 05.5 h = normal MB = 5.3
	LmV	B	25.5	D = 59.05 Az = 34.2 (USNOAA)
	LmH	B	25.6	PV(A):1.6s 55.0nm MPV(A)=5.4 LmV(B):17s 1.8/um MLV(B)=5.4 LmH(B):17.5s 2.3/um MLH(B)=5.5
22.	ePKHKP	A	01 32 03	Tonga Islands 19.04 S 172.96 W
	ePKP2	A	32 06	H = 01 12 18.2 h = normal MB = 5.1 D = 148.25 Az = 354.5 (USNOAA)

Day	Phase		h m s	Remarks
22.	eP	A	03 05 37.5	Kurile Islands 46.4 N 151.8 E H = 02 53 42.9 h = 20 km MB = 4.7 (ISC) D = 77.5
22.	+iP	A	03 20 44.5	Fox Islands, Aleutian Is. 54.18 N 165.31 W H = 03 09 05.2 h = 65 km MB = 4.8 D = 75.52 Az = 2.0 (USNOAA) PV(A):0.6s 21.1nm MPV(A)=5.2
22.	ePKIKP ePP LmV LmH	AB B B B	06 33 06 35 07 07 28.4 28.5	West Chile Rise 36.48 S 97.24 W H = 06 14 00.2 h = normal MB = 5.3 D = 128.49 Az = 50.3 (USNOAA) PKPV(A):1.6s 16.5nm LmV(B):18s 1.6/um MLV(B)=5.7 LmV(B):17s 1.3/um MLH(B)=5.7
22.	ePKP	A	20 07 25	New Hebrides Islands 19.00 S 169.32 E H = 19 48 21.9 h = 255 km MB = 4.6 D = 143.80 Az = 335.9 (USNOAA) PKPV(A):0.9s 9.7nm
23.	ePKIKP LmH LmV	A B B	11 20 34 12 15.9 19.0	West Chile Rise 36.52 S 97.21 W H = 11 01 28.4 h = normal MB = 5.5 D = 128.49 Az = 50.3 (USNOAA) PKIKPV(A):1.6s 16.5nm LmH(B):17s 0.8/um MLH(B)=5.4 LmV(B):17s 0.9/um MLV(B)=5.5
23.	iPg e(Sg)	A A	15 05 24.7 05 41.5	Bransrode / Federal Rep. Germany 51°14.0'N 9°51.6'E explosion, H = 15 05 00.63, yield 20250 kg (Hannover) D = 1.2
23.	-iP e e	AB A A	24 06 06.5 07 07 07 50	Sea of Okhotsk 48.05 N 145.49 E H = 23 55 20.1 h = 479 km MB = 5.1(USNOAA) D = 74.0 PV(A):1.7s 18.2nm MPV(A)=5.4

Day	Phase		h m s	Remarks
24.	eP	A	00 21 14.5	Alaska Peninsula 56.07 N 158.09 W
	e	A	21 21.5	H = 00 09 46.3 h = normal MB = 4.6
	e	A	21 40	D = 73.32 Az = 6.8 (USNOAA)
24.	eiP	A	15 00 14.5	Dodecanese Islands 35.75 N 27.93 E
				H = 14 55 52.7 h = normal MB = 4.4
				D = 18.99 Az = 326.7 (USNOAA)
				PV(A):1.0s 27.6nm MPV(A)=4.4
24.	eP	A	19 38 27.5	Turkey 37.03 N 29.08 E
				H = 19 34 10.9 h = 31 km MB = 4.1
				D = 18.48 Az = 322.9 (USNOAA)
				PV(A):(1.5)s 15.1nm MPV(A)=4.0
25.	e	A	05 26 37	Nicobar Islands Region 9.06 N 94.02 E
				H = 05 14 32.6 h = normal MB = 4.9
				D = 78.24 Az = 319.9 (USNOAA)
25.	eP	A	07 56 19	Nicobar Islands Region 8.97 N 93.98 E
	e	A	56 29	H = 07 44 23.7 h = normal MB = 5.1
				D = 78.29 Az = 319.9 (USNOAA)
25.	eP	A	10 17 24.5	Nicobar Islands Region 9.02 N 94.02 E
	e	A	17 32	H = 10 05 27.2 h = normal MB = 4.8
	e	A	17 49	D = 78.28 Az = 319.9 (USNOAA)
25.	eP	A	10 27 13	Nicobar Islands Region 9.11 N 94.04 E
	e	A	27 28.5	H = 10 15 15.2 h = normal MB = 5.0
				D = 78.22 Az = 319.9 (USNOAA)
				PV(A):0.8s 9.6nm MPV(A)=4.9
25.	eP	A	11 28 07	Iran-Iraq Border Region
	LmH	C	38.6	36.77 N 45.13 E
				H = 11 22 18.2 h = 19 km MB = 5.5
				D = 27.64 Az = 310.7 (USNOAA)
				PV(A):1.5s 42.7nm MPV(A)=4.9
				LmH(C):25s 2.1µm MLH(C)=4.6

Day	Phase		h m s	Remarks
25.	eP	A	12 12 45	Mid-Indian Rise 13.69 S 66.26 E
	eS	C	22 43	H = 12 00 35.2 h = 24 km MB = 5.8(USNOAA)
	eSS	C	27 28	D = 79.8
	eSSS	C	31 25	PV(A):1.5s 75.4nm MPV(A)=5.5
	LmH	C	36.0	LmH(C):60s 4.7/um MLH(C)=5.4
	LmV	C	36.8	LmV(C):48s 2.8/um MLV(C)=5.3
25.	eiP1	ABC	15 21 47	Nicobar Islands Region 9.02 N 93.87 E
	eP2	A	21 53	H = 15 09 49.4 h = normal MB = 5.5
	eP3	A	21 57.5	D = 78.18 Az = 319.9 (USNOAA)
	ePP	C	24 47	P1V(A):1.6s 94.0nm MP1V(A)=5.5
	eS1	C	31 36	P1V(B):8s 1.9/um MP1V(B)=6.2
	iS2	C	31 42	P2V(A):2.0s 196.3nm MP2V(A)=5.9
	eSS	C	36 51	P3V(A):1.7s 254.0nm MP3V(A)=6.0
	eSSS	C	40 00	LmH(B):18s 10.2/um MLH(B) =6.2
	LmH	B	16 01.7	LmV(B):17s 10.0/um MLV(B) =6.3
	LmV	B	03.6	
25.	eP	A	15 33 52	Nicobar Islands Region 8.97 N 94.00 E
	e	A	33 55	H = 15 21 49.2 h = normal MB = 5.1(USNOAA)
	e	A	34 05	D = 78.3
25.	eP	A	15 54 31	Nicobar Islands Region 8.98 N 94.08 E
				H = 15 42 31.2 h = normal MB = 4.9
				D = 78.34 Az = 319.9 (USNOAA)
25.	eP	A	22 20 13.5	Nicobar Islands Region 9.15 N 94.08 E
				H = 22 08 16.1 h = normal MB = 4.9
				D = 78.21 Az = 319.9 (USNOAA)
				PV(A):1.7s 24.2nm MPV(A)=5.0
25.	eP	A	22 32 46	Nicobar Islands Region 9.12 N 94.36 E
				H = 22 20 46.8 h = 31 km MB = 4.9
				D = 78.41 Az = 319.9 (USNOAA)
26.	ePKIKP	A	08 31 32.5	Fiji Islands Region 18.17 S 177.87 W
	+iPKHKP	A	31 33.6	H = 08 12 57.7 h = 609 km MB = 5.1
	e	A	32 53.5	D = 146.72 Az = 349.0 (USNOAA)
	e	A	34 11	PKHKPV(A):0.7s 34.5nm

Day	Phase		h m s	Remarks
26.	+iP	ABC	20 59 34.7	Greenland Sea 79.82 N 2.67 E
	Pm	A	59 55	H = 20 53 32.4 h = 32 km MB = 5.6
	e	B	21 01 20	D = 29.46 Az = 168.4 (USNOAA)
	eS	BC	04 20	P1V(A):1.1s 36.3nm MP1V(A)=5.0
	e	C	07 26	P1V(B):6s 1.6/um MP1V(B)=6.0
	eX	A	11 10	PmV(A):1.2s 183.0nm MPmV(A)=5.6
	LmH	B	13.5	SH(B):12s 3.3/um MSH(B) =5.7
	LmV	B	13.5	XV(A):1.6s 30.2nm
				LmH(B):14.5s 6.3/um MLH(B)=5.4
				LmV(B):13s 7.1/um MLV(B)=5.6
28.	ePn	A	01 15 56	Austria 47.46 N 11.17 E
	ePg	A	16 11	H = 01 15 05.0 h = 0 km
	eSn	A	16 35	D = 3.20 Az = 5 (ISC)
	eSg	A	16 51	PgV(A):1.0s, 11.8nm
28.	e	A	05 32 35	Upper Silesia / Poland
				rockburst (CLL)
28.	ePKP	A	22 46 36	Fiji Islands Region 16.39 S 177.50 W
	e	A	47 16.5	H = 22 27 01.0 h = normal MB = 4.9
				D = 145.05 Az = 349.9 (USNOAA)
29.	eiP	A	01 09 26.5	Unimak Island Region 54.17 N 164.63 W
				H = 00 57 42.8 h = 27 km MB = 5.3
				D = 75.51 Az = 2.5 (USNOAA)
				PV(A):1.0s 31.4nm MPV(A)=5.3
29.	eP	A	03 01 06.5	North Atlantic Ridge 50.83 N 29.47 W
				H = 02 55 36.8 h = normal MB = 4.5
				D = 25.77 Az = 74.2 (USNOAA)
29.	LmH	B	03 39.0	Probably Mid-Indian Rise (USNOAA)
	LmV	B	41.5	LmH(B):17s 1.6/um
				LmV(B):18s 1.3/um

Day	Phase		h m s	Remarks
29.	+eiP	A	09 23 01.5	Mid-Indian Rise 15.45 S 67.21 E
	+eiX	A	23 07.5	H = 09 10 44.0 h = normal MB = 5.4
				D = 81.80 Az = 327.9 (USNOAA)
				PV(A):1.8s 67.5nm MPV(A)=5.4
				XV(A):1.6s 57.6nm
29.	+eP	A	19 42 29.5	Kurile Islands 44.56 N 149.03 E
				H = 19 30 35.8 h = 67 km MB = 5.2
				D = 78.23 Az = 333.9 (USNOAA)
				PV(A):1.1s 42.3nm MPV(A)=4.8
29.	+iPg	A	22 47 38.5	D c. 80 km
	i	A	47 48	
	iSg	A	47 49	
29.	e(pP)	A	23 27 06.5	Honshu, Japan 35.97 N 139.93 E
				H = 23 14 42.9 h = 62 km MB = 4.9
				D = 82.49 Az = 329.7 (USNOAA)
				(pP)V(A):1.3s 10.9nm
30.	ePKHKP	A	05 34 24	Tonga Islands 18.56 S 174.73 W
				H = 05 14 52.5 h = 115 km MB = 4.5
				D = 147.59 Az = 352.5 (USNOAA)
				PKHKPV(A):1.2s 10.2nm
30.	ePKHKP	A	11 12 34	Fiji Islands Region 19.84 S 177.52 W
	ePKP2	A	12 38	H = 10 53 34.9 h = 421 km MB = 4.6
				D = 148.42 Az = 348.9 (USNOAA)
30.	eP	A	23 54 44.5	Greece-Albania Border Region
				39.95 N 20.5 E
				H = 23 51 47 h = 3 km
				D = 12.38 Az = 333 (ISC)
31.	e(P)	A	02 54 12.5	Republic of South Africa
				26.85 S 26.88 E
				H = 02 42 09.9 h = normal MB = 5.3
				D = 78.32 Az = 350.1 (USNOAA)
				(P)V(A):0.8s 11.5nm

Day	Phase		h m s	Remarks
31.	eP	A	09 20 33	Southern Sumatra 4.34 S 102.76 E
				H = 09 07 13.7 h = 13 km MB = 5.5
				D = 94.04 Az = 320.3 (USNOAA)
31.	eP	A	15 07 45	Kodiak Island Region 56.54 N 152.99 W
				H = 14 56 20.3 h = 44 km MB = 4.6
				D = 72.42 Az = 10.2 (USNOAA)
31.	eP1	A	16 10 03	Yugoslavia 42.13 N 19.32 E
	eiP2	A	10 07.5	H = 16 07 37.9 h = 23 km MB = 4.9
	eSn	A	12 11	D = 10.04 Az = 330.7 (USNOAA)
	LmH	B	15.3	P2V(A):1.3s 26.2nm
	LmV	B	15.4	LmH(B):15s 0.6 μm MLH(B)=3.6
				LmV(B):16s 1.0 μm
31.	eiP	A	16 20 37.5	Andreanof Islands, Aleutian Is.
				51.21 N 179.38 W
				H = 16 08 41.7 h = 39 km MB = 5.0
				D = 78.04 Az = 352.9 (USNOAA)
				PV(A):1.3s 26.2nm MPV(A)=5.1
31.	eP diff	C	18 08 35	Near North Coast of New Guinea
	eP diff	AB	08 38	4.93 S 145.47 E
	ePKIKP	A	11 57	H = 17 53 09.3 h = 42 km MB = 6.0
	e	A	12 10	D = 120.35 Az = 327.9 (USNOAA)
	e	B	12 56	P diff V(C):12s 0.5 μm
	e	B	13 20	LmH(B):22.5s 109.2 μm MLH(B)=7.4
	iPP	BC	13 31	LmV(B):22s 66.7 μm MLV(B)=7.2
	ePPP	B	15 55	
	eSKKS	BC	20 15	
	ei	BC	20 20	
	eS diff	B	21 08	
	ePS	C	23 00	
	eSS	C	29 44	
	LmH	B	55.9	
	LmV	B	19 03.4	
31.	eSg	A	18 26 28	Adriatic Sea 41.8 N 18.7 E
				H = 18 20 54 h = 0 km
				D = 10.2

Day	Phase		h m s	Remarks
1.	LmH	B	05 24.3	Gulf of California 23.14 N 107.92 W
	LmV	B	24.3	H = 04 31 08.5 h = 57 km MB = 4.9 (ISC)
				D = 89.3
				LmH:16s 2.6 µm MLH=5.8
				LmV:16s 2.9 µm MLV=5.8
1.	LmV	B	12 19.0	Near North Coast of New Guinea
	LmH	B	21.1	4.82 S 145.73 E
				H = 11 07 40.7 h = normal MB = 5.5
				D = 120.40 Az = 328.0 (USNOAA)
				LmH:20s 1.3 µm MLH=5.6
				LmV:20s 1.8 µm MLV=5.7
1.	eP	A	20 33 38	Honshu, Japan 39.62 N 141.12 E
				H = 20 21 41.4 h = 124 km MB = 4.8
				D = 79.81 Az = 330.1 (USNOAA)
				traces
1.	eP	A	22 43 48	Central Mid-Atlantic Ridge
	e	A	43 57.3	8.40 N 39.49 W
				H = 22 33 44.4 h = normal MB = 4.7
				D = 59.51 Az = 35.1 (USNOAA)
2.	ePn	A	02 12 50.5	Austria 47.9 N 14.3 E (VIE)
				D = 3.2
2.	eP	A	03 26 38	Central Italy 43.71 N 12.10 E
				H = 03 24 55.6 h = 0 km (ISC)
				D = 7.0
2.	ePn	A	08 43 33	Northern Italy 45.1 N 11.1 E
	ePg.	A	44 04.5	H = 08 42 12 (BCIS)
	eSn	A	44 35.5	D = 5.57 (ISC)
	eSg	A	45 10	
2.	ePKP	A	10 33 14	Fiji Islands Region 15.49 S 176.16 W
	e	A	33 23.5	H = 10 13 36.3 h = 44 km MB = 5.4
	LmH	B	11 34.2	D = 144.38 Az = 351.5 (USNOAA)

Day	Phase		h m s	Remarks
cont. 2.	LmV	B	11 37	PKPV(A):1.6s 16.5nm
				LmH(B):22s 2.6 µm MLH(B)=5.9
				LmV(B):20s 2.2 µm
2.	ePn	A	12 14 00	**Yugoslavia** 44.87 N 17.09 E
	eSn	A	15 22	H = 12 12 16.5 h = 0 km
	e	A	16 01	D = 6.86 Az = 329 (ISC)
				PnV(A):0.9s 15.6nm
3.	+iP	A	02 40 57.5	**Central Alaska** 62.03 N 151.17 W
	eipP	A	41 24.5	H = 02 30 11.4 h = 70 km MB = 5.6
	ei(sP)	A	41 34	D = 66.84 Az = 11.8 (USNOAA)
				h_{pP} = 110 km h_{sP} = 100 km
				PV(A):0.8s 77.0nm MPV(A)=5.7
3.	e(Pn)	A	08 47 23	**Belgium** 50.4 N 4.5 E
	eSn	A	48 11	H = 08 45 58 h = 0 km
	eSg	A	48 20	D = 4.73 Az = 84 (ISC)
3.	ePKP	A	09 42 42	**Easter Island Cordillera**
	ePKP2	A	42 47	49.99 S 114.42 W
	e	A	42 51.5	H = 09 23 04.5 h = normal MB = 5.2
	LmH	B	11 05	D = 146.17 Az = 67.6 (USNOAA)
	LmV	B	07	PKPV(A):0.8s 21.2nm
3.	ePKP	A	10 05 37.5	**Eastern Island Cordillera**
	epPKP	A	05 45	49.90 S 115.03 W
	epPKP2	A	05 50.5	H = 09 46 00.1 h = normal MB = 4.7
				D = 146.51 Az = 67.8 (USNOAA)
3.	ePn	A	15 14 35	**Greece** 38.6 N 20.3 E
				H = 15 11 26.7 h = 0 km
				D = 13.42 Az = 334 (ISC)
3.	eP	A	15 24 57	**Luzon, Philippine Islands**
	epP	A	25 08	18.43 N 120.88 E
	LmH	B	16 07	H = 15 12 11.6 h = 41 km MB = 5.5
	LmV	B	07	D = 87.54 Az = 323.0 (USNOAA)
				PV(A):2.0s 85.5nm MPV(A)=5.6

Day	Phase		h m s	Remarks
4.	+iP	A	06 10 43.8	Eastern Kazakh SSR 49.97 N 77.79 E
	ePn	A	12 15	H = 06 02 57.0 h = 0 km MB = 5.4
	ePP	A	12 19.5	D = 40.99 Az = 297.4 (USNOAA)
	ePcP	A	12 39.5	PV(A):0.8s 46.2nm MPV(A)=5.3
				Underground explosion MB = 6.1 (UPP)
4.	+iPKIKP	A	18 02 43.2	New Hebrides Islands 20.04 S 169.25 E
	+ipPKIKP	A	02 57.5	H = 17 43 11.1 h = 44 km MB = 5.2
				D = 144.71 Az = 335.2 (USNOAA)
				PKIKP(A):1.0s 112.0nm
				pPKIKP(A):0.9s 42.8nm
5.	eP	A	07 34 44	Andreanof Islands, Aleutian Is.
				51.47 N 179.27 W
				H = 07 22 48.7 h = 36 km MB = 4.7
				D = 77.84 Az = 352.9 (USNOAA)
5.	ePKP	A	12 27 35.5	Loyalty Islands Region 21.82 S 170.57 E
				H = 12 08 02.6 h = 91 km
				D = 146.84 Az = 335.3 (USNOAA)
				PKPV(A):1.2s 12.2nm
5.	eP	A	13 24 39	South of Panama 6.90 N 82.61 W
	e	A	24 54	H = 13 11 53.5 h = normal MB = 5.6
	e	A	25 03	D = 87.39 Az = 39.5 (USNOAA)
	eS	B	35 23	PV(A):1.2s 40.6nm MPV(A)=5.6
	ePS	B	36 23	SH(B):10s 1.0/um MSH(B)=5.8
	eSS	B	41 18	LmH:22s 1.8/um MLH =5.4
	LmV	B	56.8	LmV:22s 1.8/um MLV =5.5
	LmH	B	57	
5.	eP	A	20 43 36	North Atlantic Ridge 34.72 N 36.98 W
	esP	A	43 50	H = 20 36 11.9 h = normal MB = 5.1(USNOAA)
	LmV	B	57	D = 38.3
	LmH	B	58	LmH:18s 0.4/um MLH=4.3
				LmV:18s 1.0/um MLV=4.8

Day	Phase		h m s	Remarks
6.	eP	A	07 20 47	**Iceland Region** 63.85 N 22.73 W
	LmH	B	29.9	H = 07 15 43.8 h = 8 km MB = 4.5
	LmV	B	32.5	D = 22.45 Az = 109.8 (USNOAA)
				LmH:16s 1.2 μm MLH=4.4
				LmV:11s 1.6 μm MLV=4.8
6.	eP	A	11 30 24.5	**Iceland Region** 63.79 N 23.27 W
	LmH	B	40.2	H = 11 25 24.9 h = normal MB = 4.3
	LmV	B	41.1	D = 22.65 Az = 109.0 (USNOAA)
				PV(A):0.6s 9.6nm MPV(A)=4.4
				LmH:15s 1.0 μm MLH =4.4
				LmV:14s 1.4 μm MLV =4.6
7.	eP	A	03 35 11.5	**Kurile Islands** 48.96 N 155.20 E
				H = 03 23 26.6 h = normal MB = 4.7
				D = 75.98 Az = 337.1 (USNOAA)
				PV(A):1.1s 20.2nm MPV(A)=5.2
7.	e	A	04 36 56	**Albania** 41.8 N 19.5 E
				H = 04 34 17 (ATH)
				D = 10.45 traces
7.	ePKP2	A	08 05 24	**South of Kermadec Islands**
				34.63 S 179.71 E
				H = 07 44 57.0 h = 76 km MB = 5.1(USNOAA)
				D = 161.8
				PKP2V(A):1.9s 30.3nm
7.	eiP	A	23 31 07	**Luzon, Philippine Islands**
	e	A	31 18.5	18.51 N 120.88 E
	epP	A	31 26	H = 23 18 24.0 h = 55 km MB = 5.4
	eSKKS	C	41(30)	D = 87.48 Az = 323.0 (USNOAA)
	eS	B	41 40	PV(A):3.0s 64.1nm MPV(A)=5.5
	eSP	B	42 46	LmH:21s 1.7 μm MLH =5.4
	eSS	B	47.5	LmV:18s 1.5 μm MLV =5.5
	LmV	B	24 13.7	
	LmH	B	13.8	

Day	Phase		h m s	Remarks
8.	eP	A	00 14 43.5	<u>Eastern Caucasus</u> 41.30 N 48.92 E
				H = 00 09 03.4 h = 87 km MB = 4.7
				D = ca. 27 traces
8.	ePKHKP	A	04 54 39	<u>Tonga Islands</u> 19.75 S 174.47 W
				H = 04 34 52.5 h = normal MB = 5.0(USNOAA)
				D = 148.7
8.	eP	A	09 26 38	<u>Szechwan Province, China</u> 32.21 N 101.31 E
	LmH	B	52.0	H = 09 15 55.6 h = 38 km MB = 5.1
	LmV	B	56.8	D = 65.68 Az = 315.7 (USNOAA)
				PV(A):1.5s 22.6nm MPV(A)=5.0
				LmH:20s 2.1/um MLH =5.4
				LmV:14s 1.7/um MLV =5.4
8.	eP	ABC	15 12 30	<u>Mindanao, Philippine Islands</u>
	eSKS	BC	23 12	9.13 N 126.33 E
	eS	C	23 56	H = 14 58 53.6 h = 22 km MB = 5.7
	eSP	BC	25 20	D = 98.13 Az = 324.3 (USNOAA)
	eSS	C	30 45	PV(A):1.8s 47.3nm MPV(A)=5.7
	LmH	B	16 01.5	PV(B):11s 0.65/um MPV(B)=6.1
	LmV	B	09.1	LmH:17s 4.4/um MLH =6.0
				LmV:16s 4.0/um MLV =6.0
8.	eP diff	A	22 50 34.5	<u>West New Guinea Region</u> 3.44 S 135.63 E
	ePKIKP	AB	54 23.5	H = 22 35 46.7 h = normal MB = 6.2
	ePP	ABC	55 18	D = 113.67 Az = 324.8 (USNOAA)
	eSKS	BC	23 01 18	P diff V(A):2.0s 47.0nm
	eSKKS	B	02 20	PKIKPV(A):1.7s 36.4nm
	eiS diff	BC	02 55	PPV(B):12s 3.63/um MPPV(B)=6.8
	ePS	BC	04 40	PPH(B):12s 1.98/um MPPH(B)=6.9
	ePKKP2	A	05 20	PKKP2(A):3.4s 410.0nm
	ePPS	BC	06 00	LmH:20.5s 36.7/um MLH =7.0
	iSS	BC	11 00	LmV:18s 25.5/um MLV =6.9
	eiSSS	BC	14 50	e(A) 55 25 e(A) 55 31
	LmH	B	38.3	
	LmV	B	43.0	

Day	Phase		h m s	Remarks
9.	eiP	A	17 49 06	Southern Iran 29.52 N 56.85 E
	ipP	A	49 30	H = 17 41 42.2 h = 106 km MB = 5.5
	esP	A	49 42	D = 39.76 Az = 315.0 (USNOAA)
	ePP	A	50 04	h_{pP} = 115 km h_{sP} = 110 km
	eScP	A	54 49.5	PV(A):1.0s 25.6nm MPV(A)=5.0
10.	eP1	A	00 38 07.5	Southern Honshu, Japan 34.60 N 136.71 E
	eP2	A	38 13	H = 00 26 21.6 h = 349 km MB = 5.3
	eipP	AB	39 26.5	D = 82.30 Az = 328.3 (USNOAA)
	i	A	39 28	P1V(A):1.2s 24.4nm MP1V(A)=4.9
	ePP	A	41 15	P2V(A):1.2s 26.4nm MP2V(A)=5.0
	eSKS	B	47 48	pPV(A):1.5s 118.0nm (MpPV(A)=5.5)
	eS	B	47 52	LmH:14s 0.7 µm
	LmH	B	01 14.0	LmV:16s 1.2 µm
	LmV	B	22.7	
10.	ePKIKP	A	14 07 30	South of Kermadec Islands
	ePKP2	A	08 09	32.04 S 178.11 W
	LmV	B	15 40.5	H = 13 47 34.1 h = 37 km MB = 5.4
	LmH	B	41.0	D = 160.06 Az = 341.6 (USNOAA)
11.	eP	A	11 58 49.5	Kurile Islands 44.56 N 148.08 E
				H = 11 46 55.7 h = 50 km MB = 5.1
				D = 77.92 Az = 333.4 (USNOAA)
				PV(A):1.1s 16.1nm MPV(A)=4.9
11.	eP	AB	21 02 31.5	Dodecanese Islands 36.12 N 28.21 E
	Pm	A	02 45	H = 20 58 13.2 h = 43 km MB = 4.9
	eS	B	06 04	D = 18.81 Az = 325.7 (USNOAA)
	e	C	06 06	PV(A):1.3s 96.0nm MPV(A)=4.9
	e	B	06 10	PmV(A):2.0s 242.0nm MPmV(A)=5.1
	LmH	B	09.3	LmH:12.5s 5.8 µm MLH =5.1
	LmV	B	12.2	LmV:12s 5.0 µm MLV =5.0
12.	eP	A	04 13 09	Hokkaido, Japan Region 41.85 N 142.54 E
				H = 04 01 13.5 h = 61 km MB = 5.0
				D = 78.41 Az = 330.6 (USNOAA)
				PV(A):1.2s 18.3nm MPV(A)=4.9

Day	Phase		h m s	Remarks
12.	ePKIKP	A	06 26 06	East New Guinea Region 5.05 S 145.06 E
	epPKP	A	26 12	H = 06 67 12.4 h = 15 km MB = 5.9
	ePP	B	27 42	D = 120.24 Az = 327.7 (USNOAA)
	e	B	35 40	PKIKPV(A):2.0s 77.0nm
	ePKKP	A	36 20	PKKPV(A):1.8s 47.3nm
	eSKKP	A	40 09	PPV(B):9s 2.0 µm MPPV(B)=6.8
	ePcPPKP	A	40 20	LmH:17.5s 10.3 µm MLH =6.5
	eSS	B	44 15	LmV:16s 8.2 µm MLV =6.5
	eSSS	B	48 24	
	LmH	B	07 24.7	
	LmV	B	24.7	
12.	ePKP	A	07 51 09.5	Off East Coast of North Island, N.Z.
				36.77 S 179.42 E
				H = 07 31 02.8 h = 21 km MB = 5.1 (ISC)
				D = 163.7
12.	ePn	A	11 36 20	Albania 41.0 N 20.8 E
	eSn	A	38 40	H = 11 33 34 h = 75 km
				D = 11.52 Az = 330 (ISC)
12.	eP1	A	13 16 49.5	Iran-USSR Border Region 38.27 N 57.17 E
	eP2	A	16 51.5	H = 13 10 06.3 h = 63 km MB = 4.6
				D = 34.25 Az = 306.1 (USNOAA)
				P1V(A):1.1s 16.1nm MP1V(A)=4.9
				P2V(A):1.2s 24.4nm MP2V(A)=5.0
13.	LmH	B	01 11.5	LmH:18s 0.8 µm
	LmV	B	10.0	
13.	ePKP	A	02 32 44.5	Loyalty Islands Region 22.63 S 170.3 E
				H = 02 13 06 h = 39 km (ISC)
				D = 147.4
				PKPV(A):1.4s 18.6nm
13.	e	A	06 55 48.5	Rockburst Upper Silesia

Day	Phase		h m s	Remarks
13.	eP	A	12 49 05	Hokkaido, Japan Region 42.02 N 142.13 E
				H = 12 37 12.0 h = 80 km MB = 4.7
				D = 78.11 Az = 330.4 (USNOAA)
13.	eP1	A	14 29 42	Leyte, Philippine Islands
	epP	A	29 47	11.95 N 123.97 E
	esP	A	29 50.5	H = 14 16 18.0 h = 15 km MB = 5.4
	eP2	ABC	29 54	D = 94.50 Az = 323.8 (USNOAA)
	e	B	39 00	sPV(A):1.3s 48.0nm MsPV(A)=5.7
	eSKS	BC	40 25	P2V(A):1.9s 137.0nm MP2V(A)=6.0
	eSKKS	BC	40 46	P2V(B):7s 0.65/um MP2V(B)=6.1
	eiS	BC	40 52	LmH:25s 48.4/um MLH =6.9
	ePS	C	42 25	LmV:20s 14.0/um MLV =6.4
	eSS	C	47 28	
	LmH	B	15 07.4	
	LmV	B	18.5	
13.	iP	A	17 38 05.5	Afghanistan-USSR Border Region
	eipP	A	38 31	36.91 N 71.62 E
	esP	A	38 42	H = 17 30 06.7 h = 124 km MB = 5.3(USNOAA)
	ePP	A	39 58	D = 44.2 h = 117 km
				PV(A):2.0s 111.0nm MPV(A)=5.2
14.	eP	A	05 05 32	South of Mariana Islands
	e	A	06 17	12.64 N 143.30 E
	e	A	09 46.5	H = 04 51 37.8 h = 95 km MB = 5.5
	ePP	A	09 50	D = 104.22 Az = 330.6 (USNOAA)
	LmH	B	48.8	PV(A):1.1s 14.1nm MPV(A)=5.8
	LmV	B	58.8	PPV(A):2.0s 85.5nm MPPV(A)=6.0
				LmH:18s 3.0/um
				LmV:20s 1.5/um
14.	e	A	08 01 40	Czechoslovakia, explosion 49.50 N 14.93 E
				H = $08^h\ 00^m$ yield: 12 t (PRU)
				D = 2.4
14.	eP1	A	08 10 50	Taiwan Region 22.71 N 121.34 E
	+iP2	BC	10 51.2	H = 07 58 19.8 h = 28 km MB = 5.7
	+iPcP	A	10 55	D = 84.40 Az = 323.0 (USNOAA)

Day	Phase		h m s	Remarks
cont. 14.	e	A	08 14 00	P1V(A):1.4s 102.5nm MP1V(A)=5.7
	ePP	ABC	14 06	PmV(A):2.2s 273nm MPmV(A)=6.1
	i	A	14 09	PV(B):10.5s 3.45/um MPV(B) =6.5
	eS	BC	21 10	PPV:8s 1.63/um MPPV =6.6
	ePS	BC	22 00	SH:12.5s 3.9 /um MSH =6.4
	eSS	C	26 40	LmH:16s 38.3/um MLH =6.9
	e	C	27 32	LmV:17s 67.7/um MLV =7.1
	LmH	B	53.4	
	LmV	B	53.7	
14.	eP	A	16 02 53	South of Honshu, Japan 33.42 N 140.09 E H = 15 50 31.8 h = 110 km MB = 4.5 D = 84.75 Az = 330.0 (USNOAA)
15.	ePKIKP	A	03 32 26	South of Fiji Islands 22.98 S 177.09 W
	iPKHKP	AB	32 32.5	H = 03 12 54.8 h = 153 km MB = 5.1
	iPKP2	A	32 41.5	D = 151.57 Az = 348 (ISC)
	i	A	32 45.5	PKIKPV:1.4s 23.3nm
	esPKP	A	33 20	PKHKPV:1.5s 100.5nm
16.	ePn	A	04 26 12	Federal Republic Germany 48.38 N 11.08 E
	e	A	26 19	H = 04 25 33.1 h = 0 km
	eSn	A	26 46.5	D = 2.30 Az = 8 (ISC)
	eSg	A	27 00	
	i	A	27 01.5	
16.	eP	A	05 06 00	Northern Sinkiang Prov., China
	ei	A	06 07	43.24 N 81.20 E
	LmH	B	22.9	H = 04 57 32.9 h = 24 km MB = 5.2
	LmV	B	26.3	D = 46.48 Az = 304.6 (USNOAA) PV(A):1.5s 22.6nm MPV(A)=5.0 LmH:20s 1.9/um MLH =5.0 LmV:14.5s 1.4/um MLV =5.1
16.	ePKIKP	A	07 03 12	New Britain Region 6.08 S 148.58 E
	LmH	B	56	H = 06 44 21.4 h = 81 km MB = 5.5
	LmV	B	57	D = 122.95 Az = 328.8 (USNOAA)

Day	Phase		h m s	Remarks
16.	eX	A	21 51 45	Molucca Sea 1.03 S 126.87 E
				H = 21 32 58.8 h = normal MB = 5.4
				D = 106.58 Az = 323.1 (USNOAA)
				XV(A):2.0s 55.5nm
17.	eP	A	00 28 58.5	Turkey 36.68 N 29.58 E
	e	A	29 03.5	H = 00 24 36.8 h = 29 km MB = 4.4
	e	A	29 10.5	D = 19.00 Az = 322.9 (USNOAA)
				PV(A):1.4s 25.6nm MPV(A)=4.3
18.	LmH	B	10 56.5	Bismarck Sea 3.51 S 148.89 E
	LmV	B	55.8	H = 09 45 26.4 h = 15 km MB = 5.8(USNOAA)
				D = 120.8
				LmH:20s 3.5/um MLH=6.0
				LmV:21s 3.9/um MLV=6.0
18.	eP	A	12 30 28.5	North Atlantic Ridge 35.1 N 35.7 W
	epP	A	30 33.5	H = 12 23 18.0 h = normal MB = 5.4(USNOAA)
	ePP	B	31 56	D = 37.3
	e	B	31 58	PV(A):1.7s 18.2nm MPV(A)=4.5
	iS	B	36 22	PmV(A):3.2s 419nm MPmV(A)=5.6
	eSS	B	39 10	PPV(B):11s 1.4/um MPPV(B)=5.8
	LmH	B	44.3	SH(B):13.5s 5.06/um MSH(B)=6.2
	LmV	B	44.2	LmH:12.5s 7.7/um MLH =5.7
				LmV:15s 5.9/um MLV =5.9
18.	ePKIKP	A	17 01 53.5	South of Fiji Islands 21.89 S 175.17 E
	-iPKHKP	A	01 57.5	H = 16 43 14.1 h = 570 km MB = 5.6
	+iPKP2	A	02 04	D = 148.54 Az = 339.8 (USNOAA)
	epPKP	A	04 10	PKIKPV(A):2.0s 51.4nm
	+ipPKP2	A	04 12	PKHKPV(A):1.7s 373.0nm
18.	LmH	B	21 30.2	LmH:16s 1.4/um
	LmV	B	30.3	LmV:18s 1.9/um
19.	eP	A	05 42 02	Republic of South Africa 22.19 S 30.73 E
				H = 05 30 18.6 h = 0 km (ISC)
				D = 74.4
				PV(A):1.2s 16.3nm MPV(A)=5.0

Day	Phase		h m s	Remarks
19.	e	A	18 05 11.5	Probably rockburst in Upper Silesia
	e	A	05 32.5	(Poland)
19.	ePn	A	18 13(38)	Northern Italy 44.2 N 10.2 E
	ePg	A	13 59	H = 18 12 01 h = 0 km
	eSn	A	14 41.5	D = 6.50 Az = 8 (ISC)
	eSg	A	15 21	
19.	eP	A	18 21 10	Mid-Indian Rise 12.21 S 65.68 E
	e	A	21 16	H = 18 09 11.5 h = normal MB = 5.2
				D = 78.27 Az = 328.2 (USNOAA)
20.	eP	A	04 44 58	Andreanof Islands, Aleutian Is.
				51.30 N 179.94 W
				H = 04 33 02.6 h = 40 km MB = 5.1
				D = 77.96 Az = 352.5 (USNOAA)
20.	eP	A	12 56 17	Kurile Islands Region 50.00 N 159.26 E
				H = 12 44 31.9 h = 35 km MB = 4.6
				D = 76.00 Az = 339.4 (USNOAA)
				traces
20.	eP	A	13 59 53	Bonin Islands Region 28.12 N 142.48 E
				H = 13 46 52.8 h = normal MB = 5.4
				D = 90.36 Az = 331.2 (USNOAA)
				PV(A):1.5s 25.2nm MPV(A)=5.3
20.	+eiP	AB	14 00 22	Kurile Islands 43.5 N 146.91 E
	eipP	A	00 32	H = 13 48 23.7 h = 36 km MB = 5.7(USNOAA)
	eS	BC	10 12	D = 78.5
	eSKS	BC	10 31	PV(A):1.8s 257.0nm MPV(A)=6.0
	eSS	C	15.9	pPV(A):2.0s 171.0nm
	LmH	B	39.5	PV(B):5s 1.03 μm MPV(B)=6.2
	LmV	B	39.8	SH(B):7.5s 0.71 μm MSH(B)=5.8
				LmH:14.5s 2.1 μm MLH =5.6
				LmV:17.5s 2.8 μm MLV =5.7

Day	Phase		h m s	Remarks
21.	ePKHKP	A	00 48 03	Tonga Islands 20.41 S 174.23 W
	epPKP2	A	48 13.5	H = 00 28 14.5 h = normal MB = 5.2
	LmH	C	02 05	D = 149.48 Az = 352.7 (USNOAA)
				PKHKPV(A):1.6s 49.5nm
				pPKPV(A):1.7s 72.7nm
21.	eP	A	02 18 09	Turkey 36.96 N 28.96 E
	LmH	C	25	H = 02 13 51.3 h = normal MB = 4.3
				D = 18.48 Az = 323.2 (USNOAA)
				PV(A):1.6s 41.2nm MPV(A)=4.3
21.	eP	A	08 41 41	Eastern Caucasus 40.05 N 48.26 E
	e	A	41 42.5	H = 08 35 59.1 h = 68 km MB = 5.1
				D = 27.58 Az = 304.9 (USNOAA)
				PV(A):1.5s 27.6nm MPV(A)=4.6
21.	eP	AB	12 32 34.5	Luzon, Philippine Islands
	epP	A	32 52	14.95 N 120.10 E
	ePP	A	36 09.5	H = 12 19 39.2 h = 47 km MB = 5.5
	eSKS	BC	43 02	D = 89.85 Az = 322.9 (USNOAA)
	eS	BC	43 16	PV(A):1.2s 36.6nm MPV(A)=5.6
	ePS	BC	44 22	pPV(A):1.8s 67.5nm
	eSS	C	49.7	PV(B):8s 0.44/um MPV(B)=5.8
	LmH	B	13 15	LmH:19s 1.4/um MLH =5.4
	LmV	B	15.1	LmV:20s 1.9/um MLV =5.5
21.	eP	A	19 37 46	Alaska Peninsula 56.95 N 156.99 W
				H = 19 26 28.1 h = 90 km MB = 4.6
				D = 72.36 Az = 7.6 (USNOAA)
22.	eP	A	12 07 38.5	Mariana Islands 18.33 N 145.97 E
	e	A	11 38	H = 11 53 59.1 h = 91 km MB = 5.5
	ePP	A	11 45	D = 100.47 Az = 332.4 (USNOAA)
	e	A	12 15.5	LmH:15s 0.6/um
	LmH	B	52.4	LmV:20s 0.9/um
	LmV	B	56.6	

Day	Phase		h m s	Remarks
22.	eP	A	15 30 08	North of Ascension Island 0.14 S 16.14 W H = 15 20 32.8 h = 33 km MB = 4.5 (ISC) D = 55.8
22.	eP	A	19 07 17	West Pakistan 30.82 N 67.75 E H = 18 58 52 h = 18 km (ISC) D = 45.7 PV(A):1.6s 22.0nm MPV(A)=4.8
22.	ePn e(Sn) eSg	A A A	20 39 39 40 24.5 40 41	Austria 47.4 N 16.1 E H = 20 38 36 h = 0 km D = 4.10 Az = 322 (ISC)
23.	eP	A	00 37 44	Norwegian Sea 72.41 N 2.05 E H = 00 32 49.3 h = normal MB = 4.5 D = 22.25 Az = 163.8 (USNOAA) PV(A):1.3s 21.8nm MPV(A)=4.5
24.	ePKP e	A A	02 12 13 12 37	Tonga Islands 15.14 S 173.36 W H = 01 52 39.1 h = 25 km MB = 5.0 D = 144.35 Az = 354.6 (USNOAA) PKPV(A):0.8s 9.6nm
24.	iP eiPcP e	A A A	05 18 18.5 18 30.5 18 36.5	Kurile Islands 47.44 N 152.49 E H = 05 06 41.4 h = 136 km MB = 5.3 D = 76.65 Az = 335.6 (USNOAA) PV(A):1.1s 121.0nm MPV(A)=5.5
24.	eP epP LmH LmV	A A B B	11 54 22.5 54 28.5 12 11.0 14.1	Kazakh Sinkiang Border Region 47.55 N 84.31 E H = 11 45 59.2 h = normal MB = 4.9 D = 45.93 Az = 302.2 (USNOAA)
24.	eP	A	15 53 45	Southern Sumatra 0.06 S 99.33 E H = 15 41 01.1 h = 97 km MB = 4.9 D = 88.59 Az = 320.5 (USNOAA) PV(A):1.5s 15.1nm MPV(A)=4.9

Day	Phase		h m s	Remarks
24.	eP	A	17 25 08.5	Jan Mayen Island Region 71.82 N 2.54 W
	LmH	B	34.4	H = 17 20 13.5 h = normal MB = 4.6
	LmV	B	36.0	D = 22.19 Az = 155.7 (USNOAA)
				PV(A):1.7s 48.5nm MPV(A)=4.7
25.	eP	A	09 51 55.5	Kurile Islands 43.16 N 146.68 E
	ePcP	A	52 05.5	H = 09 39 58.9 h = 67 km MB = 4.7
				D = 78.71 Az = 332.7 (USNOAA)
				PV(A):1.8s 33.8nm MPV(A)=5.0
26.	eP	A	02 01 53	Crete 34.58 N 23.96 E
	e	A	02 38	H = 01 57 39.1 h = normal MB = 4.6
	LmH	C	10.8	D = 18.41 Az = 334.5 (USNOAA)
	LmV	C	10.9	PV(A):1.3s 34.9nm MPV(A)=4.4
26.	eiP1	A	03 23 50	Off Coast of Oregon 43.78 N 127.45 W
	isP1	A	23 55	H = 03 11 42.8 h = 14 km MB = 5.6
	eP2	A	24 07	D = 79.44 Az = 25.1 (USNOAA)
	eS	BC	33 56	MP1V(A):1.4s 39.5nm MP1V(A)=5.3
	eSS	C	39 06	MP2V(A):2.1s 183.0nm MP2V(A)=5.7
	LmH	B	04 04.0	LmH:15.5s 4.0/um MLH =5.9
	LmV	B	04.0	LmV:15s 3.8/um MLV =5.9
26.	ePKHKP	A	06 18 08	Tonga Islands 19.96 S 174.02 W
				H = 05 58 20.5 h = normal MB = 4.7
				D = 149.05 Az = 353.0 (USNOAA)
				PKHKP(A):1.3s 17.5nm
26.	ePKHKP	A	16 35 25.5	South of Fiji Islands 23.38 S 179.92 E
				H = 16 16 33.3 h = 549 km MB = 4.7
				D = 151.30 Az = 344.4 (USNOAA)
				PKHKP(A):1.3s 15.3nm

Day	Phase		h m s	Remarks
27.	eP	A	08 10 18.5	Nicobar Islands Region 6.49 N 94.70 E
	ePcP	A	10 23.5	H = 07 58 14.0 h = 85 km MB = 5.2
	epP	A	10 41.5	D = 80.63 Az = 320.2 (USNOAA)
27.	eP	AB	09 51 47	Taiwan Region 24.23 N 122.31 E
	ePP	A	55 01.5	H = 09 39 23.2 h = 57 km MB = 5.9
	eS	C	10 02 00	D = 83.72 Az = 323.2 (USNOAA)
	eSS	C	08.5	PV(A):1.5s 126.0nm MPV(A)=5.7
	LmH	B	33.7	PV(B):7s 0.65/um MPV(B)=5.8
	LmV	B	33.7	PPV(A):1.4s 51.2nm MPPV(A)=5.7
				LmH:18s 1.3/um MLH =5.4
				LmV:18s 1.7/um MLV =5.5
27.	eP	A	19 58 39.5	South Atlantic Ridge 14.34 S 13.64 W
	LmH	C	20 21	H = 19 47 39.7 h = normal MB = 4.8
				D = 68.39 Az = 17.0 (USNOAA)
				PV(A):1.4s 20.9nm MPV(A)=5.0
28.	eX	A	01 14 40	Molucca Sea 0.10 S 126.80 E
				H = 00 56 06.8 h = normal MB = 5.6
				D = 106.51 Az = 323.1 (USNOAA)
				XV(A):1.9s 37.9nm
28.	eP	A	02 35 30	Iceland Region 63 N 27.5 W
				H = 02 30 23 h = 33 km MB = 4.1 (ISC)
				D = 25.0
28.	eP	A	11 22 30	Northern Chile 20.92 S 69.83 W
	ePP	C	26 34	H = 11 08 42.5 h = 33 km MB = 6.0
	ePS	C	35.5	D = 100.67 Az = 39.8 (USNOAA)
	eSS	C	41 00	PPV(A):1.8s 60.8nm MPPV(A)=5.8
	LmH	B	12 09.3	LmH:18s 1.1/um MLH =5.4
	LmV	B	09.4	LmV:19s 0.9/um MLV =5.3
28.	eP	A	14 59 19	Northern Chile 20.92 S 69.87 W
	ePP	A	15 03 21	H = 14 45 31.7 h = 34 km MB = 5.9
	LmH	B	39.7	D = 100.70 Az = 39.9 (USNOAA)
	LmV	B	40.4	PPV(A):2.0s 59.9nm MPPV(A)=5.7
				LmH:25.5s 1.3/um MLH =5.3
				LmV:20s 0.8/um MLV =5.3

Day	Phase		h m s	Remarks
28.	ePKIKP	A	20 41 25.5	New Guinea 4.13 S 142.90 E
	e	A	42 27	H = 20 20 50.6 h = 114 km MB = 5.8
	ePP	AB	42 44	D = 118.31 Az = 327.1 (USNOAA)
	ePKKP	A	51 45	PKIKPV(A):1.1s 16.1nm
	ePS	C	52 32	PPV(B):8s 1.63 um MPPV(B)=6.2
	eSKKP	A	55 30	LmH:22s 2.8 um
	eSS	C	58 52	LmV:20s 2.2 um
	LmH	B	21 23.4	e(C) 52 36 e(C) 53 18 e(B) 53 20
	LmV	B	32.0	e(BC) 53 25 e(C) 54 28 e(B) 59 44
29.	eP	A	02 12 13	Southern Sinkiang Prov., China
	LmV	B	33.2	41.62 N 81.81 E
	LmH	B	33.3	H = 02 03 37.4 h = normal MB = 5.1
				D = 47.78 Az = 306.0 (USNOAA)
				PV(A):1.2s 12.2nm MPV(A)=4.8
29.	ePP	A	05 21 49	Afghanistan-USSR Border Region
				36.2 N 71.4 E
				H = 05 12 06.8 h = 117 km MB = 4.9
				D = 44.6 (USNOAA)
29.	eP1	A	06 12 02	Ascension Island Region 11.68 S 14.09 W
	eP2	AB	12 04.5	H = 06 01 18.7 h = normal MB = 5.3
	+ipP	A	12 09.5	D = 65.99 Az = 17.6 (USNOAA)
	esP	A	12 15	P1V(A):1.3s 17.5nm MP1V(A)=5.0
	ePPP	C	16 04	P2V(A):2.4s 110.5nm MP2V(A)=5.5
	eS	C	20 48	pPV(A):1.2s 48.7nm MpPV(A)=5.5
	ePS	C	21 14	sPV(A):2.1s 216.1nm MsPV(A)=5.9
	eiSS	C	25 00	LmH:17s 13.5 um MLH =6.2
	eSSS	C	27 55	LmV:16s 16.4 um MLV =6.4
	LmH	B	41.4	
	LmV	B	41.7	
29.	ePKP2	A	15 24 11	Tonga Islands 20.81 S 174.46 W
				H = 15 04 16.0 h = 28 km MB = 4.3
				D = 149.84 Az = 352.3 (USNOAA)

Day	Phase		h m s	Remarks
29.	eP	A	15 40 11	Southern Sinkiang Prov., China
	LmH	B	16 01.0	41.59 N 81.78 E
	LmV	B	01.2	H = 15 31 29.7 h = normal MB = 4.7
				D = 47.77 Az = 306.0 (USNOAA)
				PV(A):1.0s 11.8nm MPV(A)=4.9
				LmH:10s 0.65/um MLH =4.9
				LmV:10s 0.7/um MLV =5.0
29.	eiP	A	17 44 08.5	Turkmen SSR 39.62 N 54.64 E
	epP	A	44 20	H = 17 37 45.9 h = 53 km MB = 4.7
				D = 31.87 Az = 304.6 (USNOAA)
				PV(A):1.2s 24.4nm MPV(A)=4.8
29.	eP	A	20 13 30	Mexico - Guatemala Border Region
	LmH	B	53.0	15.29 N 92.79 W
	LmV	B	53.0	H = 20 00 54.0 h = 105 km MB = 5.0
				D = 87.18 Az = 38 (ISC)
30.	eP	A	09 52 18.5	Greece 39.07 N 21.87 E
	e	A	52 20	H = 09 49 03.2 h = 48 km MB = 4.3
				D = 13.65 Az = 331.3 (USNOAA)
30.	eP	A	17 28 09.5	Ascension Island Region 11.43 S 14.43 W
	e	A	28 17.5	H = 17 17 24.1 h = normal MB = 4.7
				D = 65.86 Az = 17.8 (USNOAA)
				LmH:18s 0.5/um MLH =4.8
				LmV:18s 0.5/um MLV =4.8
30.	ePKHKP	A	20 36 46	Fiji Islands Region 21.81 S 178.62 W
	ePKP2	A	36 55	H = 20 17 56.7 h = 555 km MB = 4.7
				D = 150.12 Az = 346.9 (USNOAA)

Day	Phase		h m s	Remarks
1.	eP	A	01 06 00.5	<u>Tunisia</u> 36.94 N 9.73 E
	e	A	06 06.5	H = 01 02 44.2 h = 24 km MB = 5.1
	LmH	B	11.5	D = 13.77 Az = 5.1 (USNOAA)
	LmV	B	12.5	PV(A):1.8s 40.5nm
				LmH(B):17s 1.8 μm MLH(B)=4.2
				LmV(B):12s 1.8 μm
1.	ePn	A	10 49 50	**<u>Federal Rep. Germany</u>** 50.30 N 7.79 E
	i	A	49 52	H = 10 49 09.3 h = 18 km MB = 3.9
	iPg	A	49 56.7	D = 2.47 Az = 80.5 (USNOAA)
	iSn	A	50 18.7	
	iSg	A	50 27.2	
1.	eP1	AB	12 02 23.5	<u>Turkey</u> 39.89 N 38.85 E
	eP2	A	02 27	H = 11 57 29.1 h = 21 km MB = 4.7
	eSS	B	06 32	D = 21.85 Az = 308.5 (USNOAA)
	LmH	B	13.2	P1V(A):1.4s 30.3nm MP1V(A)=4.5
	LmV	B	13.3	P2V(A):1.3s 94.0nm MP2V(A)=5.0
				LmH(B):16s 1.7 μm MLH(B)= 4.6
				LmV(B):15.5s 2.0 μm MLV(B)= 4.8
1.	ePKHKP	A	13 08 13	<u>New Britain Region</u> 6.00 S 152.60 E
				H = 12 49 19.0 h = 34 km MB = 5.4
				D = 124.90 Az = 330.7 (USNOAA)
1.	ePKIKP	ABC	18 34 13	<u>Solomon Islands</u> 10.97 S 163.39 E
	ePP	AB	36 43	H = 18 14 38.6 h = normal MB = 5.5(USNOAA)
	eSS	C	54 10	D = 134.2
	eSS	B	54 12	PPV(B):8s 1.2 μm MPPV(B)=6.1
	e	C	56 20	LmV(B):16s 1.2 μm MLV(B)= 5.7
	eSSS	BC	59 12	LmH(B):19s 2.9 μm MLH(B)= 6.0
	LmV	B	19 37.3	
	LmH	B	37.9	
1.	+iP	A	19 41 06.2	<u>Fox Islands, Aleutian Is.</u>
				52.42 N 169.08 W
				H = 19 29 16.6 h = 60 km MB = 5.2
				D = 77.31 Az = 359.5 (USNOAA)
				PV(A):1.0s 35.4nm MPV(A)=5.3

Day	Phase		h m s	Remarks
1.	+eiP	ABC	21 21 34	Andreanof Islands, Aleutian Is.
	ePP	B	24 30	51.40 N 175.30 W
	e	C	24 40	H = 21 09 37.2 h = 36 km MB = 5.6
	ePPP	B	26 22	D = 78.16 Az = 355.5 (USNOAA)
	ePPP	C	26 24	PV(A):1.0s 51.2nm MPV(A)=5.5
	eS	C	31 30	PV(A):2.5s 184.3nm MPV(A)=5.8
	eSKS	B	31 36	PV(A):superposition of two periods
	eSKS	C	31 40	PV(B):10s 1.4/um MPV(B)=6.0
	ePPS	B	32 16	PPV(B):8s 0.8/um MPPV(B)=5.9
	e(SS)	C	36.7	LmH(B):16s 4.7/um MLH(B)=5.9
	e(SS)	B	36 52	
	LmH	B	22 03.4	
	LmV	B	06.4	
1.	ePKP	A	22 21 28	Tonga Islands 15.56 S 173.18 W
				H = 22 01 54.1 h = normal MB = 5.4
				D = 144.79 Az = 354.7 (USNOAA)
				PKPV(A):1.4s 23.3nm
2.	eP	ABC	02 46 54	Andreanof Islands, Aleutian Is.
	e	A	47 31.5	51.44 N 175.25 W
	e	A	47 49	H = 02 34 59.5 h = 57 km MB = 5.4
	ePP	B	49 54	D = 78.12 Az = 355.5 (USNOAA)
	eS	C	56 48	PV(A):1.3s 52.4nm MPV(A)=5.3
	e(SS)	C	03 02 16	PV(B):8.5s 0.6/um MPV(B)=5.7
	LmH	B	29.0	LmH(B):16.5s 2.1/um MLH(B)=5.5
	LmV	B	31.0	
2.	eP	AB	09 15 09	Andreanof Islands, Aleutian Is.
	LmH	B	59.4	51.43 N 175.24 W
	LmV	B	10 00.0	H = 09 03 14.6 h = 52 km MB = 5.2
				D = 78.14 Az = 355.5 (USNOAA)
				PV(A):(1.5)s 25.1nm MPV(A)=4.9
				LmH(B):16s 1.5/um MLH(B)=5.4
				LmV(B):16s 1.3/um MLV(B)=5.4
2.	e	B	15 45 24	Solomon Islands 10.86 S 163.46 E
	ePP	B	47 38	H = 15 25 46.7 h = 36 km MB = 5.3

Day	Phase		h m s	Remarks
cont. 2.	e	B	15 48 00	D = 134.09 Az = 335.3 (USNOAA)
	ePKS	B	48 28	
	ePPP	B	50 19	
	ePKKP	B	52 34	
2.	ePKIKP	A	16 13 37	<u>Solomon Islands</u> 10.97 S 163.29 E
	e	B	14 12	H = 15 54 19.9 h = normal MB = 5.8
	ePP	B	16 15	D = 134.11 Az = 335.1 (USNOAA)
	e	B	16 34	LmH(B):20s 35.9/um MLH(B)=7.1
	ePKS	B	17 22	LmV(B):20s 30.3/um MLV(B)=7.1
	eSKS	B	20 31	
	e	B	23 12	
	e	B	23 31	
	ePPS	B	27 57	
	eSKSPKP	B	29 00	
	eSS	B	33 44	
	e	B	34 08	
	LmH	B	17 16.0	
	LmV	B	16.2	
2.	eP	A	19 23 36	<u>Kansu Province, China</u> 35.90 N 105.49 E
				H = 19 12 53.8 h = 28 km MB = 5.3
				D = 65.46 Az = 315.7 (USNOAA)
				PV(A):1.8s 54.1nm MPV(A)=5.4
3.	eP	A	05 12 16	<u>Northern Colombia</u> 7.36 N 76.04 W
	LmH	B	50.2	H = 04 59 53.4 h = 38 km MB = 5.7
	LmV	B	50.2	D = 82.91 Az = 39.9 (USNOAA)
				PV(A):1.9s 68.2nm MPV(A)=5.4
				LmH(B):20s 1.7/um MLH(B)=5.4
				LmV(B):20s 1.8/um MLV(B)=5.5
3.	eP	A	07 21 33	<u>Molucca Sea</u> 1.06 S 126.80 E
	LmH	B	08 11.4	H = 07 02 46.6 h = 26 km MB = 5.8
	LmV	B	11.4	D = 106.56 Az = 323.1 (USNOAA)
				PV(A):2.0s 51.2nm

Day	Phase		h m s	Remarks
3.	ePKP	A	15 34 12.5	Tonga Islands 16.27 S 173.99 W
				H = 15 14 43.1 h = 85 km MB = 4.6
				D = 145.41 Az = 353.7 (USNOAA)
				PKPV(A):1.2s 20.3nm
4.	eP	ABC	02 04 02	Western Caucasus 43.78 N 39.13 E
	eS	BC	07 45	H = 01 59 29.1 h = normal MB = 4.9
	LmV	B	15.2	D = 19.83 Az = 299.9 (USNOAA)
	LmH	B	16.9	PV(A):1.5s 40.2nm MPV(A)=4.4
				PV(B):6s 1.3 μm MPV(B)=5.3
				LmV(B):12s 3.9 μm MLV(B)=5.1
				LmH(B):10s 3.3 μm MLH(B)=5.0
4.	eP	A	10 03 44	Panama 9.78 N 79.68 W
	LmH	B	37.7	H = 09 51 16.1 h = 20 km MB = 5.3
	LmV	B	37.7	D = 83.34 Az = 39.9 (USNOAA)
				LmH(B):18s 1.9 μm MLH(B)=5.5
				LmV(B):18s 1.9 μm MLV(B)=5.6
4.	eP	AC	17 22 45	Near Coast of Northern Chile
	e	A	22 58.5	23.13 S 70.11 W
	e	A	23 03	H = 17 08 48.7 h = 36 km MB = 5.9
	ePP	ABC	26 52	D = 102.53 Az = 40.2 (USNOAA)
	e	A	27 06.5	PV(A):1.6s 38.5nm MPV(A)=5.8
	e	A	27 19	LmH(B):20s 9.9 μm MLH(B)=6.3
	e	A	27 24	LmV(B):20s 9.2 μm MLV(B)=6.3
	eSKS	BC	33 22	
	eiPS	BC	36 04	
	eSS	C	41 32	
	e	C	41 44	
	LmH	B	18 05.6	
	LmV	B	05.6	
5.	eiPKP	A	22 18 41	Tonga Islands 18.12 S 175.38 W
	epPKP	A	19 47	H = 21 59 25.3 h = 241 km MB = 5.0
				D = 147.07 Az = 351.8 (USNOAA)
				PKPV(A):1.5s 45.2nm

Day	Phase		h m s	Remarks
6.	ePKP	A	02 51 53	Fiji Islands Region 17.26 S 176.33 W H = 02 32 19.9 h = 48 km MB = 4.9 D = 146.09 Az = 350.9 (USNOAA) PKPV(A):1.9s 37.9nm
6.	eP	A	04 29 02	Banda Sea 6.27 S 130.14 E H = 04 10 37.8 h = 118 km MB = 5.7 D = 112.70 Az = 322.7 (USNOAA)
6.	e	A	12 59 07	Kurile Islands 43.58 N 147.03 E H = 12 47 31.4 h = 40 km MB = 4.9 D = 78.45 Az = 332.9 (USNOAA)
6.	eP	A	17 45 42	Ascension Island Region 7.95 S 13.41 W H = 17 35 20.1 h = normal MB = 4.2 D = 62.26 Az = 17.7 (USNOAA)
6.	eP	A	17 53 43	Ascension Island Region 7.95 S 13.50 W H = 17 43 21.4 h = normal MB = 4.4 D = 62.29 Az = 17.8 (USNOAA)
6.	+iP	AB	20 32 51.5	Hokkaido, Japan Region 41.80 N 143.51 E
	Pmax	A	32 54	H = 20 20 52.2 h = 48 km MB = 5.7
	+i	A	33 03	D = 78.80 Az = 331.1 (USNOAA)
	ePP	BC	35 52	PmaxV(A):1.2s 118.0nm MPmaxV(A)=5.7
	eS	BC	42 44	PV(B):11s 2.6/um MPV(B)=6.2
	eSS	C	48 45	PPV(B):10s 1.35/um MPPV(B)=6.0
	LmH	B	21 06.9	LmH(B):17s 23.6/um MLH(B)=6.6
	LmV	B	10.7	LmV(B):19.5s 20.1/um MLV(B)=6.5
7.	eiP	A	12 26 50.5	Panama-Colombia Border Region 6.98 N 77.68 W H = 12 14 21.6 h = 25 km MB = 4.8 D = 84.21 Az = 39.8 (USNOAA)
7.	-eiP1	ABC	21 47 50.5	South of Honshu, Japan 29.69 N 140.01 E
	iP2	AB	47 56.5	H = 21 35 21.4 h = 179 km MB = 5.9
	epP1	B	48 44	D = 87.96 Az = 330.0 (USNOAA)

Day	Phase		h	m	s	Remarks
cont. 7.	+ipP2	C	21	48	50	P1V(A):1.5s 178.0nm MP1V(A)=5.8
	ePP1	C		51	14	P2V(A):2.0s 940.0nm MP2V(A)=6.4
	eiPP2	BC		51	21	LmH(B):16.5s 10.6μm
	eipPP	C		52	21.5	LmV(B):13s 6.8μm
	eiSKS	C		57	56.5	
	eiSKS	B		58	00	
	eiS	B		58	22	
	iSP	BC		59	16	
	ei	C		59	24	
	ei	BC		59	35	
	i	C	22	03	30	
	eiSS	C		04	08	
	eSS	B		04	14	
	e	C		04	58	
	LmH	B		23.3		
	LmV	B		37.3		
8.	eP	A	09	11	30	Off W. Coast of Northern Sumatra
	e	A		11	38.5	2.39 N 94.88 E
						H = 08 59 00.2 h = 21 km MB = 5.3
						D = 83.88 Az = 320.5 (USNOAA)
						PV(A):1.4s 18.6nm MPV(A)=5.1
8.	ePKP	A	11	28	52	New Hebrides Islands 18.87 S 169.25 E
						H = 11 09 45.2 h = 226 km MB = 4.6
						D = 143.65 Az = 335.9 (USNOAA)
8.	eP diff	B	19	44	25	Near Coast of Central Chile
	e	A		44	52.5	30.70 S 71.21 W
	ePKIKP	A		48	25	H = 19 30 06.7 h = 50 km MB = 5.8
	ePP	C		48	54.5	D = 108.86 Az = 41.9 (USNOAA)
	iPP	B		48	56	PdiffV(B):10s 1.0μm MPdiffV(B)=7.0
	ePP	A		48	57	PKIKPV(A):2.3s 61.0nm
	i	BC		49	00	PPV(B):9s 2.4μm MPPV(B)=6.8
	eSKS	B		55	32	LmH(B):19.5s 12.1μm MLH(B) =6.5
	iSKS	C		55	36	LmV(B):20s 16.7μm MLH(B) =6.6
	e	C		58	16	
	ei	BC		58	18	
	e	C		58	23	

Day	Phase		h m s	Remarks
cont. 8.	e	B	19 58 30	
	e	A	59 38.5	
	eSS	C	20 04 29	
	ei	C	04 37	
	+i	B	04 43	
	eSSS	C	08 06	
	ei	B	08 21	
	LmH	B	33.9	
	LmV	B	33.9	
9.	ePn	A	00 46 48.5	Peißenberg, FRG 47.7 N 11.2 E
	eiPb	A	46 54.5	H = 00 46 04 (BCIS)
	iPg	A	46 59.2	D = 2.6
	iSn	A	47 24.2	
	iSg	A	47 38.2	
9.	eP	A	08 15 43	Near Coast of Guerrero, Mexico
	e	A	15 50	16.15 N 99.41 W
	LmH	C	57.0	H = 08 02 43.3 h = 34 km MB = 5.5
	LmV	C	57.5	D = 90.37 Az = 36.5 (USNOAA)
9.	+iPKP2	A	08 32 02.3	Tonga Islands 18.41 S 174.80 W
				H = 08 12 31.1 h = 124 km MB = 4.9
				D = 147.43 Az = 352.4 (USNOAA)
				PKPV(A):1.3s 19.6nm
9.	ePKP2	A	12 06 00	Loyalty Islands Region 22.64 S 172.16 E
				H = 11 46 16.9 h = 44 km MB = 4.9
				D = 148.19 Az = 336.3 (USNOAA)
9.	eP	A	16 41 18	Ecuador 1.40 S 77.66 W
	e	A	42 02.5	H = 16 28 33.4 h = 178 km MB = 5.0
				D = 90.61 Az = 39.5 (USNOAA)
				PV(A):1.6s 27.5nm
9.	eP	A	19 45 02	Kyushu, Japan 31.53 N 130.19 E
				H = 19 33 01.2 h = 175 km MB = 4.7
				D = 81.88 Az = 325.6 (USNOAA)
				PV(A):1.2s 18.3nm MPV(A)=4.7

Day	Phase		h m s	Remarks
10.	+eiP1	ABC	04 47 58.5	Peru-Ecuador Border Region
	eiP2	A	48 03.5	3.99 S 80.72 W
	iP3	A	48 08.8	H = 04 34 38.9 h = 25 km MB = 6.3
	iPP	B	51 41	D = 94.54 Az = 39.7 (USNOAA)
	iPP	C	52 00	P1V(A):1.6s 154.0nm MP1V(A)=6.2
	iSKS	BC	58 32	P2V(A):2.0s 666.0nm MP2V(A)=6.6
	iS	BC	59 26	P3V(A):1.6s 824.0nm MP3V(A)=6.9
	-iPKKP	A	05 05 01.5	PKKPV(A):2.0s 115.0nm
	eP'P'1	A	13 11	P'P'1V(A):2.2s 120.0nm
	eP'P'2	A	13 15	P'P'2V(A):2.0s 132.0nm
	eP'P'3	A	13 19	P'P'3V(A):3.0s 842.0nm
	LmH	B	27.7	LmH(B):20s 152.0/um MLH(B)=7.5
	LmV	B	27.9	LmV(B):20s 174.8/um MLV(B)=7.5
10.	eP	A	05 46 08.5	Peru-Ecuador Border Region
				3.95 S 80.73 W
				H = 05 32 49.7 h = 33 km MB = 5.4
				D = 94.52 Az = 39.7 (USNOAA)
10.	eP1	A	10 26 55	Fox Islands, Aleutian Islands
	eP2	A	26 58.5	53.10 N 169.85 W
	LmH	B	11 11.6	H = 10 15 07.2 h = 48 km MB = 5.5(USNOAA)
	LmV	B	11.6	D = 76.5
				P2V(A):1.8s 33.8nm MP2V(A)=5.0
				LmH(B):15s 1.0/um MLH(B) =5.6
				LmV(B):16s 0.8/um MLV(B) =5.2
10.	eP	A	11 51 04	Peru-Ecuador Border Region
	e	A	51 12	3.98 S 80.79 W
	LmH	B	12 31.5	H = 11 37 42.4 h = 32 km MB = 5.4(USNOAA)
	LmV	B	31.8	D = 94.6
				LmH(B):22s 1.5/um MLH(B)=5.4
				LmV(B):22s 1.6/um MLV(B)=5.5
10.	eP	A	20 04 03	Peru-Ecuador Border Region
				4.11 S 80.75 W
				H = 19 50 43.1 h = normal MB = 5.1
				D = 94.65 Az = 39.7 (USNOAA)

Day	Phase		h m s	Remarks
10.	eP	A	22 00 00	Peru, Ecuador Border Region
				3.94 S 80.81 W
				H = 21 46 41.7 h = 34 km MB = 5.0
				D = 94.56 Az = 39.7 (USNOAA)
				PV(A):1.9s 22.7nm MPV(A)=5.3
11.	e	A	07 40 49	North Atlantic Ridge 43.82 N 28.42 W
	e	A	40 56.5	H = 07 34 54.5 h = normal MB = 5.2
	LmH	B	51.2.	D = 27.76 Az = 61.6 (USNOAA)
	LmV	B	51.3	PV(A):2.6s 60.6nm
				LmH(B):17s 2.0/um MLH(B)=4.8
				LmV(B):18.5s 3.4/um MLV(B)=5.1
11.	eP1	A	10 37 54.5	Peru-Ecuador Border Region
	eP2	A	37 56	3.95 S 80.72 W
	e	B	38 00	H = 10 24 36.2 h = 37 km MB = 5.7
	eSKS	B	48 24	D = 94.51 Az = 39.7 (USNOAA)
	e	C	48 30	P2V(A):2.0s 42.8nm MPV(A)=5.5
	e	BC	49 08	LmH(B):19s 2.7/um MLH(B)=5.7
	LmH	B	11 17.7	LmV(B):22s 3.5/um MLV(B)=5.8
	LmV	B	18.6	
11.	iPg	A	11 44 20	ČSSR, explosion, yield 10.7 to
	iSg	A	44 40.5	50°33.8'N 14°00.4'E
				(PRU)
				D c. 1.6
12.	ePKIKP	A	01 29 39	Fiji Islands Region 20.81 S 177.99 W
	eiPKHKP	A	29 43.5	H = 01 10 41.2 h = 411 km MB = 5.5
	ePKP2	A	29 48.5	D = 149.28 Az = 348.0 (USNOAA)
	epPKP	A	31 25	h = 425 km
	ePP	A	33 14	PKIKPV(A):1.1s 16.1nm
				PKHKPV(A):1.7s 96.9nm
				PKP2V(A):1.7s 121.0nm
12.	+iP	ABC	07 07 06.8	Western Kazakh SSR 43.85 N 54.77 E
	e	A	07 33	H = 07 00 57.3 h = 0 km MB = 6.1
	LmH	B	17.9	D = 29.73 Az = 298.6 (USNOAA)

Day	Phase		h m s	Remarks
cont. 12.	LmV	B	07 19.1	PV(A):1.4s 125.5nm MPV(A)=5.7 LmH(B):16s 3.3/um MLH(B)=5.1 LmV(B):13s 2.3/um MLV(B)=5.1
12.	ePKP2	A	10 42 50	<u>Fiji Islands</u> 19.67 S 177.80 W H = 10 24 08.3 h = 573 km MB = 4.2 D = 148.20 Az = 348.6 (USNOAA)
12.	eP	A	11 22 53	<u>Iran-USSR Border Region</u> 37.25 N 55.43 E H = 11 16 13.4 h = 46 km MB = 4.5 D = 33.76 Az = 307.5 (USNOAA)
13.	e	A	01 04 38	<u>Near Coast of Northern Peru</u> 4.2 S 81.2 W H = 00 50 51 h = 9 km MB = 4.8 (ISC) D = 95.1
13.	-eP LmH	A C	04 15 49 47.9	<u>Near West Coast of Honshu, Japan</u> 39.81 N 139.38 E H = 04 03 42.5 h = 14 km MB = 5.4 D = 78.97 Az = 329.2 (USNOAA) PV(A):1.4s 27.9nm MPV(A)=5.1 LmH(C):18s 0.45/um MLH(C)=4.8
13.	eP e	A A	13 00 08.5 00 33	<u>Southern Sumatra</u> 4.25 S 103.37 E H = 12 47 00.3 h = 119 km MB = 5.4 D = 94.37 Az = 320.3 (USNOAA) PV(A):1.2s 14.2nm MPV(A)=5.2
14.	eP ePP eSKS eS ePS LmH LmV	ABC B BC BC BC B B	07 46 04 49 43 56 38 57 08 58 14 08 23.2 25.3	<u>Near Coast of Ecuador</u> 1.31 S 80.91 W H = 07 32 52.5 h = normal MB = 5.4 D = 92.61 Az = 39.5 (USNOAA) PV(A):1.8s 33.8nm MPV(A)=5.5 LmH(B):21s 2.4/um MLH(B)=5.6 LmV(B):20s 2.5/um MLV(B)=5.7

Day	Phase		h m s	Remarks
14.	eP	A	15 00 00	Fox Islands, Aleutian Is.
	LmH	B	41.0	53.03 N 169.93 W
				H = 14 48 11.8 h = 50 km MB = 5.3
				D = 76.69 Az = 359.0 (USNOAA)
				LmH(B):18s 0.5/um MLH(B)=4.9
14.	eP	A	21 23 27	Fox Islands, Aleutian Is.
	LmH	B	22 02.5	53.01 N 170.02 W
	LmV	B	08.6	H = 21 11 39.1 h = 54 km MB = 5.2
				D = 76.71 Az = 358.9 (USNOAA)
				PV(A):2.0s 47.0nm MPV(A)=5.1
				LmH(B):20s 1.9/um MLH(B)=5.4
				LmV(B):16s 1.5/um MLV(B)=5.4
15.	eP	A	02 17 30	Off W. Coast of Northern Sumatra
				2.92 N 95.47 E
				H = 02 05 00.3 h = normal MB = 4.9
				D = 83.85 Az = 320.5 (USNOAA)
15.	eP	A	08 07 39.5	Near Coast of Chiapas, Mexico
	LmH	B	48.8	14.39 N 93.06 W
	LmV	B	51.2	H = 07 54 51.6 h = normal MB = 5.1
				D = 88.04 Az = 38.0 (USNOAA)
				PV(A):0.7s 9.6nm MPV(A)=5.2
				LmH(B):18s 1.6/um MLH(B)=5.5
				LmV(B):16s 1.5/um MLV(B)=5.5
15.	ePKJKP	A	15 43 34	New Hebrides Islands 14.39 S 167.34 E
	ePKHKP	A	43 37	H = 15 24 37.4 h = 182 km MB = 5.8
	ePKIKP	A	43 42	D = 138.83 Az = 336.6 (USNOAA)
	e	A	46 21	PKJKPV(A):0.6s 11.5nm
	e	A	47 03.5	PKIKPV(A):1.5s 75.5nm
15.	ePKP2	A	16 26 08.5	Tonga Islands 18.41 S 175.62 W
				H = 16 06 53.1 h = 252 km MB = 4.6
				D = 147.33 Az = 351.5 (USNOAA)
				PV(A):1.6s 30.2nm

Day	Phase		h m s	Remarks
15.	eP	A	21 28 41	Peru-Ecuador Border Region 4.03 S 80.86 W H = 21 15 20.2 h = 32 km MB = 5.2 D = 94.66 Az = 39.7 (USNOAA)
16.	eiP	ABC	01 13 22.5	New West Coast of Colombia 6.04 N 77.54 W H = 01 00 46.9 h = 14 km MB = 5.6 D = 84.84 Az = 39.7 (USNOAA) PV(A):1.6s 85.2nm MPV(A)=5.7 LmH(B):19s 1.0/um MLH(B)=5.2 LmV(B):20s 1.0/um MLV(B)=5.2
	eSKS	B	23 55	
	LmH	B	50.2	
	LmV	B	50.2	
16.	e(P)	A	08 57 03	Costa Rica 8.71 N 83.15 W H = 08 44 22.0 h = 64 km MB = 5.1 D = 86.35 Az = 39.5 (USNOAA) LmH(B):21s 0.7/um LmV(B):22s 0.5/um
	LmH	B	09 29.4	
	LmV	B	29.4	
16.	eP	A	16 12 18.5	Southern Nevada 37.10 N 116.01 W H = 16 00 00.1 h = 0 km MB = 5.1 D = 81.23 Az = 30.7 (USNOAA) PV(A):1.2s 12.2nm MPV(A)=5.0
16.	eP	A	16 36 32	Andreanof Islands, Aleutian Is. 51.28 N 174.89 W H = 16 24 34.8 h = 37 km MB = 4.9 D = 78.30 Az = 355.8 (USNOAA) PV(A):2.0s 34.2nm MPV(A)=5.0
16.	e(PKP)	A	21 06 33	Fiji Islands Region 17.59 S 178.54 W H = 20 47 50.2 h = 544 km MB = 4.6 D = 146.03 Az = 348.4 (USNOAA)
17.	+iP	A	07 08 46.8	Eastern Kazakh SSR 49.73 N 78.13 E H = 07 00 57.4 h = 0 km MB = 5.5 D = 41.29 Az = 297.8 (USNOAA) PV(A):0.8s 80.7nm MPV(A)=5.5
	e	A	08 53.5	
	ePn	A	10 19.5	

Day	Phase		h m s	Remarks
17.	e(P)	A	07 52 23	Near West Coast of Colombia 6.48 N 77.54 W H = 07 39 53.8 h = 18 km MB = 5.1 D = 84.50 Az = 39.8 (USNOAA)
17.	eP	A	09 28 58.5	North of Severnaya Zemlya 83.38 N 115.38 E
	e	A	29 06.5	H = 09 21 13.0 h = normal MB = 4.8
	LmH	B	43.6	D = 41.55 Az = 291.2 (USNOAA)
	LmV	B	44.0	PV(A):1.6s 33.0nm MPV(A)=4.8
17.	ePg	A	11 05 04	ČSSR, explosion, yield 28 to
	iSg	A	05 24.8	50°35.2'N 14°03.2'E (PRU)
17.	ePKP	A	15 10 45.5	Fiji Region 18.15 S 177.89 W H = 14 52 07.3 h = 570 km MB = 4.7 (ISC) D = 146.8
17.	+iP	A	16 17 18.0	Southern Nevada 37.13 N 116.08 W
	ePP	A	20 22	H = 16 05 00.2 h = 0 km MB = 5.7 D = 81.24 Az = 30.6 (USNOAA) 37°06'00.6"N 116°00'28.6"W "Artesia" Nevada test site (USAEC) PV(A):1.2s 69.1nm MPV(A)=5.7
18.	eP	A	15 40 52.5	South of Alaska 53.30 N 160.78 W
	ePcP	A	41 03.5	H = 15 29 05.7 h = normal MB = 5.1 D = 76.23 Az = 5.0 (USNOAA) PV(A):1.6s 41.2nm MPV(A)=5.2 PcPV(A):1.2s 28.4nm
18.	eP	A	15 42 19	Southern Nevada 37.17 N 116.10 W H = 15 30 00.2 h = 0 km MB = 5.2 D = 81.21 Az = 30.6 (USNOAA) 37°10'23.3"N 116°05'55.9"W Nevada test site "Baneberry" (USAEC)

Day	Phase		h m s	Remarks
18.	ePKP	A	23 14 47.5	Loyalty Islands Region 21.9 S 169.6 E
				H = 22 55 09 h = 36 km
				D = 146.53 Az = 334 (ISC)
18.	eP1	A	24 03 02.5	Mindanao, Philippine Islands
	-iP2	ABC	03 06.5	5.12 N 123.55 E
	epP	A	04 59	H = 23 50 12.3 h = 511 km MB = 5.5
	e	A	05 07	D = 99.70 Az = 323.2 (USNOAA)
	ePP	C	07 08	P1V(A):1.3s 52.4nm MP1V(A)=5.8
	e(PPP)	C	09 48	P2V(A):1.2s 150.3nm MP2V(A)=6.3
	eSKS	B	13 48	LmH(B):18s 1.7 µm
	ePS	C	16 09	LmV(B):18s 1.8 µm
	e	C	17 00	
	e	BC	17 22	
	e	BC	19 54	
	LmH	B	52.0	
	LmV	B	52.3	
19.	ePKIKP	A	02 47 50	New Britain Region 5.63 S 151.80 E
	LmH	B	03 41.1	H = 02 28 56.0 h = 56 km MB = 5.6
	LmV	B	47.4	D = 124.23 Az = 330.5 (USNOAA)
				LmH(B):18s 1.8 µm MLH(B)=5.8
				LmV(B):20s 2.2 µm MLV(B)=5.8
19.	ePn	A	03 01 18	Yugoslavia 46.42 N 16.40 E
	e(Sn)	A	02 18.5	H = 02 59 58.1 h = 16 km MB = 4.2
	e	A	02 46	D = 5.29 Az = 324.8 (USNOAA)
	e	B	02 48	
19.	eP1	ABC	10 51 02.5	Southern Sumatra 1.63 S 99.90 E
	+iP2	A	51 06.5	H = 10 38 05.2 h = 46 km MB = 5.8
	ipP2	AB	51 17	D = 90.16 Az = 320.5 (USNOAA)
	ePP	BC	54 40	P1V(A):1.5s 37.7nm MP1V(A)=5.5
	eiPPP	C	56 48	P1V(B):5s 1.6 µm MP1V(B)=6.5
	eSKS	B	11 01 30	P2V(A):1.9s 356.0nm MP2V(A)=6.4
	eSKS	C	01 32	PPV(B):8s 1.2 µm MPPV(B)=6.4
	eiS	B	01 52	SH(B):12s 1.95 µm MSH(B) =6.4
	eiS	C	01 54	LmH(B):16.5s 4.5 µm MLH(B) =6.0
	ePS	B	03 08	LmV(B):22s 4.8 µm MLV(B) =5.9

Day	Phase		h m s	Remarks
cont. 19.	ei	C	11 03 16	
	e	C	07 36	
	eSS	B	08 00	
	eSSS	C	11 31	
	e	C	14 36	
	LmH	B	39.6	
	LmV	B	39.7	
19.	eiP	A	12 21 32.5	United Arab Republic 27.51 N 33.78 E H = 12 15 34.1 h = 20 km MB = 4.6 D = 28.60 Az = 329.9 (USNOAA) PV(A):0.7s 11.5nm MPV(A)=4.7
19.	e	A	22 50 13	United Arab Republic 27.51 N 33.88 E H = 22 44 09.3 h = 23 km MB = 4.6 D = 28.64 Az = 329.8 (USNOAA)
20.	ePKP	A	00 51 22	Loyalty Islands Region 21.7 S 169.8 E H = 00 31 06 h = 24 km (ISC) D = 146.4
20.	ePKP	A	03 00 36	Loyalty Islands Region 21.9 S 169.6 E H = 02 40 55 h = 0 km (ISC) D = 146.9
20.	+iP	A	06 12 09	Central Alaska 63.09 N 151.45 W H = 06 01 36.1 h = 130 km MB = 5.3
	e	A	12 42	D = 65.83 Az = 11.7 (USNOAA) PV(A):1.4s 37.2nm MPV(A)=5.1
20.	ePKP	A	10 59 43	Near North Coast of New Guinea 4.84 S 144.14 E H = 10 41 03.1 h = 104 km MB = 5.5 D = 119.57 Az = 327.4 (USNOAA) PKPV(A):1.1s 14.1nm

Day	Phase		h m s	Remarks
20.	eP1	AB	11 05 40	Turkey 39.40 N 29.24 E
	eP2	AC	05 43	H = 11 01 48.0 h = 38 km MB = 5.0
	P2max	A	05 50	D = 16.73 Az = 318.0 (USNOAA)
	eSS	C	08 55	P2V(A):1.4s 27.9nm MP2V(A)=4.2
	e	B	09 11	P2maxV(A):1.8s 111.6nm MP2maxV(A)=4.7
				LmH(B):14s 8.8/um MLH(B) =5.2
				LmV(B):10s 7.0/um MLV(B) =5.4
20.	eiPKP	A	12 14 27.5	Fiji Islands Region 17.83 S 179.96 W
				H = 11 55 56.7 h = 635 km MB = 4.9
				D = 145.98 Az = 346.8 (USNOAA)
				PV(A):1.4s 39.6nm
21.	eP	A	07 46 35.5	Peru-Ecuador Border Region
				3.96 S 80.81 W
				H = 07 33 15.5 h = 32 km MB = 5.1
				D = 94.58 Az = 39.7 (USNOAA)
				PV(A):1.6s 27.5nm MPV(A)=5.5
21.	eP	A	11 04 59	Kurile Islands Region 43.83 N 151.03 E
				H = 10 52 54.9 h = 38 km MB = 5.1
				D = 79.50 Az = 335.1 (USNOAA)
				PV(A):1.8s 27.0nm MPV(A)=5.0
21.	LmH	B	13 56.3	Probably Philippine Islands (USNOAA)
	LmV	B	59.0	LmH(B):16s 1.3/um
				LmV(B):16s 1.2/um
21.	eX	A	14 59 13	Sumbawa Island Region 9.12 S 116.42 E
	e	A	59 16	H = 14 40 45.0 h = 92 km MB = 5.9
	LmH	C	15 45.5	D = 106.38 Az = 320.1 (USNOAA)
	LmV	C	48.6	XV(A):1.8s 47.3nm
				LmH(C):23s 0.9/um
				LmV(C):21s 0.6/um
22.	eP	A	02 01 49	Peru-Ecuador Border Region
	LmH	C	42.0	3.96 S 80.82 W

Day	Phase		h m s	Remarks
cont. 22.	LmV	C	02 42.4	H = 01 48 30.2 h = 33 km MB = 5.5
				D = 94.58 Az = 39.7 (USNOAA)
				PV(A):1.4s 23.3nm MPV(A)=5.5
				LmV(C):19s 0.5/um MLV(C)=5.0
22.	ePKP2	A	02 49 51	<u>Fiji Islands Region</u> 19.49 S 178.90 W
				H = 02 31 13.9 h = 608 km MB = 4.6
				D = 147.81 Az = 347.4 (USNOAA)
22.	ePKIKP	A	05 57 13	<u>New Hebrides Islands</u> 20.81 S 109.84 E
	+iPKHKP	A	57 14.5	H = 05 37 47.9 h = 119 km MB = 5.6
	eX1	A	57 43	D = 145.63 Az = 335.3 (USNOAA)
	eX2	A	57 58	PKIKPV(A):0.8s 15.4nm
				PKHKPV(A):1.8s 87.8nm
				X1V(A):1.5s 40.2nm
				X2V(A):1.2s 106.0nm
22.	+iP	A	20 59 46.5	<u>North Atlantic Ridge</u> 28.30 N 43.91 W
	e	A	59 52.5	H = 20 51 16.2 h = normal MB = 5.3
	LmV	B	21 18.5	D = 47.04 Az = 45.8 (USNOAA)
	LmH	B	18.6	PV(A):1.9s 83.4nm MPV(A)=5.4
				LmV(B):18s 1.6/um MLV(B)=5.1
				LmH(B):16s 1.2/um MLH(B)=5.0
22.	iP1	AB	21 01 34	<u>North Atlantic Ridge</u> 28.29 N 43.94 W
	eP2	A	01 40	H = 20 53 04.3 h = normal MB = 5.4
	ePcP	C	06(50)	D = 47.07 Az = 45.8 (USNOAA)
	eS	BC	08 36	P1V(A):1.7s 79.0nm MP1V(A)=5.5
	LmV	B	18.5	P2V(A):1.8s 128.3nm MP2V(A)=5.8
	LmH	B	18.6	LmV(B):18s 1.6/um MLV(B) =5.1
				LmH(B):16s 1.2/um MLH(B) =5.0
				Superposition of surface waves
22.	iPn	A	22 14 16	<u>Peißenberg, Federal Rep. Germany</u>
	e(Pg)	A	14 24	47.8 N 11.02 E
	i	A	14 47.5	H = 22 13 31 (BCIS)
	i(Sn)	A	14 51.5	D = 2.9
	i	A	15 07.5	

Day	Phase		h m s	Remarks
23.	+iP	A	07 07 07.5	Western Kazakh SSR 43.83 N 54.85 E
	ePn	A	07 35	H = 07 00 57.3 h = 0 km MB = 6.1
	e	A	07 36	D = 29.79 Az = 298.6 (USNOAA)
	e(PP)	A	08 09.5	PV(A):1.7s 242.0nm MPV(A)=5.8
	i	A	08 48	LmH(B):10s 0.9 μm MLH(B)=4.7
	+eiPcP	A	10 10	LmV(B):10s 0.9 μm MLV(B)=4.8
	LmH	B	22.4	
	LmV	B	25.3	
23.	LmH	B	12 40.6	Probably South of Mariana Islands
	LmV	B	46.4	(USNOAA)
				LmH(B):16s 0.6 μm
				LmV(B):20s 0.6 μm
23.	eP1	A	15 36 17	Near Coast of Chiapas, Mexico
	eP2	A	36 18	15.90 N 93.82 W
	eS	BC	46 50	H = 15 23 39.1 h = 90 km MB = 5.3
	LmH	C	16 06.0	D = 87.31 Az = 37.9 (USNOAA)
	LmV	C	10.0	P1V(A):1.1s 12.1nm MP1V(A)=4.9
				P2V(A):1.4s 55.8nm MP2V(A)=5.4
				LmV(C):24s 0.5 μm
24.	e(PKP2)	A	07 45 31.5	Fiji Islands Region 20.30 S 178.66 W
				H = 07 26 51.2 h = 610 km MB = 4.3
				D = 148.65 Az = 347.4 (USNOAA)
				(PKP2)V(A):0.8s 15.4nm
24.	eP	ABC	08 13 11	Chiapas, Mexico 16.10 N 93.56 W
	ePP	A	16 24	H = 08 00 37.6 h = 116 km MB = 5.6
	e	C	17 06	D = 87.00 Az = 38.0 (USNOAA)
	e	B	17 09	PV(A):1.5s 181.0nm MPV(A)=5.9
	eSKS	BC	23 32	LmH(C):48s 1.2 μm
	eS	BC	23 42	LmV(C):40s 2.2 μm
	LmH	C	42.5	
	LmV	C	42.7	

Day	Phase		h m s	Remarks
24.	eP	A	08 34 14	Andreanof Islands, Aleutian Is.
				51.47 N 178.31 W
				H = 08 22 20.8 h = 53 km MB = 5.3
				D = 77.91 Az = 353.6 (USNOAA)
				PV(A):1.6s 38.5nm MPV(A)=5.1
25.	e(PKP)	A	04 25 36	Tonga 21.3 S 175.4 W
				H = 04 05 50 h = 22 km MB = 4.5 (ISC)
				D = 150.3
25.	+eiP	ABC	13 03 22.5	Central Mid-Atlantic Ridge
	ePP	BC	05 30	0.32 S 19.17 W
	ePPP	C	06 37	H = 12 53 37.4 h = normal MB = 5.5
	ePPP	B	06 38	D = 57.13 Az = 22.8 (USNOAA)
	eiS	BC	11 13	PV(A):1.2s 95.5nm MPV(A)=5.7
	LmH	B	26.6	PV(B):10s 0.9 μm MPV(B)=5.7
	LmV	B	30.3	LmH(B):17s 5.8 μm MLH(B)=5.7
				LmV(B):16s 3.6 μm MLV(B)=5.6
25.	ePKP	A	20 07 56	New Hebrides Islands Region
				21.85 S 174.40 E
				H = 19 48 09.7 h = 13 km MB = 5.0
				D = 148.24 Az = 339.0 (USNOAA)
26.	e(PKP)	A	06 45 28	Fiji Islands Region 20.22 S 176.39 W
	eX	A	45 50	H = 06 26 32.9 h = 270 km MB = 4.8
				D = 148.99 Az = 350.1 (USNOAA)
				XV(A):0.9s 17.5nm
26.	eP	AB	10 14 43	Nicobar Islands Region
	eS	B	24 28	9.28 N 94.06 E
	ePS	B	25 04	H = 10 02 47.9 h = 47 km MB = 5.4
	LmH	B	53.6	D = 78.10 Az = 319.8 (USNOAA)
	LmV	B	54.1	PV(A):1.1s 14.1nm MPV(A)=4.8
				LmH(B):19.5s 1.5 μm MLH(B)=5.3
				LmV(B):20s 1.3 μm MLV(B)=5.3

Day	Phase		h m s	Remarks
26.	eP	A	10 27 27.5	Nicobar Islands Region
	e	A	27 46	9.38 N 94.04 E
				H = 10 15 49.3 h = normal MB = 4.8
				D = 78.02 Az = 319.8 (USNOAA)
26.	e(PKP)	A	14 35 29.5	Fiji Islands Region 21.63 S 176.85 W
				H = 14 16 06.4 h = 251 km MB = 4.1
				D = 150.28 Az = 349.1 (USNOAA)
26.	ePKP	A	19 25 11.5	Fiji Islands 15.98 S 178.22 E
	LmV	B	20 21.6	H = 19 05 46.9 h = 86 km MB = 5.3
	LmH	B	22.7	D = 143.77 Az = 345.5 (USNOAA)
				PKPV(A):2.8s 107.0nm
26.	ePKP	A	21 59 18.5	Tonga Islands 16.35 S 174.43 W
				H = 21 39 40.6 h = 9 km MB = 4.7
				D = 145.44 Az = 353.2 (USNOAA)
26.	eX	A	22 06 14.5	West of Macquarie Islands
				56.50 S 146.85 E
				H = 21 46 17.3 h = normal MB = -
				D = 153.17 Az = 276.6 (USNOAA)
				XV(A):1.8s 23.6nm
27.	ePKP	A	02 26 21.5	Fiji Islands 16.20 S 178.79 E
				H = 02 06 45.1 h = 29 km MB = 5.1
				D = 144.12 Az = 346.0 (USNOAA)
				PKPV(A):1.4s 18.6nm
27.	ePKP	A	10 32 28	Fiji Islands Region 20.40 S 178.52 W
				H = 10 13 42.5 h = 557 km MB = 4.5
				D = 148.77 Az = 347.5 (USNOAA)
				PKPV(A):1.2s 24.4nm
27.	eP	A	15 32 25.5	Near East Coast of Kamchatka
				54.00 N 160.06 E
				H = 15 21 10.7 h = 123 km MB = 4.4
				D = 72.41 Az = 339.5 (USNOAA)
				PV(A):0.9s 15.6nm MPV(A)=4.8

Day	Phase		h m s	Remarks
27.	eP	A	19 26 48	Off East Coast of Kamchatka
				52.57 N 158.96 E
				H = 19 15 14.4 h = 18 km MB = 4.7
				D = 73.51 Az = 339.0 (USNOAA)
27.	eP	A	20 56 46	Kurile Islands Region 44.90 N 150.65 E
	e	A	56 59	H = 20 44 48.9 h = 48 km MB = 5.2
	LmH	C	21 34.0	D = 78.42 Az = 334.8 (USNOAA)
				PV(A):1.4s 51.1nm MPV(A)=5.3
				LmH(C):20s 0.5 µm MLH(C)=4.8
28.	eP	A	03 46 35	Turkey 37.05 N 28.93 E
				H = 03 42 15.5 h = 11 km MB = 4.5
				D = 18.39 Az = 323.1 (USNOAA)
				PV(A):1.6s 27.5nm MPV(A)=4.2
28.	eP	A	11 34 12.5	Hokkaido, Japan 41.30 N 142.7 E
				H = 11 22 20 h = 28 km (ISC)
				D = 78.8
28.	-iP	A	17 05 08	Dodecanese Islands 36.06 N 28.25 E
	LmV	B	14.8	H = 17 00 47.9 h = 46 km MB = 4.5
	LmH	B	14.9	D = 18.38 Az = 325.7 (USNOAA)
				PV(A):1.5s 60.3nm MPV(A)=4.6
				LmV(B):12s 0.9 µm MLV(B)=4.5
				LmH(B):10s 0.6 µm MLH(B)=4.2
28.	ePKIKP1	AB	20 22 20	New Ireland Region 5.16 S 153.62 E
	iPKIKP2	A	22 21.5	H = 20 03 25.1 h = 61 km MB = 6.0
	ePP	BC	24 08	D = 124.65 Az = 331.5 (USNOAA)
	ePPP	B	26 44	PKIKP1V(A):1.2s 16.3nm
	eiPS	B	33 58	PKIKP2V(A):2.0s 103.0nm
	iPPS	B	35 34	LmH(B):20.5s 12.1 µm MLH(B)=6.5
	iSS	B	41 00	LmV(B):18.5s 11.3 µm MLV(B)=6.6
	eSSS	B	45 10	
	LmH	B	21 20.4	
	LmV	B	21.0	

Day	Phase		h m s	Remarks
29.	eP	A	00 53 11	Crete 34.98 N 23.23 E
	eX	A	53 16	H = 00 49 04.8 h = 53 km MB = 4.5
	ePP	A	53 26	D = 17.79 Az = 335.2 (USNOAA)
				XV(A):1.7s 27.2nm
29.	ePKIKP	A	02 45 23.5	Solomon Islands 10.55 S 161.40 E
	ePP	B	47 44	H = 02 26 12.2 h = 72 km MB = 6.1
	iPKS	B	48 44	D = 132.94 Az = 334.0 (USNOAA)
	ePS	B	58 00	LmH(B):20s 13.3/um
	ePPS	B	59 56	LmV(B):20s 9.1/um
	iSS	B	03 05 14	
	eiSSP	B	05 44	
	eSSS	B	10 16	
	LmH	B	45.8	
	LmV	B	47.1	
29.	eP	A	08 15 18	Peru-Ecuador Border Region
	eX	A	15 26	3.90 S 80.90 W
	LmH	B	55.0	H = 08 01 59.3 h = 47 km MB = 5.8
	LmV	B	56.1	D = 94.59 Az = 39.6 (USNOAA)
				PV(A):1.2s 24.4nm MPV(A)=5.6
				XV(A):1.4s 32.6nm
				LmH(B):20s 1.3/um MLH(B)=5.4
				LmV(B):20s 1.6/um MLV(B)=5.5
29.	eP	A	10 36 35	Fox Islands, Aleutian Is.
				51.24 N 168.44 W
				H = 10 24 31.4 h = 9 km MB = 5.0
				D = 78.49 Az = 360.0 (USNOAA)
29.	eP	A	12 51 14	Crete 35.02 N 23.25 E
				H = 12 47 11.4 h = 63 km MB = 4.7
				D = 17.76 Az = 335.1 (USNOAA)
29.	eP	A	15 51 33	Kurile Islands 44.59 N 147.60 E
				H = 15 39 37.5 h = normal MB = 4.7
				D = 77.75 Az = 333.1 (USNOAA)
				PV(A):1.0s 15.8nm MPV(A)=5.0

Day	Phase		h m s	Remarks
29.	eP	A	17 29 39.5	Rat Islands, Aleutian Is. 51.56 N 177.65 E H = 17 17 51.4 h = 58 km MB = 4.4 D = 77.49 Az = 350.9 (USNOAA)
29.	eP	A	20 38 30	Eastern, Mediterranean Sea 35.93 N 28.11 E H = 20 34 07.4 h = 16 km MB = 4.1 D = 18.93 Az = 326.1 (USNOAA)
29.	eP	A	21 08 00	Dodecanese Islands 36.00 N 28.21 E H = 21 03 37.9 h = 22 km MB = 4.4 D = 18.91 Az = 325.9 (USNOAA) PV(A):1.6s 30.2nm MPV(A)=4.2
30.	e	A	02 21 50.5	Northern Italy 44.35 N 8.17 E
	ePg	A	22 17.5	H = 02 20 05.8 h = normal MB = 4.0
	e	A	22 47.5	D = 6.70 Az = 19.1 (USNOAA)
	e	A	23 08.5	LmH(E):10s 1.3/um MLH(B)=3.9
	eSg	A	23 47	
	LmH	B	24.3	
30.	eX	A	04 00 16	New Ireland Region 3.26 S 152.52 E
	ePP	A	01 46.5	H = 03 41 11.9 h = 29 km MB = 5.2
	LmH	B	55.8	D = 122.47 Az = 331.6 (USNOAA)
	LmV	B	58.1	XV(A):2.0s 72.7nm LmH(B):20s 2.2/um MLH(B)=5.8 LmV(B):19s 2.2/um MLV(B)=5.8
30.	eP	A	08 23 48	Northern Sumatra 1.44 N 99.09 E
	epP	A	24 11.5	H = 08 11 05.1 h = 86 km MB = 5.5 D = 87.29 Az = 320.4 (USNOAA) PV(A):1.6s 44.0nm MPV(A)=5.3 pPV(A):1.4s 46.5nm
30.	eP	A	09 43 23	Kirgiz SSR 40.26 N 72.73 E H = 09 35 25.9 h = normal MB = 5.0 D = 42.97 Az = 305.1 (USNOAA)

Day	Phase		h m s	Remarks
30.	eP	A	11 37 10	Dodekanese Islands 36.05 N 28.34 E
				H = 11 32 46.5 h = 36 km (ISC)
				D = 18.9
30.	eP	A	18 59 02	Turkey 37.08 N 28.94 E
				H = 18 54 43.8 h = normal MB = -
				D = 18.38 Az = 323.0 (USNOAA)
30.	-eP1	A	21 02 37.5	North Atlantic Ocean 37.21 N 15.05 W
	iP2	AB	02 38.8	H = 20 57 30.6 h = normal MB = 5.1
	i	BC	02 42	D = 23.28 Az = 46.3 (USNOAA)
	-iX	B	02 45	P1V(A):0.9s 23.4nm MP1V(A)=4.7
	eiS	BC	06 50	P2V(A):1.4s 123.0nm MP2V(A)=5.2
	Smax	B	07 10	XV(A):1.6s 274.0nm
	LmH	B	10.2	SmaxH(B):11s 4.0 µm MSmaxH(B)=5.8
	LmV	B	12.9	LmH(B):18s 9.5 µm MLH(B)=5.3
				LmV(B):14s 4.4 µm MLV(B)=5.2
31.	eP	A	05 46 02	Off Coast of Washington 47.77 N 128.76 W
				H = 05 34 13.5 h = normal MB = 5.2
				D = 76.20 Az = 24.7 (USNOAA)
31.	eP	A	06 04 53.5	Kurile Islands 46.03 N 153.07 E
				H = 05 52 57.3 h = 44 km MB = 4.7
				D = 78.10 Az = 336.1 (USNOAA)
31.	eP	A	08 13 29	North of Svalbard 80.32 N 0.34 E
				H = 08 07 20.4 h = normal MB = 4.5
				D = 30.04 Az = 165.6 (USNOAA)
31.	ePn	A	22 06 25	Northern Italy 44.39 N 8.46 E
	e	A	06 58.5	H = 22 04 47.9 h = normal MB = 4.1
	iSn	A	07 44.5	D = 6.61 Az = 17.7 (USNOAA)
	i	A	07 56	LmH(B):11s 1.4 µm MLH(B)=3.8
	iSg	A	08 38	LmV(B):8s 0.5 µm
	LmH	B	09.0	
	LmV	B	09.7	

Day	Phase		h m s	Remarks
31.	eP	A	23 45 24	Kurile Islands 44.56 N 148.03 E H = 23 33 31.7 h = 67 km MB = 5.1 D = 77.90 Az = 333.3 (USNOAA)

Distribution of Azimuth and Epicentral Distance of Earthquakes Relative to Moxa (MOX)

by
HORST NEUNHÖFER

The knowledge how the azimuth A and epicentral distance D of earthquakes are distributed relative to a seismic station is useful for estimating the possibility to carry out special investigations by using the seismograms of that station. For instance, the distribution of the azimuths can be helpful for an effective arrangement of a seismic network to be established for the investigation of seismic surface waves. Sometimes, only earthquakes in a distanct epicentral distances give the sought information on a depth interval of the Earth. Than the distribution of the epicentral distances helps us to decide whether a planned investigation is possible with the records of the station or not. Therefore, the distribution of the azimuth and epicentral distance of the earthquakes is characteristic of a station.

The distribution density $p(D)$ and $p(A)$ of the epicentral distance and the azimuth relative to Moxa are given in Figs. 1 and 2, respectively. In both cases the compilation was done by grouping into the intervals i one degree wide with the midvalues 0.5, 1.5, 2.5, etc. The relations

$$\sum_{i=1}^{180} p_i(D) = 1$$

and

$$\sum_{i=1}^{360} p_i(A) = 1$$

are valid. For compiling, the punched cards of the U.S. Geological Survey National Earthquake Information Service (formerly NOAA and USCGS) and the computer routine 'AZENT' developed by the author are used. All earthquakes occurred during the period 1st January, 1968 to 23rd March, 1973 with the magnitude $m_B \geqq 5.0$ are considered.

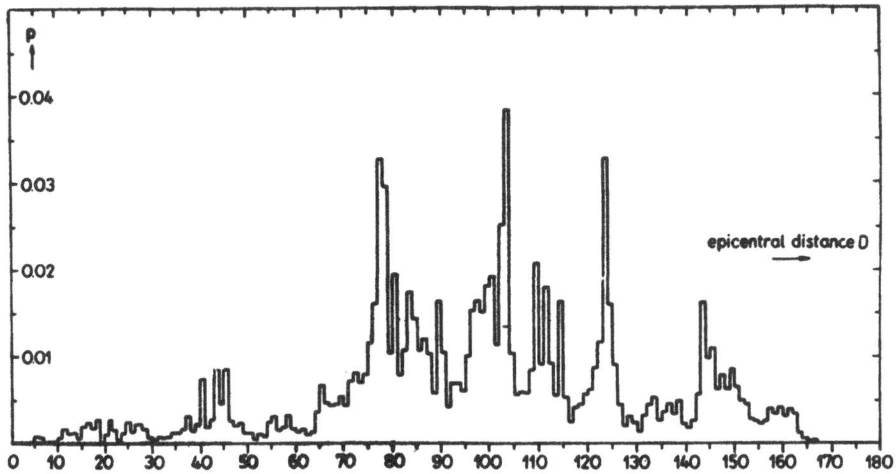

Fig. 1. Distribution density of the epicentral distance relative to Moxa

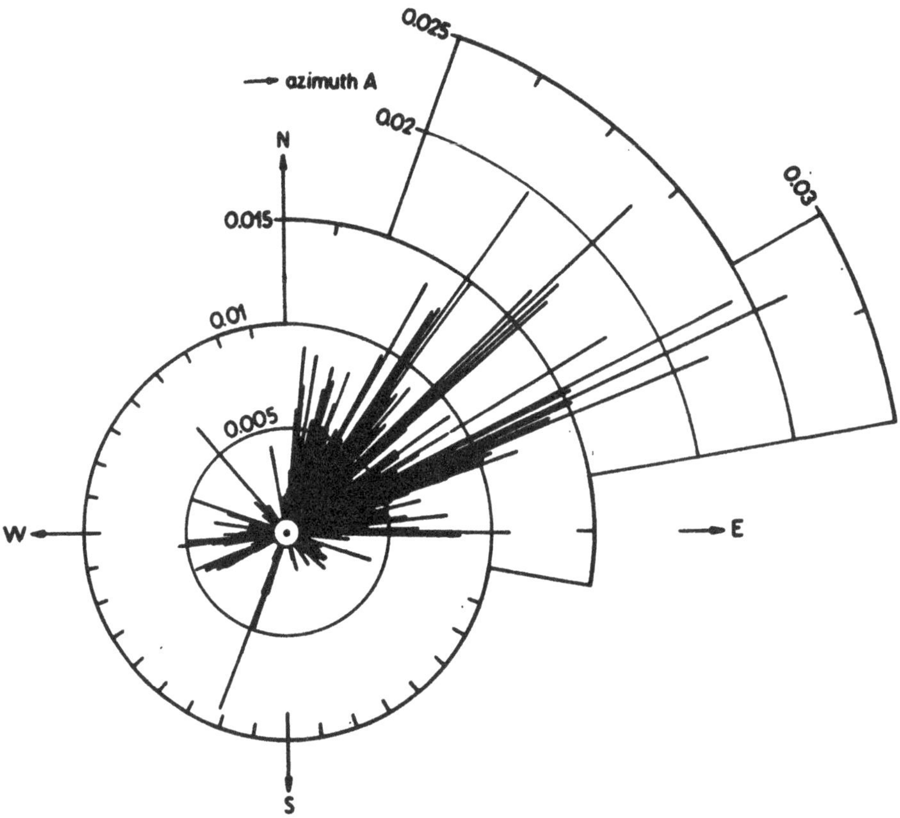

Fig. 2. Distribution density of the azimuth relative to Moxa

Bei Fragen zur Produktsicherheit wenden Sie sich bitte an:
If you have any questions regarding product safety,
please contact:

Walter de Gruyter GmbH
Genthiner Straße 13
10785 Berlin
productsafety@degruyterbrill.com